QK861 . S97

KU-405-412

0222803 -3

SWAIN, TONY
COMPARATIVE PHYTOCHEMISTRY
000222803

QK861.S97

WITHDRAWN FROM STOCK
The University of Liverpool

2

21

WITHDRAWN FROM STOCK
University of Liverpool

COMPARATIVE
PHYTOCHEMISTRY

COMPARATIVE PHYTOCHEMISTRY

Edited by

T. SWAIN

Low Temperature Research Station, Cambridge, England

 1966

ACADEMIC PRESS

LONDON AND NEW YORK

ACADEMIC PRESS INC. (LONDON) LTD
BERKELEY SQUARE HOUSE
BERKELEY SQUARE
LONDON, W.1

U.S. Edition published by
ACADEMIC PRESS INC.
111 FIFTH AVENUE
NEW YORK, NEW YORK 10003

Copyright © 1966   By Academic Press Inc. (London) Ltd

All rights reserved

NO PART OF THIS BOOK MAY BE REPRODUCED IN ANY FORM BY PHOTOSTAT, MICROFILM,
OR ANY OTHER MEANS, WITHOUT WRITTEN PERMISSION FROM THE PUBLISHERS

Library of Congress Catalog Card Number: 66–16700

PRINTED IN GREAT BRITAIN BY
SPOTTISWOODE BALLANTYNE & CO. LTD.
LONDON AND COLCHESTER

List of Contributors

ALSTON, R. E., *The Cell Research Institute and The Department of Botany, The University of Texas, Austin, Texas, U.S.A.*

BATE-SMITH, E. C., *Low Temperature Research Station, Downing Street, Cambridge, England*

BELL, E. A., *Department of Biochemistry, King's College, London, England*

BU'LOCK, J. D., *Department of Chemistry, The University, Manchester, England*

DOUGLAS, A. G., *Chemistry Department, The University, Glasgow, Scotland*

EGLINTON, G., *Chemistry Department, The University, Glasgow, Scotland*

GOODWIN, T. W., *Department of Biochemistry and Agricultural Biochemistry, University College of Wales, Aberystwyth, Wales*

HARBORNE, J. B., *John Innes Institute, Hertford, Hertfordshire, England*

HEGNAUER, R., *Laboratorium voor Experimentele Plantensystematiek, Leiden, Netherlands*

HEYWOOD, V. H., *The Hartley Botanical Laboratories, University of Liverpool, England*

KJÆR, ANDERS, *Department of Organic Chemistry, Royal Veterinary and Agricultural College, Copenhagen, Denmark*

MABRY, T. J., *Department of Botany, The University of Texas, Austin, Texas, U.S.A.*

MATHIS, C., *Department of Chemistry, University of Strasbourg, France*

MENTZER, C., *Muséum National d'Histoire Naturelle, Paris, France*

PERCIVAL, ELIZABETH, *Royal Holloway College, University of London, Englefield Green, Surrey, England*

RUIJGROK, H. W. L., *Laboratorium voor Experimentele Plantensystematiek Leiden, Netherlands*

SWAIN, T., *Low Temperature Research Station, Downing Street, Cambridge, England*

WAGNER, H., *Institut für Pharmazeutische Arzneimittellehre, Munich, Germany*

WEISSMANN, G., *Federal Research Organization for Forestry and Forest Products, Institute for Wood Chemistry and Chemical Technology of Wood, Hamburg, Germany*

WILLIAMS, A. H., *Long Ashton Research Station, University of Bristol, England*

Preface

During the last three years, there has been an explosive growth in the number of books and papers that have appeared on the use of chemical compounds as taxonomic characters. It can be confidently asserted that comparative phyto-chemistry, dealing as it does with both the distribution and biogenesis of chemical substances in plants, is now well established and has proved to be useful not only as an addition to the armoury of plant taxonomists, but also by indicating interesting taxa and biosynthetic sequences to natural product chemists.

This conclusion is reinforced by the fact that in the last year both the International Union of Pure and Applied Chemistry and the International Association for Plant Taxonomy have set up official committees on chemical plant taxonomy.

This book can therefore be regarded not as an advertisement for a new branch of science, but as a necessary up-to-date report showing the present state of the art, the progress made in the field in the last year or so, and providing, moreover, a useful source book both to botanists and to chemists for years to come.

What of the future? The interesting phylogenetic implications of recent work on the differences in amino acid sequences in haemoglobin from different animal species indicates that similar investigations on plant proteins and on base-sequences in plant nucleic acids will undoubtably be equally rewarding. However, until suitable techniques have been developed for such work, we may expect that the results of greatest interest will come from biosynthetic investigations which should give a better insight into the interrelationships between compounds, and help to answer questions about parallelism and diversification, convergence and divergence as applied to chemical characters.

Almost equal importance can be attached to work that leads to an understanding of the biological function of the so-called secondary plant products and the relationship between such functions and other characters of taxonomic importance. It is gratifying to note that, as can be seen from the contributions in this book, such problems are already being tackled with vigour.

The chapters in this book are based on papers presented at a meeting of the Phytochemical Group in Cambridge in April 1965, which was generously supported by a grant from the Scientific Affairs Division of NATO, and all those interested in chemotaxonomy must be grateful for the continued support both organizations have so readily given over the last few years.

As editor, I am also especially grateful to the contributors who have all responded generously and promptly to the various demands I made on them, and I hope they will forgive me for any faults in their chapters due to an oversight on my part. I would also like to record my debt to Professor Holger Erdtman for his continued wise guidance and inspiration, and, finally, I wish to thank the staff of Academic Press for their invaluable assistance in preparing the book for publication.

December, 1965   T. SWAIN

Contents

Contents

CHAPTER 1

Phytochemistry and Taxonomy

V. H. HEYWOOD

The Hartley Botanical Laboratories, University of Liverpool, England

task

I. INTRODUCTION

The role of a taxonomist writing the opening chapter in a book on chemical plant taxonomy is a difficult and somewhat embarrassing one. It might appear straightforward at first glance—basically what has to be attempted is to set out the ways in which taxonomic principles should be applied to chemical data. The difficulty stems from the fact that there is currently a ferment of debate about which principles taxonomists themselves should follow and even wide disagreement as to what taxonomists actually do. Moreover, taxonomists are often prone to say one thing and do another. This explains, perhaps, the embarrassment of my predicament. I shall attempt, in this chapter, to sketch out the major principles of taxonomic procedure about which there is a wide measure of agreement, and in those instances where currently held attitudes are widely divergent I shall endeavour to give a fair account of the alternative positions.

It is universally agreed that biological classification is an intellectual procedure which allows us to come to grips with, and organize the vast amount of information which we can perceive in the natural world. It involves the formation and description of taxonomic groups (species) and the grouping of these into a limited number of more inclusive groups (genera, families, orders, etc.) based upon degrees of similarity in respect of greater or lesser numbers of the characters possessed by the organisms concerned. A classification may, therefore, be regarded as a series of theoretical generalizations built up by taxonomists about the pattern of nature (Grant, 1963). These generalizations are improved upon by successive generations of taxonomists. This aspect of

1

taxonomy is sometimes termed "data-processing for biology." We are, how-ever, free to base our classifications on those principles that appear most appropriate to our purpose (Mackerras, 1964) and while the data-processing approach may serve a wide variety of purposes (and if broadly based may be termed "general purpose") other approaches are widely championed, notably the evolutionary or phylogenetic one. Such special classifications are designed to serve only a single or few purposes although it should be pointed out that evolutionary classifications are frequently believed to be the most generally useful kind (cf. Fosberg and Sachet, 1961).

If we consider classifications in the light of general purpose data-processing it can be easily seen that the principles that govern their construction are simply an extension of those that apply to inanimate objects. The groups or classes are formed by placing together those individuals which share the maximum number of resemblances; such groups store large amounts of information and have a high predictivity. They form an information retrieval system.

The likening of the classification of organisms to that of inanimate objects such as motor-cars has led to a great deal of controversy (cf. Constance, 1964). Clearly plants are not motor-cars, but for certain purposes we can treat them as though they were. The plants we classify have arisen as the result of historical evolution and reproduce themselves, but it is their evolutionary persistence at a certain level of time (stasigenesis) that allows us to classify them. The kind of classification I have been discussing so far is often referred to as *empirical* since it is based on the degree of overall similarities between organisms without explicit attention to the means by which such resemblances have been achieved. In other words it is non-evolutionary to the extent that it does not take into account or deliberately reflect the evolutionary history of the organisms which are grouped together and makes no claim about the evolutionary nature of the groups. It is, of course, as a result of evolution that groups can be recognized. The pejorative term "static" is also applied to such classifications, although it is difficult to see how a useable classification could be otherwise.

Before considering the role of chemical data in the construction of empirical classifications the nature of evolutionary or phylogenetic classifications with which they are contrasted will be considered briefly. An evolutionary classifica-tion is one that in some ways reflects or is at least consistent with what is known of the evolutionary history of the groups it contains (Simpson, 1961). It is based on an assessment of relative propinquity of descent, or as Edwards and Cavalli-Sforza (1964) put it, it is an attempt to estimate the form of an evolutionary tree and therefore not primarily a classification at all. As I shall show later, the only way to achieve an evolutionary classification is to start with the basis of an empirical classification based on overall resemblances which is then modified to reflect what factual or highly probable evidence of evolution is available. A fully phylogenetic classification is a misnomer—it is really an arrangement or a system.

These two alternatives—empirical or evolutionary—were the choices available in classification until the last decade, even though the epithets might not have been accepted by many practising taxonomists at that time, or the distinctions between them recognized, for that matter. The situation has been greatly complicated by three developments: (1) the formation of a school of numerical taxonomy or taximetrics with an associated type of classification known as phenetic or neo-Adansonian; (2) arising out of the previous development, a detailed consideration of the nature of taxonomic characters and the premise of "equal weighting", which has led many taxonomists to the third development; (3) a realization that in practice many or perhaps most classifications are based on an as yet unanalysed system of selection and weighting of characters, and that they are, therefore, neither strictly empirical, nor based on all possible characters equally weighted, nor are they evolutionary in the sense outlined above. New terms have been proposed to meet this situation—omnispective (Blackwelder, 1964), holotaxonomic (Stafleu, 1965).

What terms we employ for the description of our classifications is unimportant so long as we can recognize that they refer to different kinds of activity and that we distinguish between them in practice. It is relevant here to draw attention to an extraordinary situation which arose at the 10th International Botanical Congress held at Edinburgh in 1964. There, after ten days of intensive discussion by taxonomists from all over the world, it was realized, to use the words of Constance (1964) that "A phylogenetic classification means quite different things to some British and some American botanists. The British, doubtless as a consequence of childhood efforts to trace the lineage of the English kings supplemented by a later interest in the pedigree of dogs and horses, tend to interpret phylogeny in terms of strict genealogy. Many Americans, on the contrary, seek a self-reinforcing classification based upon maximum correlation of characters, and believe that its explanation can only be an evolutionary or phylogenetic one. An awareness of this difference in viewpoint can alleviate considerable mutual irritation." You might well think, as I do, that this gives the game away—that Constance's definition of a phylogenetic classification would fit equally well that of a general classification which we started off with and that it was just a matter of viewpoint whether one prefers to believe that its explanation is evolutionary or not. Constance further comments that those of us who emit warnings about confusing conjecture with reality are wasting our breath in trying to cure people of the "naive view that our classifications have some meaning over and above the simple-minded pigeon-holing of facts." The point at issue, however, is to establish just what evolutionary interpretation or meaning our groups have and how we can find this out. Evolutionary classification as interpreted by other distinguished exponents such as Simpson (1961) means something radically different from what Constance outlined, at least in theory. I shall attempt to resolve these difficulties later in this chapter.

II. TAXONOMIC CHARACTERS

In the construction of taxonomic groups we are concerned with *characters*. Characters are any attributes referring to form, structure or behaviour which the taxonomist separates from the whole organism for the purpose of comparison or interpretation (Davis and Heywood, 1963). Each organism affords the taxonomist thousands of potential characters, and although ideally all of them should be employed when constructing a classification, it is clear that a selection has to be made for purely practical reasons. A choice, therefore, has to be made: some characters are selected, some are rejected, some are overlooked, some are not understood, and so on.

Since different selections of characters from the same set of material may produce different classifications, the problem of making a *relevant* choice is most important. What criteria should be employed in making the selection of characters? Some of them are (1) the general convenience of the scientist, (2) availability or accessibility, (3) traditional usage, (4) experience, (5) biological commonsense.

Strictly speaking characters are abstract, and it is their expressions or states which we employ. A taxonomic character must be divisible into two or more states, and it is these states which we score. Some characters appear to be single, some are clearly compound; it is clearly necessary to distinguish between these in making comparisons, since their information content will be different.

Recently, and particularly in numerical taxonomy, attempts have been made to define single or unit characters. The definition given by Sneath and Sokal (1962) is "a taxonomic character of two or more states, which within the study at hand cannot be further subdivided." The aim is that each unit character will, as far as possible, contribute a new item of information (Sneath, 1964). There is, however, a great deal of variation in the skill and perception of different taxonomists in making the logical division of characters into units.

In the construction of taxonomic groups we are concerned with the correlation of characters, but different kinds of correlations exist. For the greatest predictive value we are normally concerned with the correlations of those characters that can also vary independently, but there are various reasons why characters may be *necessarily* correlated. There are mathematical correlates, single-cause correlates (Harrison, 1964), functional correlates, ecological correlates and so on. In the case of intrinsically correlated features, care must be taken not to score the same thing twice by assessing it in a different way. As Harrison has pointed out, one should not count the DNA code, then the RNA code, then the enzyme that this produces, then the products, and so on to the observable phenotype effect. On the other hand one cannot ignore the correlated characters in a functional complex when assessing overall resemblances between groups. Yet as Mayr (1964) has pointed out, taxonomists tend to shy away from characters with a low information content, and the more one breaks down a functional complex into unit characters, the lower the information content of any single characters of the complex. In the construction of monophyletic groups, however, necessarily correlated characters must be scored once only since if one of them arises through convergence, the others will arise as well.

There are great difficulties in recognizing necessary correlates and they may not become apparent until the groups have been constructed. The problems raised by the association of characters in suites or pleiades have recently been discussed by Olson (1964).

Having established that a selection of characters has to be made, we now have to consider two further points: (1) what classes of characters and how many should we select? (2) when we have chosen the characters, how do we treat them?

A. SOURCES OF CHARACTERS

Despite apparent disagreements over the purpose and interpretation of classifications, it is widely accepted that the most useful assessment of the overall relationships of organisms is obtained by using the largest possible number of similarities and differences. In the original definition of *phenetic* classification (Cain and Harrison, 1960) it was made clear that it was based on all available, observable characters (including genetic data), not just morphological ones. Thus phytochemical characters are included along with cytological, anatomical, palynological and other attributes.

Man is a visually gifted animal and he tends therefore to employ those characters for classification which are easily observed. This has led to an emphasis on the use of morphological characters, especially in higher plants. Apart from ease of observation, morphological characters can be seen in combination and the taxonomist is able, in many cases, to assess at a glance a wide range of features and character complexes. This mental assessment is virtually impossible to analyse but involves not only selection of some characters but rejection of others, all as part of a complex trial and error process. Recent studies have attempted to rationalize these various steps so that they can be analysed and distinguished and therefore repeated by other people. Attempts to define unit characters are precisely concerned with this.

Although it is undeniably convenient to be able to express classifications in morphological terms, any other kinds of character may be used in constructing a classification, and no attempt should be made to claim any special role for morphology on other than practical grounds. When dealing with non-visual characters we are unable to make mental assessments of the correlations that exist to anything like the same extent.

As a general principle every attempt should be made to make as random a selection of characters as possible—from different organ systems, from different sources, exomorphic and endomorphic, conspicuous and cryptic—so that they reflect as far as possible a random selection of the genotype. One advantage of choosing characters which are nearer the direct action of genes is that we are more likely to avoid scoring the same character with the same causation more than once, and also of avoiding features which are susceptible to environmental modification (Harrison, 1964).

It can be easily shown that the taxonomic groups that are produced by using a particular selection of characters from the plants under study may be quite different from those groups that we would arrive at by using a different selection of characters. Or, put another way, taxa that are very similar in respect of one set of characters may be quite different from each other in respect of other features. A further complication is that morphological or other conspicuous characters may not be correlated with endomorphic or cryptic characters so that one may well get different groupings by using different "kinds" of characters. One of the commonest uses of phytochemical data today is to see if they fit in with taxonomic groups based on other features. In many cases they do, in others they cut across them. If, however, we were able to start our classification from scratch, it is clear that by using a large number of chemical characters we might get a classification which was not expressible morphologically. The fact is that no attempts have yet been made to construct a phenetic classification using a wide range of characters from all possible sources. We might well be surprised at the results when we attempt this, and we are proposing to undertake a study along these lines at Liverpool.

The reaction of the taxonomist to new classes of data has been somewhat sceptical. "Better the devil we know . . ." is an attitude for which there is much to be said. But before we can continue this discussion of character selection it is necessary to consider how we are going to treat characters.

B. CHARACTER WEIGHTING

Cain and Harrison's original definition of phenetic included the phrase "based on all available characters *without any weighting*" (my italics). Equal weighting of characters has stirred up one of the greatest controversies of taxonomy in recent years and must be considered in the context of weighting as a whole.

To weight a character means to give it greater or lesser importance than other characters. In other words some characters are more important than others in the various activities of taxonomy. Even if characters are treated as of equal value when constructing a taxonomic group, some of them can be shown *after* the group is constructed to be of different value for, e.g., separating the group from others. Likewise, we can say that certain of the many characters employed to make a group are markers, and if they alone were used, the same groups would have resulted. Again, the very fact that we select characters means that we are giving them greater importance than those which we have consciously or unconsciously rejected. It follows that before we can decide which characters should be weighted, we must consider the various forms of weighting that are possible. The phrase equal weighting suggests that characters are used mechanically or indiscriminately, but as Mayr (1964) correctly points out, the masters of the taxonomic craft (as opposed to the bunglers) are virtuosi in the art of weighting.

There are two major kinds of weighting—*a priori* and *a posteriori* (Davis and Heywood, 1963; Mayr, 1964). *A priori* weighting is undertaken "because the character is believed to be of particular importance for reasons other than those deduced by correlation studies—very often because of its known or inferred phylogenetic importance, or its recognition as a 'conservative character'" (Davis and Heywood, 1963: p. 49). This is a deductive procedure and has a long history stemming from Aristotelian logic as practised by most taxonomists of the pre-Darwinian period and occasionally by most of us to the present day. Today few people would claim to know the relative physiological or functional importance of characters. Yet there are numerous examples of *a priori* weighting remaining in classifications which we have inherited from pre-Darwinian times.

Moreover, insistence on the primacy of morphological features in classification may sometimes stem from a belief that they are in some way more important rather than merely more practical. However, the commonest source of *a priori* weighting today is from newer lines of evidence, particularly cytology, anatomy and of course chemistry and serology. There is an innate tendency to half-believe that the more "internal" it is, the more importance it possesses. But what is new is not necessarily better. The classic case is that of chromosome number which in the early days of cytotaxonomy was hailed as a much more important attribute than, say, a morphological feature. There were numerous attempts to use chromosome numbers as an *a priori* basis for constructing groups—those plants possessing a particular chromosome number were placed in a separate group from those with a different number, often irrespective of the other features of the resultant groups. Frequently this was a successful procedure, but then it was, in fact, an example of the second type of weighting—*a posteriori*—as is discussed below. What happened usually was that after construction of the groups delimited by different chromosome numbers it was found that chromosome number showed a high correlation with the other features of the groups and was therefore a good marker character. Unfortunately there are many cases where this has been found *not* to be true and there are no serious grounds for believing that, *a priori*, chromosome numbers will make good marker characters. Similar considerations apply to any other class of taxonomic character.[1]

In the same way, phytochemical characters are frequently considered as more basic, fundamental or privileged than other classes of characters; there has, consequently, been a tendency in several instances to give them *a priori* weighting. This point is discussed at length by Davis and Heywood (1963: p. 232 *et seq.*). In general the value of chemical data in taxonomy stems from their stability, lack of ambiguity and resistance to change; that is, when these characteristics can be established! On the other hand, there is no reason to believe that such features are not susceptible to parallelism, convergence,

---

[1] Certainly there are reasons why cytological data are of special value in taxonomy—for purposes of comparison, chromosome number is one of the most constant features of plants available to the taxonomist. Most individuals of species have the same number although there are numerous exceptions. Because of this high degree of stability and because they can be shown, *a posteriori*, to correlate with other features in phenetic groups, chromosome numbers may be of great value in suggesting to us *where* to look for discontinuity of variation. From an interpretative viewpoint, chromosome number may be of exceptional use since the chromosomes are more closely connected with the hereditary mechanism than most other classes of character. We must not, however, confuse the construction of groups with their interpretation in phyletic or evolutionary terms.

inconsistency and all the other "defects" which the more traditional taxonomic characters show (cf. Erdtman, 1963).

At the level of what has recently been termed the "informational molecules", that is, the proteins and nucleic acids, there is some justification for considering chemical information as more basic. The selection of characters, as I mentioned previously, should attempt to reflect a selection of the genotype, and clearly the nearer one can get to directly sampling the genetic material, the firmer our attempts at selection should be. Since proteins are the immediate secondary products of the genetic material, and since the amino-acid sequence of a protein is a direct translation of the genetic information in the DNA which determined the sequence, it is not surprising that such prophets as F. C. Crick envisage a protein taxonomy—"the study of the amino-acid sequences of the proteins of organisms and the comparison of them between different species." This takes us into the field of future developments which I shall consider in the final section of this paper.

To summarize, there seem to be no grounds at the moment for treating any class of data as specially privileged and therefore deserving *a priori* weighting. Classifications based on such principles may by lucky chance prove satisfactory, but one certainly cannot generalize from individual successes, and we cannot proceed on such a hit or miss basis.

Characters may be given *a posteriori* or *correlation* weighting because it is found that they show the highest correlation with others in a group. Clearly we cannot know which these characters are until the groups have been made—correlation weighting is a form of being wise after the event. These correlations may be made by numerical means using computers, or by the taxonomist subconsciously. As Mayr (1964) says no electronic computer has yet been able to surpass the integrating capacity of an intelligent and experienced taxonomist. *A posteriori* weighting has been practised since Linnaean times, although botanists then usually professed to be employing *a priori* methods (Cain, 1959); it is, I believe, the only justifiable form of weighting. Unfortunately it does not constitute a procedure for the construction of taxonomic groups but tells us what characters are "good" ones in the sense that they help to diagnose or mark off the groups.

The experienced taxonomist often knows or can judge what characters are going to be useful to construct a group. In practice this means that in the light of his general knowledge of character distribution in other groups, of the known or assumed variation in certain types of character (through environmental modification, etc.), he makes a tentative delimitation of groups based on one or a few selected characters. If the groups so delimited are coherent, i.e. show good character correlations, the procedure has been successful and the characters initially chosen are given *a posteriori* weighting. This means attempting to decide *a priori* which characters are going to be *a posteriori* weighted! In a discussion of this problem Mayr (1964) notes that among the considerations on which the taxonomist bases his decisions are giving high weighting to characters that are rather constantly present in a given group and rather constantly absent from other groups. This is a principle which has been familiar to taxonomists for nearly two centuries—working out what characters are constant in a group so as to define it. But as Cain (1962) has pointed out, if characters have been singled out as important in natural (correlation) groups, there is no point in proclaiming that they must be important on principle, since the principle is seldom more than a restatement of the fact that they have already been found to be important.

To make the system work we have to consider *a posteriori* weighting as a form of model testing, since the value of characters, that is, their constancy, can only be tested after groups have been tentatively delimited (Mayr, 1964).

Model testing, one might say, is all very well but how does one set up the model? Clearly we cannot just proceed by the system of blind groping which de Candolle (1813) castigated, hoping that we will be lucky in our choice of characters. The answer is one that involves the principle of equal weighting. Equal weighting, as I have said, is a procedure which has provoked much argument (cf. Burtt, 1964, Meeuse, 1964, Kendrick, 1965). Mayr (1964) considers its claims along with *a priori* and *a posteriori* weighting and finds it unsatisfactory. He terms it "non-weighting" and suggests that the idea behind it is that "weighting of characters is such a high art and the majority of taxonomists so inept that it is better not to let any taxonomist do any weighting but entrust the whole job for determining the degree of similarity to an electronic computer." This is a fair example of the way in which the whole idea of equal weighting has been misrepresented although some numerical taxonomists have no doubt given grounds for such a sharp reaction.

The first point that has to be made is that equal weighting is undertaken only after certain other procedures have been carried out and that still other procedures are undertaken after similarity groups have been made. Equal weighting is regarded as a characteristic of neo-Adansonian taxonomy but, as I shall show, it also plays a vital role in non-mathematical approaches to classification. The best way to explain its role is to break down the whole procedure of group making into a series of steps:

(1) Because of the virtually unlimited number of characters that are in theory available, a selection of a limited number is made according to criteria mentioned earlier—ease of observation, availability, experience, etc. This may be considered *selection weighting*. At the same time certain other characters are rejected because they are known to be susceptible to environmental modification or otherwise judged to be unreliable. This form of *rejection* or *residual weighting* has the effect of giving greater importance to those characters which are chosen. Throughout this first stage, full use is made of our biological knowledge of the organisms and in no way can the choice of characters be considered automatic or objective. Deliberate attempts are made to see that the selection of characters is as representative as possible. Even if the ultimate treatment of the characters chosen is to be by numerical taxonomic means, this first step is undertaken by the taxonomist using all his skill and perception.

(2) If *a priori* weighting is rejected and if we accept, as we must, that *a*

posteriori weighting can only be given after groups have been constructed we have no choice but to treat all those characters which we have selected for use in step one as having equal weight when we attempt to work out which groups show maximum co-variation or similarity of characters.

If we attempt to make our groups neurally without the aid of machines, it is not entirely clear what the mental processes involved are. It is probably certain that we do not compare every character with every other simultaneously as Sokal (1963) has pointed out, and it seems that we engage in a complex series of model-testing with a great deal of back-tracking so that the groups we eventually come up with satisfy us as to their naturalness (i.e. co-variation). This process probably involves testing what kind of results we get by using certain features as markers, i.e. seeing what happens if we weight them *a posteriori*. A certain amount of irrational *a priori* weighting is also involved and the whole method is somewhat imprecise and can only be tested by the results.

If we employ numerical methods for this purpose, we are using the computer to do these sums for us, but more precisely and objectively. This will give us an assessment of overall similarity, in respect of the characters employed, in numerical terms. Of the various numerical approaches the CABOD method of Silvestri *et al.* (1963) appears to approximate closest to the mental operations carried out by the traditional taxonomist. Silvestri and Hill (1964) summarize the CABOD method as follows: " Each successive repartition of the individuals into groups is made by determining that character, or group of characters, with the highest content of information, which determination is carried out *ex novo* by the computer at each iteration. Thus, not only do the characters have different weights, but also each character changes its weight at each iteration. "

The phrase unweighted or equal weighted characters is somewhat misleading since, as Silvestri (1964) points out, if characters really had equal weight in a matrix no classification at all would result, or "expressed in a geometrical model, the elements would be dispersed in a homogeneous manner in a hyper-sphere with no possibility of drawing partitions between them." Silvestri suggests the expression "unweighted *a priori*" or that characters are used in an unprejudiced manner. This is a valid point as it draws attention to the fact that after the initial selection, the characters are not loaded before we work out the correlations between them. The correlation process, whether performed by the taxonomist or by the computer, works out possible weightings for us, allowing us to give *a posteriori* weighting to characters.

The reason for giving characters equal weighting is mainly that we simply do not know beforehand which characters should be given greater or less value and no logical system has been devised as a basis for so doing (Sneath and Sokal, 1962). Most of the objections to equal weighting appear either to confuse *a priori* and *a posteriori* weighting or are directed at the phylogenetic or evolutionary value of characters. Thus Mayr (1964) states that "the most serious criticism of the method of non-weighting is that it throws a great deal of information away and is therefore inefficient." By this he apparently means information about the "goodness" or "badness" of characters or about their evolutionary significance. But as I have already indicated, the former information ("goodness" or "badness") is taken into account when the initial selection of characters is made or is derived from groups *a posteriori*, whereas the latter (evolutionary significance) involves a special kind of *phyletic a priori* weighting which is out of place in a phenetic classification. Heslop-Harrison (1962) makes the point that phyletic weighting, such as Mayr appears to be recommending, has the paradoxical effect of making taxonomists adopt a more artificial basis for the delimitation of groups since it implies deliberately rejecting some of the information we possess about organisms! The explanation of this paradox is that phyletic or evolutionary classification is a separate activity from phenetic classification and depends on being able to assess what characters are *a priori* important in indicating various kinds of evolutionary relationship so that these characters can then be weighted in the construction or modification of a classification. This is considered further in the next section.

(3) When a classification has been obtained we should regard it as an hypothesis to be verified experimentally if possible (Silvestri, 1964). In practice this means, particularly in the case of numerical classifications, that the taxonomist should evaluate the results, consider if they make sense, go back to the organisms in nature and see if the proposed classification can be supported by further observation. One of the common procedures today is to test the classification by applying data from other sources which were not used in the construction of the classification. Comparative phytochemical data are being increasingly used for this purpose. When they support a classification there is general satisfaction—both the original work and the chemical approach appear to be vindicated. There may, however, be another explanation, depending on the nature of the chemical data involved. It is quite possible that the chemical characters chosen may simply reflect or be logically correlated with the diagnostic or marker features used to delimit the group. An example of such a possibility can be taken from the work by Bate-Smith and Metcalfe (1957) on the distribution of tannins in Angiosperms. As could have been predicted most tanniniferous families are largely woody or related to woody families while most of the non-tanniniferous families are herbaceous. Also, in 83 per cent of cases, the presence of tannins was found to coincide with the presence of leuco-anthocyanins. It was naturally to be considered whether the histological reaction for tannins was usually due to the presence of leuco-anthocyanins (Gibbs, 1958), and although the coincidence with woody habit is not exact, it seems that the tannins and leuco-anthocyanins *reflect* woodiness and cannot therefore be taken as a confirmation of Hutchinson's division of the Dicotyledones into woody and herbaceous stocks.

When phytochemical data conflict with an established classification, the taxonomist may well be advised to consider whether the pattern of distribution of the compounds in question indicates that a more satisfactory taxonomic

V. H. Heywood

grouping could be made. It is certainly not mandatory and usually the taxonomist would seek some further source of data which would reinforce the chemical
characters before considering an alteration of the classification.

III. EVOLUTIONARY CLASSIFICATION

The avowed aim of perhaps the majority of taxonomists is the production of
classifications that have an evolutionary meaning or content. "Phylogeny
provides intellectual vitality to taxonomy" (Alston and Turner, 1963). It is
only natural that biologists should expect their classifications to have more
significance than the mere accumulation of facts. The majority of classifications
do, but it is up to the taxonomist to find out what evolutionary interpretations
can be applied to them.

A fully evolutionary classification is incapable of achievement since no
classification can encompass all the dimensions that are involved. There are two
evolutionary relationships in classification that are often not distinguished:
patristic and *cladistic*. Patristic relationship refers to community of ancestry,
while cladistic relationship expresses the actual pathways by which the groups
evolved in time (Fig. 1). An important point to consider is that both these
evolutionary components as well as others that are involved in classification
such as rate of change, parallelism, convergence, degree of advancement or
primitiveness, are assessed by reference to phenetic (usually morphological or
anatomical) change. Fossil evidence which provides factual phylogenetic data
has to be evaluated in terms of phenetic resemblances and differences. The
extent to which a classification can be constructed following phylogenetic as
opposed to phenetic principles is extremely restricted. Features which are
known, or at least assumed, to indicate patristic relationship can be selected
as *a priori* marker characters but this is a procedure fraught with danger and
normally these phylogenetic markers are deduced from a previously formed
phenetic classification which is then modified in accord with this form of
weighting. It is probably true that the more avowedly evolutionary the
construction of a classification is, the less likely it is to be a generally satisfactory one. It is also, paradoxically enough, less likely to be as good a reflection
of evolutionary relationships as a phenetic classification. Likewise other
evolutionary criteria such as degree of advancement, rate of change, etc. are
calculated from within a classification.

It is highly probable that the resemblances shown by many of our taxonomic
groups are the consequence of the common ancestry of their component
members. If our groups have been well established and based on the correlation
of numerous features, a common genetic and evolutionary background is the
most likely explanation. All we have said so far is that we *believe* that phenetic
groups are in some way evolutionary. Evolutionary change is not always
divergent, however, but may be parallel or convergent, and in order to satisfy
ourselves that similarity is due to common ancestry we must remove from our
groups those features of similarity that are the result of such parallelism or

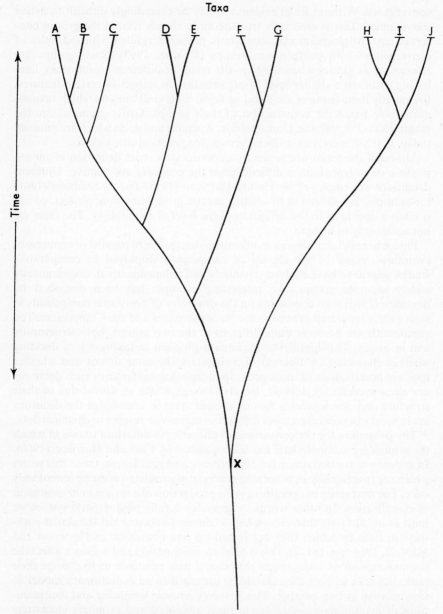

FIG. 1. Taxa A-J are all *patristically* related—they have a common ancestry through the stem at X. Each taxon shows a different degree of *cladistic* relationship with each other. The pairs A & B, D & E, F & G, H & I are separated by only one divergence in time before retracing common ancestry. Between A and G there are five divergences before they meet in a common stock.

convergence. Without fossil evidence it may be exceedingly difficult to detect convergence. This is especially true the more closely related the groups concerned are. Angiosperm classification is probably riddled with examples of convergence—with polyphyletic groups (Thorne, 1963). Such groups are referred to as grades when they are the result of different evolutionary lines having achieved a similar level of organization in respect of certain features. Frequently these features are floral or floral biological ones on which taxonomists have based the construction of their groups. Gross examples are the major taxa—Polypetalae, Gamopetalae, Apetalae and so on which are retained today, if at all, merely as artificial groups for practical convenience.

Although the terms are generally understood, a strict definition of monophyletic or polyphyletic is difficult since the concepts are relative. Different definitions are employed by Heslop-Harrison (1958) and by Simpson (1961) for example. The debate as to whether certain groups are monophyletic or not is often a debate as to the definition of the level of monophyly. The facts are not necessarily in dispute.

Phytochemical characters are, of course, susceptible to parallel or convergent evolution—many of the classes of compounds employed in comparative studies appear to have evolved separately and independently in taxonomically widely separate groups. An interesting example has been described by Sørensen (1963) who comments on the discovery of acetylenic compounds in such widely separated groups as the Basidiomycetes and the Compositae. The compounds are however quite different in the two groups, both structurally and in origin. This highlights a common problem in taxonomy of deciding whether characters in different organisms are the same or not and whether they are homologous or analogous. In morphological features such decisions are often exceedingly difficult, largely through a lack of knowledge of their structure and development. At our present state of knowledge the decisions are in most instances even more difficult to make with respect to chemical data.

The procedure for the conversion of phenetic classification to one in which the groups are monophyletic has been outlined by Cain and Harrison (1960). In addition to the correction for convergence and parallelism, those characters which are functionally or otherwise necessarily correlated must be scored only once. The next series of operations is the production of a *system* or *arrangement* of classification. In other words to approach a fully phylogenetic system we must know not only that groups had a common ancestor but the actual pathways in time by which they are joined up (see discussion in Heywood and McNeill, 1964: pp. 161–2). This is *cladistic* information and does not affect the circumscription of our groups although it may persuade us to change their rank, but tells us how they should be arranged in an evolutionary model or tree, relative to one another. This involves *phyletic* weighting and considerations of evolutionary sequences in time, advanced and primitive characters and a host of related problems which have been discussed by many authors (e.g. Cain and Harrison, 1960; Simpson, 1961; Sporne, 1956). Chemical characters, when we learn to understand them, may be of special value in estimating the probability of evolutionary sequences and patterns. They may be useful in *negating* certain proposed sequences. We are faced, however, with enormous problems, not only in obtaining chemical data and knowing what it is we are attempting to measure, but in working out biochemical or biogenetic pathways. The recent volume by Mentzer and Fatianoff (1964) is a valuable contribution in this area since it attempts to arrange chemical plant constituents according to established biosynthetic pathways. It is widely stated that it is the pathways that are more significant taxonomically than the distribution of the compounds concerned. This may be true in that it will permit us to understand these compounds as taxonomic characters, but I think that the main argument is that the chemical data are phylogenetically important. As Swain (1963) puts it, the biochemistry of evolutionary processes may be deduced from present-day organisms. It is quite clear that the main interest of phytochemists is to contribute to the construction of an evolutionary classification and that they are not interested simply in providing fodder for pigeon-holing in a "static" phenetic classification. Nevertheless, it is paradoxical that chemists who are engaged by their profession in a rigorous scientific discipline should wish their data to contribute to that aspect of classification which is most speculative! I would plead that they think carefully what is meant by "a truly phyletic" classification, to cite a commonly used phrase in phytochemical papers, and consider whether science is not better served by the improvement of existing phenetic classifications. After all, as indicated above, the greatest evolutionary content is probably to be found in well constructed phenetic classifications. Two quotations are pertinent here : "One must also be realistic enough to admit that the choice of phylogenetic markers, even when based on a lifetime of experience, is a matter of personal opinion" (Joysey, 1964). "Why cannot phylogenics, even phylogenetic trees, be considered under the heading of poetry, or visual aids, or metaphors, or analogies, or hypotheses, all devices designed to lead us to a better understanding? To condemn a given phylogeny as speculative is as inappropriate as to damn a poet or composer for being imaginative" (Constance, 1964). I think that these are fair comments, and it may well be as Burtt (1964) says, that those of us who misguidedly labour with the false ideal of an ideal phylogenetic classification before us may even be getting the best results. What we have to be sure about is that we *are* getting the best results and not diverting too much of our effort to following our phylogenetic fancies at the expense of sound phenetic work.

IV. Systems of Classification of Angiosperms

Current systems of classification of the Angiosperms provide phytochemists with a framework and background against which to operate. Frequent reference is made to some particular piece of chemical evidence confirming or supporting a particular phylogenetic arrangement of some groups. I feel that I ought to make it clear that by and large taxonomists do not take the conflicting claims of the different systems of classification too seriously on phylogenetic grounds. In practice we are much more concerned with problems of specific delimitation or, at the higher levels of classification, with the naturalness of groups which means whether they have been constructed on the basis of overall similarity and not arrived at by use of one or a few *a priori* weighted characters. We are not too concerned with whether the Paeoniaceae has been derived from one group or another, or whether one places it next to the Ranunculaceae or the Dilleniaceae, so much as with the decision whether it ought to be a separate family or not. Even if a new system of classification were to be proposed which satisfied most taxonomists as to its plausibility, its value would be limited. It would be useful for teaching, for mental recreation, but not for the way in which our collections should be arranged or the sequence in which our Floras should be written. Nor would it be of much use for predictive purposes if the groups were not soundly based on phenetic principles.

In practice many so-called phylogenetic arrangements are largely the taxonomist's expression of his impressions, derived from various sources, of overall relationship between groups, as much as the result of the application of genuinely evolutionary dicta. Indeed it is when some line of evidence is over-stressed as an evolutionary principle (such as the woody versus herbaceous habit by Hutchinson) at the expense of other evidence that the resulting system is in many ways absurd.

It should be remembered that most of the higher taxa in our systems have been inherited from former generations and that many of them date from pre-evolutionary times. We are not able for practical reasons to ignore our inheritance and start the whole process from scratch again; and even if we did there is no doubt in my mind that we would aim at a phenetic system, perhaps not much different in many cases from those we already possess. Curiously enough, were we to be faced with such a situation, the sheer bulk of information would force us to use some numerical means of assessment. We are not otherwise equipped for the task. As it is, our task is to revise, modify and improve our classifications so that they not only store a large amount of information but form a good starting point for evolutionary interpretation.

V. Communication and Terminology

In the taxonomic employment of comparative phytochemistry there must be a two-way communication of information and ideas. The phytochemist must familiarize himself with some of the special jargon of the taxonomist and vice versa. Although the chemist may often complain about taxonomic terminology and nomenclature, it is only fair to point out that taxonomy is largely designed for use by other people, and in fact a great deal of effort is spent in making taxonomy usable and comprehensible by various classes of consumer, from school children upwards. On practical grounds, an internationally agreed code of nomenclature is employed by taxonomists, but this does not guarantee the use of correct nomenclature in a particular instance which involves us in the interpretation of synonymy.

Mirov (1963) has written an amusing paper on this topic and cites several examples of the failure by chemists to employ comprehensible nomenclature for their samples, so that it is impossible to tell what it was that they worked on.

As a taxonomist may I enter a plea for greater consideration from the chemist. Already we are faced with having to understand the ideas, concepts and terms of the anatomist, cytologist, geneticist, ecologist, palynologist, statistician, evolutionist and so on as well as mastering our own descriptive and conceptual vocabulary; but it is a new hazard to find ourselves faced with the abbreviations and other special shorthand of chemistry. Could we not, perhaps, devise an acceptable chemical classification of compounds, involving some form of hierarchical system, so that the taxonomist has a fighting chance of orientating himself before delving into the detail of papers? Otherwise I fear we may be at school till we are forty before we are sufficiently trained to understand each other. Metcalfe (1963) has discussed the problems that result from the development of new languages by specialists. He points out that often a specialist in one field may not be willing to admit his ignorance of technical terms when discussing a problem in a field outside his province—leading to a lack of mutual understanding of the issues involved.

VI. PHYTOCHEMISTRY AND TAXONOMY: TODAY AND TOMORROW

Most of this paper has been concerned with the principles which should be followed when employing various classes of data, and in particular phyto-chemical ones, in classification. I have not attempted to discuss detailed examples of chemotaxonomy or assess the present-day standing of such studies. A few years ago such a survey would have been possible in a single paper but the expansion of the field since then has been spectacular, and already several textbooks on the subject are available. A few observations seem, however, worth making.

It is perhaps unfortunate, although inevitable, that descriptions such as chemotaxonomy or chemical taxonomy should be employed since they give the impression that there is an approach to taxonomy in which chemical data are more important than are other classes of data. Similar objections apply to the use of the term cytotaxonomy. It is perfectly true that individual workers specialize in the production of chemical, cytological, or other kinds of data for use in classification, but as I have tried to show above, when it comes to actual construction of the classification it is the sum total of characters from all sources that has to be considered and there are quite complex processes to be applied to decide upon the weight to be given to individual characters, irrespective of their source. Moreover, the limited and sporadic nature of chemical data available today normally means that they can seldom be employed except by application to an existing classification arrived at by the use of other data. We must look forward to the time when chemical data can be more regularly used and this involves not only much further analytical work but collation of what work has already been done. Hegnauer's *Chemotaxonomie der Pflanzen* (1962–1966) is an important contribution in this direction and both taxono-mists and chemists will long be grateful to him for this outstanding work.

Obviously an enormous labour is required to work out the distribution of the various kinds of compounds throughout the plant kingdom (cf. Metcalfe,

1963, on the situation in respect of comparative plant anatomy). It is highly unlikely that more than a small fraction of this will ever be achieved in the light of scientific policy decisions if for no other reason. At the moment what is studied often depends on the development of techniques which are successful with certain kinds of plant material or on the selection of products which have economic or pharmacological importance.

Apart from such general considerations, it has to be remembered that phytochemical data will be judged by taxonomists in relation to the depth of knowledge of other kinds of data, particularly morphological. The latter are normally confirmed and tested hundreds or thousands of times by direct observations in the herbarium, field or experimental garden: their variability is fairly well known in a high proportion of cases although a great deal of the taxonomist's work is devoted to extending his knowledge of the variation and variability of morphological characters through which his classifications find expression.

In very broad terms morphological characters of angiosperms are fairly well known, at least from a descriptive point of view, although there are still surprising gaps in our knowledge (cf. Davis and Heywood, 1963: chapter 5). Karyology is still in its infancy—literally millions of chromosome counts are needed to establish a sound foundation in the angiosperms although the general pattern of chromosome number variation has become clear and much is known about the other ways in which chromosomes vary. Similarly in fields such as anatomy or palynology the outlines are clear enough and the details are known in an increasing number of cases. But in comparative phyto-chemistry we do not yet know all the major features let alone the details.

Much phytochemical work is needlessly wasted through inadequate attention to taxonomic requirements. For example identification of the material studied may be uncertain and the nomenclature employed incomprehensible. It cannot be stressed too often that accurate identification is essential; voucher specimens should be kept and an indication given as to the source of identification. Also the plants studied should be collected in the wild or at least raised from spontaneous seed and not of botanic garden origin, since such material is often of unknown provenance and hybridization in cultivation frequently occurs. Both the phytochemist and the taxonomists have their own sets of criteria that must be satisfied. Close co-operation between them is needed and the best results are likely to be derived from research teams composed of specialists from various disciplines.

REFERENCES

Alston, R. E. and Turner, B. L. (1963). "Biochemical Systematics". Prentice Hall, New Jersey.
Bate-Smith, E. C. and Metcalfe, C. R. (1957). *J. Linn. Soc. (Bot.)* **55**, 669.
Blackwelder, R. E. (1964). *In* "Phenetic and Phylogenetic Classification" (V. H. Heywood and J. McNeill, eds.) (Syst. Ass. Publ. No. 6), p. 17. London.

Burtt, B. L. (1964). *In* "Phenetic and Phylogenetic Classification" (V. H. Heywood and J. McNeill, eds.) (Sysst. Ass. Publ. No. 6), p. 5. London.

Cain, A. J. (1959) *Proc. Linn. Soc.* **170**, 233.

Cain, A. J. (1962). *Symp. Soc. gen. Microbiol.* **12**, 1.

Cain, A. J. and Harrison, G. A. (1960). *Proc. zool. Soc.* (*Lond.*) **135**, 1.

Constance, L. (1964). *Taxon* **13**, 257.

Davis, P. H. and Heywood, V. H. (1963). "Principles of Angiosperm Taxonomy", Oliver & Boyd. Edinburgh and London.

Edwards, A. W. F. and Cavalli-Sforza, L. L. (1964). *In* "Phenetic and Phylogenetic Classification" (V. H. Heywood and J. McNeill, eds.) (Syst. Ass. Publ. No. 6), p. 67. London.

Erdtman, H. (1963). *Pure appl. Chem.* **6**, 679.

Fosberg, F. R. and Sachet, M.-H. (1961). *Regn. Veget.* **39**, 1961.

Gibbs, R. D. (1958). *Proc. Linn. Soc.* **169**, 216.

Grant, V. (1963). "The Origin of Adaptations". Columbia Univ. Press, New York.

Harrison, G. A. (1964). *In* "Phenetic and Phylogenetic Classification" (V. H. Heywood and J. McNeill, eds.) (Syst. Ass. Publ. No. 6), p. 161. London.

Hegnauer, R. (1962–1966). "Chemotaxanomie der Pflanzen". Vols. I–IV, Birkhauser, Basel.

Heslop-Harrison, J. (1958). *Phytomorphology*, **8**, 177.

Heslop-Harrison, J. (1962) *Symp. Soc. gen. Microbiol.* **12**, 14.

Heywood, V. H. and McNeill, J. (1964) (eds.). "Phenetic and Phylogenetic Classification" (Syst. Ass. Publ. No. 6). London.

Joysey, K. A. (1964). *In* "Phenetic and Phylogenetic Classification" (V. H. Heywood and J. McNeill, eds.) (Syst. Ass. Publ. No. 6), p. 37. London.

Kendrick, W. B. (1965). *Taxon* **14**, 141.

Mackerras, I. M. (1964). *Proc. Linn. Soc. N.S.W.* **88**, 324.

Mayr, E. (1964). *In* "Taxonomic Biochemistry and Serology" (C. A. Leone, ed.). New York.

Meeuse, A. D. J. (1964). *In* "Phenetic and Phylogenetic Classification" (V. H. Heywood and J. McNeill, eds.) (Syst. Ass. Publ. No. 6), p. 115. London.

Mentzer, C. H. and Fatianoff, O. (1964). "Actualités de Phytochimie Fondamentale". Paris.

Metcalfe, C. R. (1963). *Advanc. bot. Res.* **1**, 101.

Michener, C. D. (1963). *Syst-Zool.* **12**, 151.

Mirov, N. T. (1963). *Lloydia* **26**, 117.

Olson, E. C. (1964). *In* "Phenetic and Phylogenetic Classification" (V. H. Heywood and J. McNeill, eds.) (Syst. Ass. Publ. No. 6), p. 123. London.

Silvestri, L. G. (1964). *In* "Phenetic and Phylogenetic Classification" (V. H. Heywood and J. McNeill, eds.) (Syst. Ass. Publ. No. 6), p. 81. London.

Silvestri, L. G. and Hill, L. R. (1964). *In* "Phenetic and Phylogenetic Classification" (V. H. Heywood and J. McNeill, eds.) (Syst. Ass. Publ. No. 6), p. 87. London.

Silvestri, L. G., Hill, L. R. and Möller, F. (1963). *Conv. naz. biofis.* (*Roma*) **1**.

Simpson, G. G. (1961). "Principles of Animal Taxonomy". New York and London.

Sneath, P. H. A. (1964). *In* "Phenetic and Phylogenetic Classification" (V. H. Heywood and J. McNeill, eds.) (Syst. Ass. Publ. No. 6), p. 43. London.

Sneath, P. H. A. and Sokal, R. R. (1962). *Nature, Lond.* **193**, 855.

Sokal, R. R. (1963). *Taxon* **12**, 190.

20 *V. H. Heywood*

Sørensen, N. A. (1963). *In* "Chemical Plant Taxonomy". (T. Swain, ed.). Academic Press, London and New York.
Sporne, K. R. (1956). *Biol. Rev.* **31**, 1.
Stafleu, F. (1965). *Taxon* **14**, 69.
Swain, T. (1963). *In* "Chemical Plant Taxonomy". (T Swain, ed.). Academic Press, New York and London.
Thorne, R.F. (1963). *Amer. Nat.* **97**, 287

CHAPTER 2

Biogenetic Classification of Plant Constituents

C. MENTZER

Muséum National d'Histoire Naturelle, Paris

I. INTRODUCTION

The chemistry of natural products is one of the oldest scientific disciplines. It has been completely revolutionized, however, by the most recent discoveries in the fields of physics and biology, and its present rate of progress is so rapid that it has become necessary to classify new findings as they are published into a logical system which can promote further development. The problem of classification arises above all in connection with compounds described for the first time; the rate of discovery of such products is now twice as fast as it was five years ago.

It should be pointed out that all workers who are interested in plants do not have the same needs. Some are guided mainly by chemotaxonomic considerations, and for them, the accepted botanical classification may be of most use. This is the basis of the well-known book of Wehmer (1928), as well as the more recent works by Hegnauer (1962–64) and Kariyone (1959, *et seq.*). Pharmacists, and in general all those who are concerned with the applications of drugs, prefer a physiological or cytological classification. The organic chemist, to whom the plant is a laboratory capable of synthesizing what are to him elaborate new molecular structures, will be satisfied with a classification according to chemical function. A fourth method of classification, known as "biogenetic classification", has enjoyed a particularly rapid development during the past few years. This method is based on the natural relationships between the various plant constituents, and its progress is closely associated with the biogenetic theories built up around the basic ideas advanced by Robinson, Ruzicka, Birch, and Davis.

21

This generalized biogenetic system of classification was first proposed in 1954 (Mentzer, 1954). Since then, ideas in this field have become increasingly definite, finally culminating in the system summarized in the first volume of "Actualités de Phytochimie Fondamentale" (Mentzer and Fatianoff, 1964). The aim of the following discussion is to emphasize a number of advantages of biogenetic classification, and to show, with the aid of some recently published results, how it can add to our understanding of phytochemistry as a whole.

II. CLASSIFICATION ON THE BASIS OF CHEMICAL STRUCTURE

A work which has been of the greatest service to chemists for a number of years is undoubtedly the outstanding treatise by Karrer (1958), which covers nearly every known organic substance discovered in the plant kingdom before 1956, with the exception of the alkaloids. The only criticism which, in my opinion, can be levelled at this work concerns the manner in which the substances described are classified. On the whole, the classification is predominately based on chemical structure. Nevertheless, certain chapters devoted to phyto-hormones, vitamins, and so on, conform to a physiological classification, while others are obviously based on the most recent biogenetic data; e.g. in the case of the tropolones, the terpenes, and the steroids.

The lack of uniformity in the methods of classification sometimes leads to unexpected results: thus in the chapter on "Alcohols, phenols, naphthols" (Karrer, 1958), solanesol (I), with a $(C_5)_n$ molecule, appears alongside dodecandiol (II), which belongs to the $(C_2)_n$ group.

Similarly the aurones, which should be included in the $(C_6C_3C_6)$ family together with the flavones, and other related compounds, are dealt with in

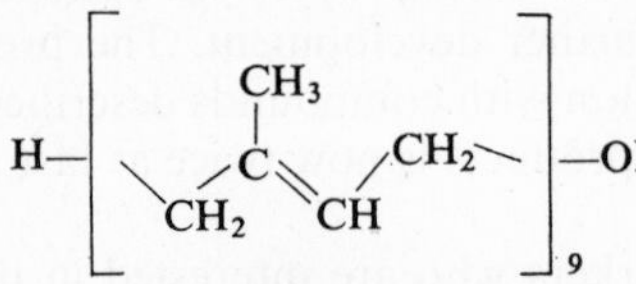

I. Solanesol

II. 1,12-Dodecandiol

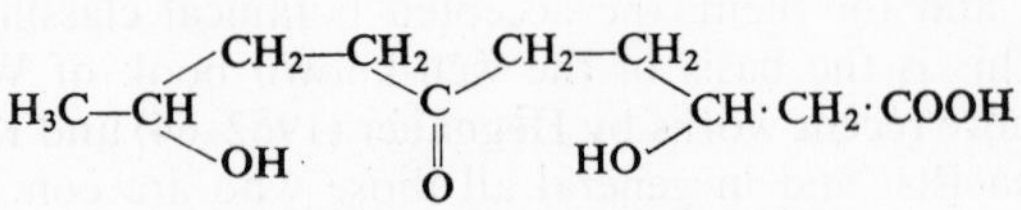

III. 3,9-Dihydroxy-6-ketodecanoic acid

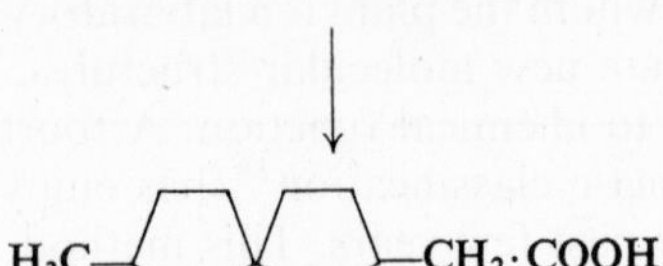

IV. Exogonic acid

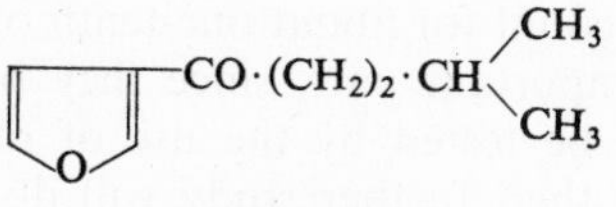

$$-CO \cdot (CH_2)_2 \cdot CH \begin{smallmatrix} CH_3 \\ CH_3 \end{smallmatrix}$$

V. Perillacetone

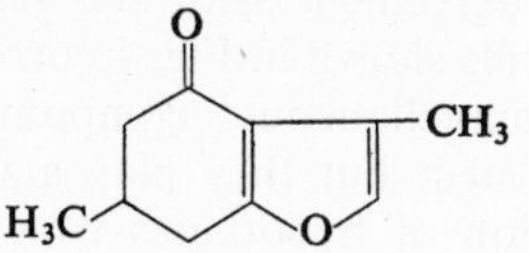

VI. Evodone

the chapter devoted to furan derivatives. The furan ring in itself, however, is of little interest either in chemotaxonomy or in the plan of biogenetic relationships, since it is found not only in shikimic acid derived compounds, but also in $(C_2)_n$ molecules such as exogonic acid (IV), which is derived from the hydrate of 3,9-dihydroxy-6-keto-decanoic acid (III) by double dehydration. It can also be formed from $(C_5)_n$ molecules, as in perillacetone (V) and evodone (VI).

III. BIOGENETIC CLASSIFICATION

To avoid disadvantages of this type, I suggest that phytochemists should whenever possible classify compounds according to their biogenetic relationship. Such a classification, moreover, offers advantages not only to the biologist who needs to understand the origin of plant products, but also to the chemist whose job it is to elucidate their structure.

At present, for example, it is possible to distinguish three large classes of substances: the primary or basic constituents, the secondary constituents, and "miscellaneous" substances. These classes consist of various groups which can themselves be divided into families, sub-families, sections and so on.

At present, the alkaloids do not readily fit into this classification, but they could fairly easily be included at a later stage: thus, as was pointed out some time ago by Robinson, some are derived from basic constituents (e.g. amino acids), while others are derived from $(C_5)_n$ molecules (steroidal alkaloids) and still others from the shikimic acid group $(C_6C_3)_n$, etc.

Macromolecules fall mainly into the group of basic constituents (proteins, nucleic acid derivatives, chlorophylls, and polysaccharides) although some, like the lignins, are derived from the shikimic acid group, or more particularly from the $(C_6C_3)_n$ family of compounds.

The formulae of any compounds which cannot be explained in the light of the current biogenetic theory, and which appear at present in the "miscellaneous" class, should *a priori* be regarded with suspicion. In many cases their structures are inaccurate and here the application of our knowledge of the natural relationships may set the research worker on the path to the true structure. Thus the perinaphthenone pigments do not correspond to the structural type (X), as suggested by Barton *et al.* (1959), but are related to the skeleton

(IX), in agreement with the results obtained by Paul *et al.* (1962). The mechanisms shown in Fig. 1 correspond to these views.

The "miscellaneous" compounds account for about one-tenth of the substances listed, but they play a very important part, since they permit the formulation of hypotheses which can be tested by the use of radioactive precursors. It is likely, therefore, that their further study will do much to promote progress in phytochemistry.

It should not, however, be imagined that biogenetic classification has no disadvantages. Several perfectly justifiable criticisms have been directed at it.

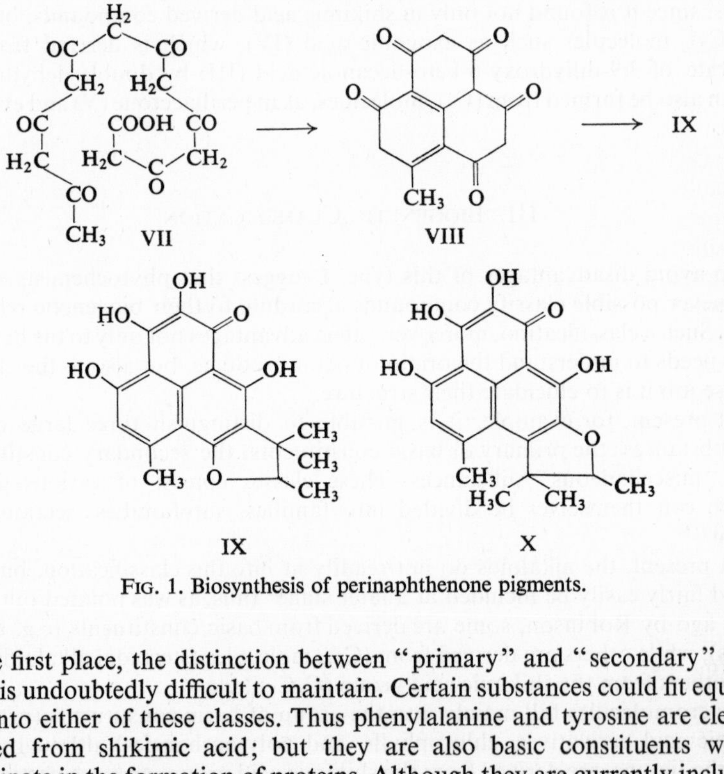

FIG. 1. Biosynthesis of perinaphthenone pigments.

In the first place, the distinction between "primary" and "secondary" products is undoubtedly difficult to maintain. Certain substances could fit equally well into either of these classes. Thus phenylalanine and tyrosine are clearly derived from shikimic acid; but they are also basic constituents which participate in the formation of proteins. Although they are currently included in the group of amino acids (Mentzer and Fatianoff, 1964), this is purely a matter of convention, since we did not wish to break up this group which forms a fairly homogeneous unit in the plan of dynamic biochemistry. The $(C_2)_n$ molecules present similar problems, since some of these (in particular the fatty acids) are also basic constituents.

Certain groups of substances may be derived from two or more different precursors, depending on the organism in question. Thus the benzoic acid derivatives may be formed either from shikimic acid, cinnamic acids or acetic acid, but the nature and position of substituents on the aromatic nucleus frequently permit the elimination of all but one pathway. In several cases, (that of orsellinic acid (XI) for example) the question has finally been answered by the use of ^{14}C acetate. The precursor is a poly-β-ketonic acid, which cyclizes in accordance with the scheme shown in Fig. 2.

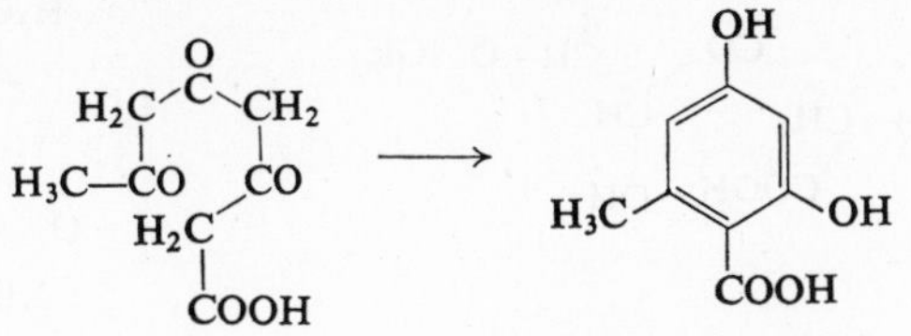

XI. Orsellinic acid

FIG. 2. Biogenesis of orsellinic acid.

Several different biochemical mechanisms are sometimes involved in the formation of certain compounds. These compounds should in future be classed in separate sub-groups. They are hybrid molecules. Thus the $(C_2)_n$ group contains substances derived by "crossing" a $(C_2)_n$ with a $(C_3)_n$; and even by crossing a $(C_2)_n$ with a $(C_m)_n$ (where $m > 3$); these are mainly branched chain compounds, very often found in micro-organisms. In the (C_6C_3), $(C_6C_3C_6)$, etc. families, (C_5) units are often grafted onto either a ring carbon or an oxygen. Many coumarins, chromones, flavones, and xanthones are the result of such hybridizations.

IV. RECENT EXAMPLES OF BIOGENETIC CLASSIFICATION

Let us now consider a number of structures which have recently been elucidated, and which fit readily into a biogenetic classification.

A. PRIMARY CONSTITUENTS

Phorbic acid (XII), which was described by Bernatek *et al.* (1963), is formed by a mechanism related to the tricarboxylic acid cycle. The aldolization by

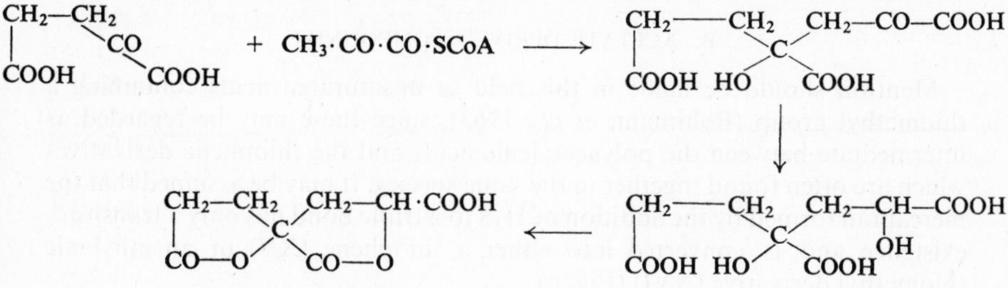

XII. Phorbic acid

FIG. 3. Biosynthesis of phorbic acid.

which its skeleton is formed is of the citric acid type, as shown in Fig. 3. The structure of piptoside (XIII) is related to that of phorbic acid, but Riggs and Stevens (1963) believe that it is formed by a Michael type of reaction (Fig. 4). In this case the problem will remain unanswered until experimental

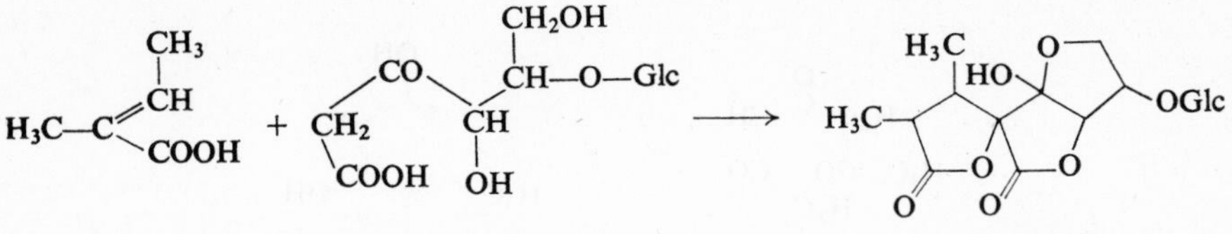

XIII. Piptoside

FIG. 4. Biosynthesis of piptoside.

evidence can be found by the use of ^{14}C precursors. Avenaciolide (XIV), which is an antifungal lactone synthesized by *Aspergillus avenaceus* (Brookes *et al.*, 1963), probably owes its formation to a chain of reactions starting with the condensation between a molecule of acetyl coenzyme A and a β-keto acid (Fig. 5).

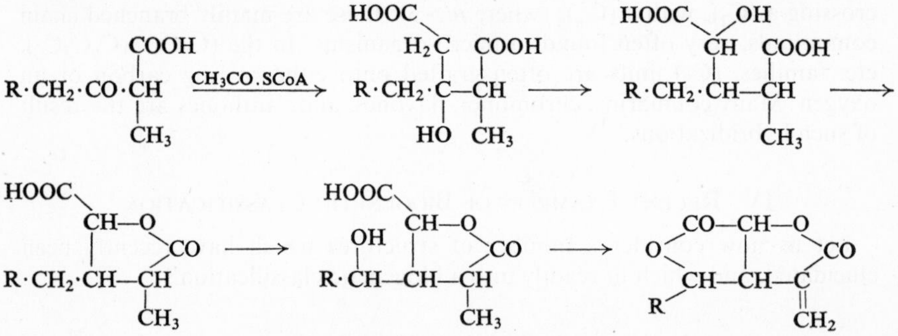

XIV. Avenaciolide

FIG. 5. Biosynthesis of avenaciolide.

B. ACETATE-DERIVED COMPOUNDS

Mention should be made in this field of unsaturated acids containing a thiomethyl group (Bohlmann *et al.*, 1963), since these may be regarded as intermediate between the polyacetylenic acids and the thiophene derivatives which are often found together in the same species. It may be assumed that the mercaptan formed by the addition of H_2S to a triple bond has only a transitory existence and is converted into either a thiophene (XV) or an ethylenic thiomethyl derivative (XVI) (Fig. 6).

Ramulosine (XVII), which is isolated from *Pestallozzia ramulosa* (Benjamin and Stodola, 1960), is one of the few natural isocoumarins in which the ring A

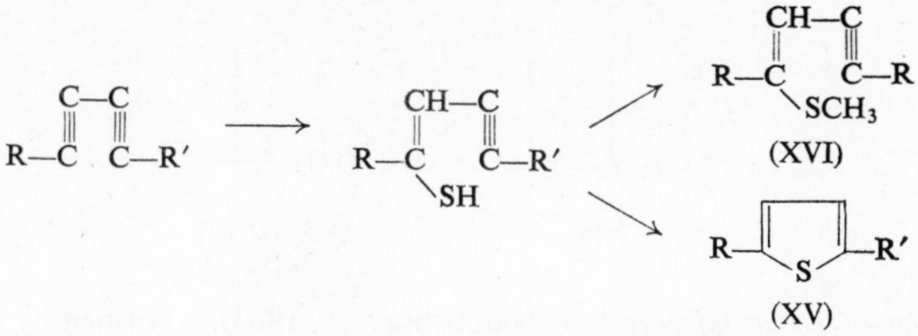

CH—C
‖ ‖
R—C   C—R
\SCH₃

(XVI)

C—C   CH—C
‖ ‖   ⟶   ‖ ‖
R—C C—R′   R—C   C—R′
                 \SH

R—⟋‾‾⟍—R′
   \S⁄

(XV)

FIG. 6. Biosynthesis of thiophene acetylenic acids.

has been almost completely hydrogenated. Citromycetin (XVIII), according to Birch, results from the fusion of two poly-β-ketone chains (Birch *et al.*, 1964), so that this compound can now be easily assigned to the heterocyclic $(C_2)_n$ group (Fig. 7).

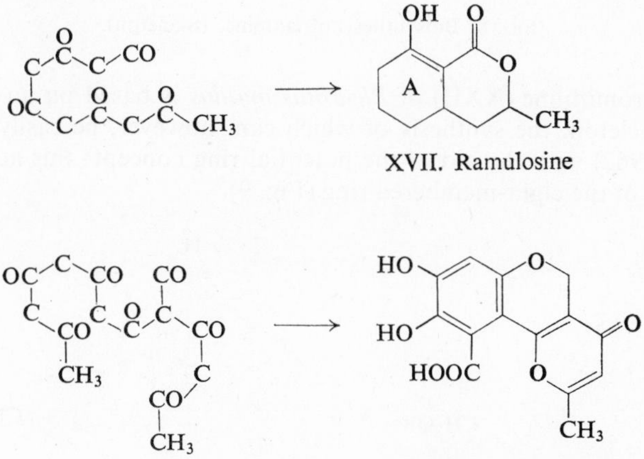

XVII. Ramulosine

XXVIII. Citromycetin

FIG. 7. Biosynthesis of ramulosine and citromycetin.

C. ISOPRENOID COMPOUNDS

Certain substances in the terpene series present difficulties either because of abnormalities in the linkage of the structural units, or because the number of carbon atoms is not a multiple of five. This is the case for example in a C_{14} compound (XIX) obtained from a species of Thymelaeaceae of the genus *Aquilaria*. Since Bhattacharyya (Maheshwari *et al.*, 1963) has isolated the sesquiterpene (XX) from the same species, it may be assumed that the molecule (XIX) results from the oxidation of (XX) by a mechanism similar to that used, for instance, to explain the formation of estrone by the cleavage of a side chain attached to C_{17} in a steroid. Taxinine (XXI), a diterpene from

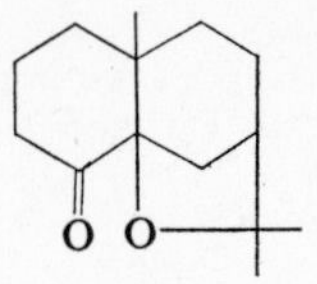

XIX 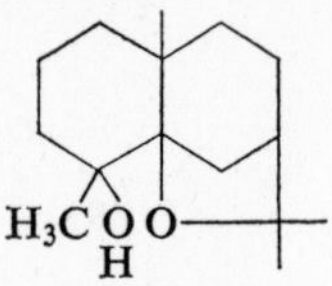   XX

the leaves of *Taxus baccata* L. (Nakanishi *et al.*, 1963), is formed by a rather unexpected rearrangement of four C_5 units which do not normally link up in the manner shown in Fig. 8.

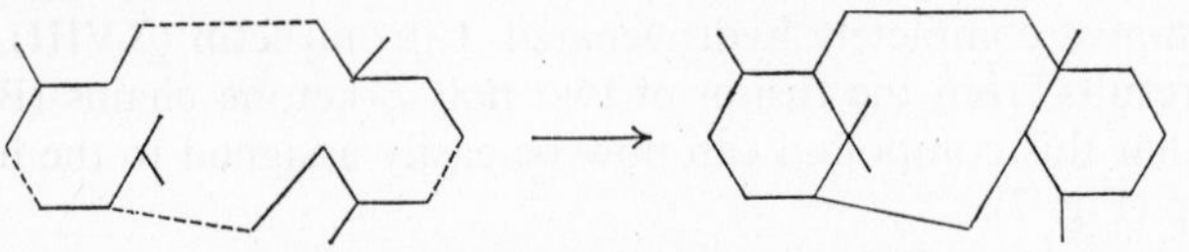

XXI

FIG. 8. Biosynthesis of taxinine. (Skeleton).

The pleuromutiline (XXII) of *Pleurotus mutilus* is based on an even more complex skeleton, the synthesis of which can, however, be easily explained (Arigoni, 1962) with the aid of the potential ring concept; this accounts for the closure of the eight-membered ring (Fig. 9).

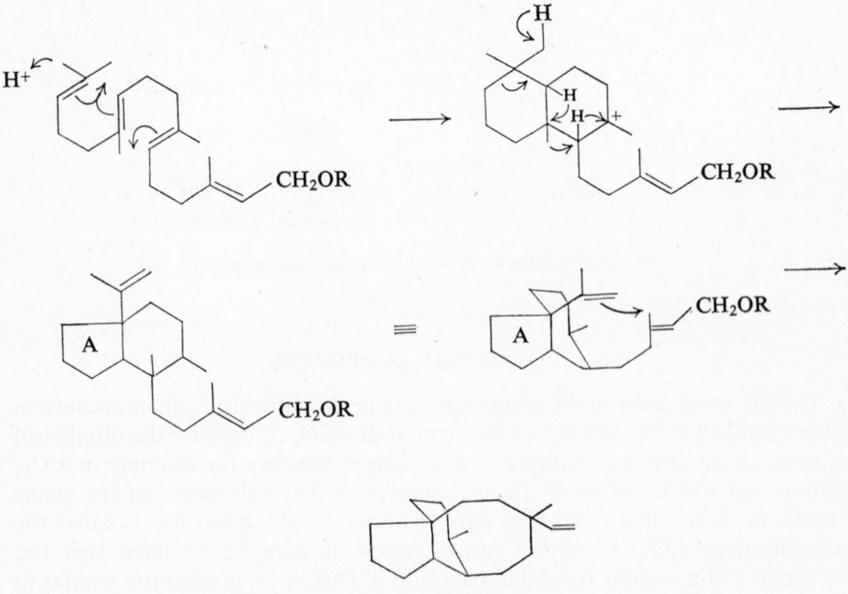

XXII

FIG. 9. Biosynthesis of the pleuromutiline skeleton.

D. COMPOUNDS DERIVED FROM SHIKIMIC ACID

The origin of the 5-membered ring in the furanocoumarins has until now been difficult to explain, but, as a result of the elucidation of the structure of calcicolin (XXIII) by Nakabayashi (1962), the synthetic chain shown in Fig. 10 can now be proposed. Since calcicolin (XXIII) can be converted *in*

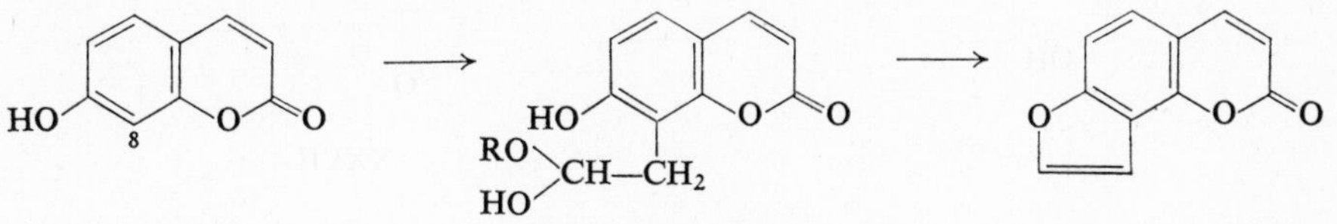

XXIII. Calcicolin   XXIV. Angelicin

FIG. 10. Biosynthesis of the furanocoumarins.

vitro into angelicin (XXIV), which is found in the same plant (*Angelica saxicola*), it may be assumed that a similar transformation is possible *in vivo*. The acetal chain in (XXIII) could result from an attack by serine in position 8 of the coumarin, followed by deamination, decarboxylation, and finally acetal formation.

Mention should also be made of the relationships between the benzylidenepyrones, the stilbenes, the 2-arylbenzofurans, the isocoumarins, and the flavone derivatives. These families may be regarded as hybrids of (C_6C_3) molecules with (C_2) units. The difference between the various families lies in the number of (C_2) units added to the (C_6C_3) precursor, and sometimes also in the form of the pseudocyclic precursor of the skeleton in question.

The benzylidenepyrones (XXV) are of the type ($C_6C_3 + 2C_2$) (Fig. 11). The type ($C_6C_3 + 3C_2$), on the other hand, accounts for the formation of the

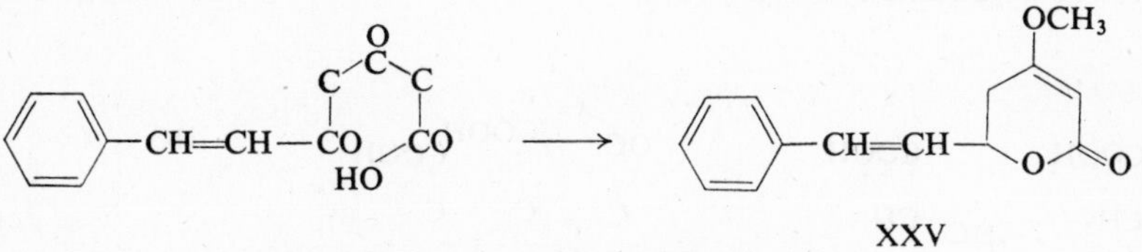

XXV

FIG. 11. Biosynthesis of benzylidenepyrones.

stilbenes (XXVI), the 2-arylbenzofurans (XXVII), the 3-phenylisocoumarins (XXVIII), and the flavone derivatives (XXIX) (Fig. 12).

Finally, the origin of dalbergin (XXXI) and its analogues, which are 4-arylcoumarins (Ahluwalia *et al.*, 1959), is still doubtful. It may, however, be

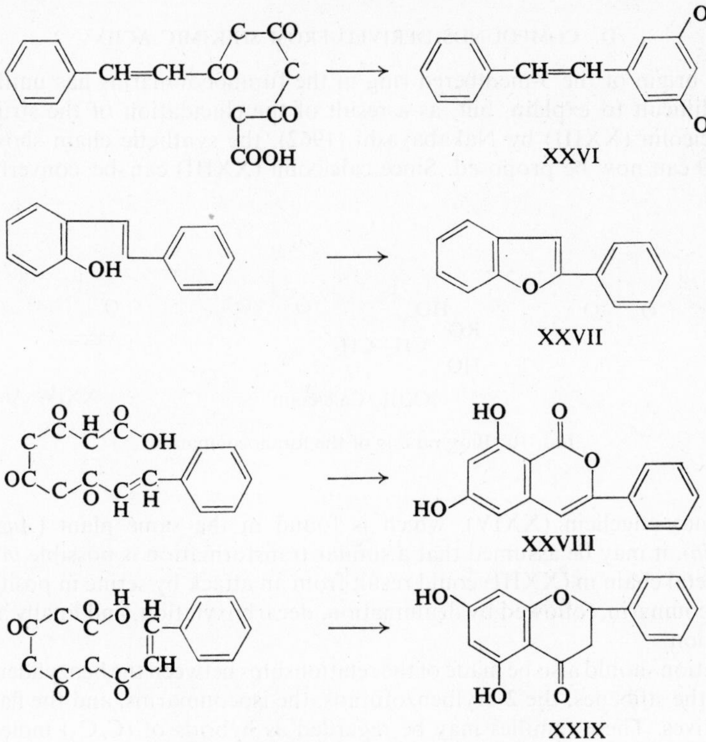

Fig. 12. Biosynthesis of C-15 compounds.

tentatively assumed that they result from a Michael type of reaction involving the addition of malonic acid to a cinnamic acid. The precursor (XXX) could be formed by decarboxylation and extension of the poly-β-ketone chain, followed by double cyclodehydration to form (XXXI) (Fig. 13).

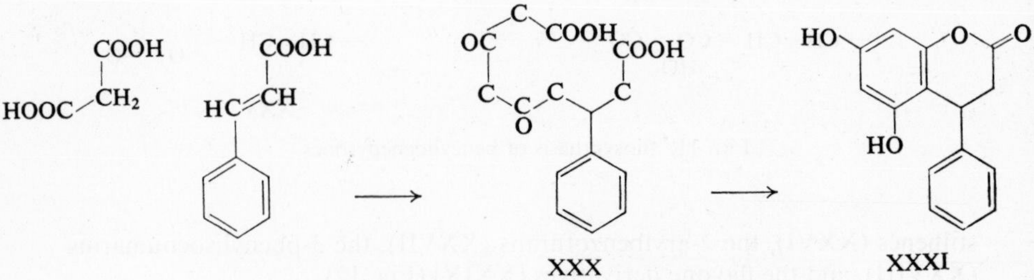

Fig. 13. Suggested mode of biosynthesis of 4-phenylcoumarins.

REFERENCES

Ahluwalia, V. K., Rustogi, C. L. and Seshadri, T. R. (1959). *Proc. Indian Acad. Sci.* **49A**, 104.

Arigoni, D. (1962). *Gazz. chim.* **92**, 884.

Barton, D. H. R., de Mayo, P., Morrison, G. A. and Raistrick, H. (1959). *Tetrahedron* **6**, 48.

Benjamin, C. R. and Stodola, F. H. (1960). *Nature, Lond.* **188**, 662.

Bernatek, E., Nordal, A. and Ogner, G. (1963). *Medd. Norsk Farm. Selskap* **25**, 77 (*Chem. Abstr.* (1963), 15, 602e.

Birch, A. J., Hussain, S. F. and Rickards, R. W. (1964). *J. chem. Soc.* 3494.

Bohlmann, F., Arndt, C., Bornowski, H. and Kleine, K. M. (1963). *Chem. Ber.* **96**, 1485.

Brookes, D., Tidd, B. K. and Turner, W. B. (1963). *J. chem. Soc.* 5385.

Cunningham, J., Haslam, E. and Haworth, R. D. (1963). *J. chem. Soc.* 2875.

Hegnauer, R. (1962, 1963 and 1964). "Chemotaxonomie der Pflanzen", 3 Vols. Birkhäuser, Basel and Stuttgart.

Kariyone, T. (1959 *et seq.*). "Annual Index of the Reports on Plant Chemistry", Hirokawa Publishing Company, Tokyo.

Karrer, W. (1958). Konstitution und Vorkommen der Organischen Pflanzenstoffe", p. 1207. Birkhäuser, Basel.

Maheshwari, M. L., Varma, K. R. and Bhattacharyya, S. C. (1963). **19**, 1519.

Mentzer, C. (1954). *Bull. Soc. Chim. Biol.* **36**, 1225.

Mentzer, C. and Fatianoff, O. (1964). "Actualités de Phytochimie Fondamentale". Masson, Paris.

Nakabayashi, T. (1962). Nippon Kagaku Zasshi **83**, 182.

Nakanishi, K., Kurono, M. and Bhacca, N. S. (1963). *Tetrahedron Letters* 30, 2161.

Paul, I. C., Sim, G. A. and Morrison, G. A. (1962). *Proc. chem. Soc.* 352.

Riggs, N. V. and Stevens, J. D. (1963). *Tetrahedron Letters* **24**, 1615.

Wehmer, C. (1928). "Die Pflanzenstoffe". G. Fischer, Jena.

CHAPTER 3

Chemotaxonomy or Biochemical Systematics?

R. E. ALSTON[1]

The Cell Research Institute and The Department of Botany
The University of Texas, Austin, Texas, U.S.A.

I. INTRODUCTION

The study of secondary compounds of plants has occupied chemists, pharmacologists and biologists for many generations, and the practical benefits of these investigations are well known. By virtue of the existence of thousands of secondary compounds among plants, the chemist was challenged to isolate these substances as pure entities, sometimes from incredibly complex mixtures (e.g. the alkaloids of *Vinca*); and to ascertain their structures. This latter task, undertaken by some of the most renowned chemists, led to remarkable triumphs of structure analysis. One must only read the colorful description by Hendrickson (1960) of the assault upon the strychnine molecule to assimilate the dramatic element in these conquests.[2]

[1] National Institutes of Health Grant GM–11111–02 is acknowledged with appreciation.
[2] "The classical period (1910–1932) may be said to have started with the entrance into the battle of the English school, brilliantly headed by Sir Robert Robinson, and the forces of Hermann Leuchs in Berlin, who contributed the massive total of 125 papers of outstanding experimental work. The efforts of this period were largely devoted to oxidative incursions into the underside of the molecule. By 1932 the escalade had successfully taken the outer wall of the molecule's defences, leaving only the inaccessible and silent heart, which, with its singular intricacy of interlaced hydrocarbon rings, was to require sixteen more years of concentrated effort for solution" Hendrickson, J. B. (1960). (p. 180).

Like the chemist, the pharmacologist discovered, especially in the alkaloids, a virtually inexhaustible source of challenge. For of these thousands of compounds residing benignly in the living plant, nearly all are capable of the most violent effect upon the physiology of higher animals, *Homo sapiens* not excepted. From an era of purely empirical methods of drug usage, we are now beginning to expose the mechanisms of some of their physiological effects.

As a biologist, the author is interested to know the prerogatives of the biologists with respect to general scientific inquiries regarding these thousands of secondary compounds. Although an oversimplification, it suits the present purpose to assume that the biologist is more properly concerned with the compounds *in the plants themselves*, although, in the contexts of natural selection, ecology and perhaps other areas, the biologist is also concerned, *inter alia*, with the effect of a plant product upon other organisms, e.g., parasites. Having established this single prerogative it is now relatively easy to raise a number of rhetorical questions:

1. How does a certain compound come into existence in the individual cell? (a biochemical, physiological and genetical question).
2. How did the compound come to be present in the species? (an evolutionary and, hence, a genetical question).
3. How do internal regulatory mechanisms control the time and place of synthesis, as well as the rate of synthesis and final amount, of a particular compound? (One might describe this as chemogenesis, analogous to morphogenesis and probably not fundamentally distinguishable from it, since form, structure, and chemistry are related much like a symphony to its musical notes.)
4. Needless to say, the *raison d'être* (Fraenkel, 1959) of the secondary compounds, individually and collectively, is a basic biological question, though regrettably it has not inspired much profound analysis.

It is desirable to touch upon all of these somewhat obvious questions in the course of this discursive treatment and, if possible, to formulate additional ideas that may elicit further questions. The four points raised above will not be discussed in sequence, or always directly, but hopefully they may pervade the discussion.

II. COMPARATIVE BIOCHEMISTRY OF SECONDARY COMPOUNDS

The tides of biological interests ebb and flow, often through the vagaries of technical advances, and occasionally through a more abstract form of creativity. Currently, biological interest in secondary compounds favors a comparative approach, with taxonomic, or preferably (to the author) evolutionary, concepts being the prevailing interest. This emphasis is manifest through the occurrence of a number of symposia and the existence of several books on comparative chemistry. Its causes are not wholly determinable, but contributing factors must have been the development of rapid and efficient screening techniques for secondary compounds, and the eventual recognition that the chemical data indeed have systematic utility.

A. PHYLOGENETIC IMPLICATIONS

In recent years, there has been increasing emphasis upon taxonomically-oriented surveys of secondary compounds. Discounting, for the present discussion, studies in which the compounds are not actually identified, these contributions provide valuable information about the occurrences of "related" compounds within a taxonomically delimited group, e.g. the occurrences of anthraquinones among certain angiosperm families. Incompleteness of a survey within the group, inadequate population sampling, the possibility of intra-individual differences and the failure to detect a minimal concentration are often cited as possible danger spots in such surveys. Yet if only strictly taxonomic use of the data is intended, one can employ them practically without regard for many biological considerations. The chemical data thus serve merely to increase the descriptive criteria which collectively circumscribe the taxonomic entity. This is not true if one is interested in the evolutionary or phylogenetic implications of such comparative chemical data. Yet, many chemotaxonomic studies have been evaluated both taxonomically and with respect to their evolutionary implications without due regard for the total biological facts. All such phylogenetic speculation is discouraged by many taxonomists who are of the opinion that the nearly total inadequacy of the fossil record, among other sound biological limitations, makes it impossible for us to achieve any significant progress in analysis of relationship by descent using objective scientific means. That is, one may produce many cheaply won hypotheses but few convincing proofs. In contrast, other taxonomists agree in principle with Constance (1964) who wrote "It is a well worn cliché to remark that a plan of classification can be termed phylogenetic only if it is based on fossils. The implication can then be drawn that the availability of fossil evidence will ultimately lead us to an indisputably evolutionary arrangement of plants, and since that is the future prospect, we would be well advised to eschew phylogeny and to content ourselves with producing such useful, if mindless, devices for naming and identifying plants as we can." It is unlikely that either group favours dogmatic assertions of relationships based on a limited number of correlations, or that either group considers it necessary to stop *thinking* about phylogeny or even writing about phylogeny.

A practice which may rightly offend the taxonomist is the assertion that the coexistence of some chemical components implies a "relationship" between two families. The obvious fallacy here is that from a total biological perspective the particular correlation simply cannot be measured equitably against another correlation (favouring an alternative relationship); nor can one gauge effectively the possibility of parallel or convergent evolution. Although the chemist is often in a better position to appreciate the probable significance of the chemical data than is the biologist, sometimes taxonomic correlations are enlightening to the former. When the possible biosynthetic relationships between isoflavones and rotenones was unrecognized, there would have been no significance, either taxonomic or phylogenetic, in correlations among these two groups of substances. However, as noted by Grisebach and Ollis (1961) multiple correlations in the systematic distributions of these compounds led to a recognition of plausible interconversions to rotenones from appropriate isoflavones. The correlation still has only limited taxonomic significance, but a phylogenetic implication is inherent in the taxonomic correlations. The lure of phylogeny, more than taxonomy, will then perhaps encourage some taxonomists to participate in further intensive searches for additional correlations at other levels of molecular structure among these compounds.

B. BIOSYNTHETIC MECHANISMS

Intensive comparative studies may give access to new experimental approaches by the chemist. For example, one reasonable mechanism of synthesis of the glycoflavonoid compounds, vitexin, saponaretin (and vicenin,

I. Vitexin   II. Saponaretin

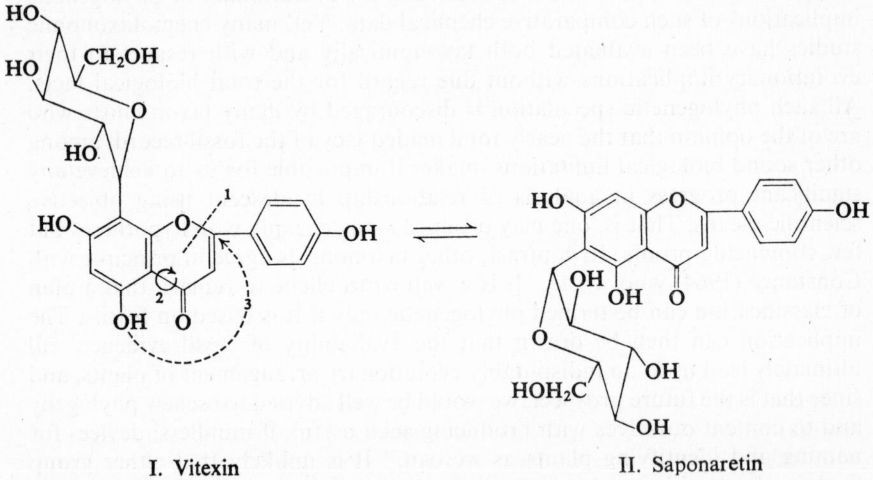

FIG. 1. Possible mechanism of interconversion between vitexin and saponaretin: (1) opening of ring C; (2) rotation of ring A; (3) closing of ring C in the alternate position.

based on its probable structure), involves opening of ring C, and closure in the alternate position thus equilibrating a substituent between the C-6 and C-8 positions (Fig. 1). It is not known, however, whether vitexin (I) and saponaretin (II) are synthesized independently or sequentially in the living plant. The availability of certain species of duckweeds which make vitexin, saponaretin, and vicenin, in various combinations enables greater flexibility in the experimental approach to this problem, and this knowledge was acquired originally for systematic purposes (McClure, 1964).

C. TAXONOMIC CORRELATIONS

Although most single chemical correlations can hardly be said to provide convincing evidence of enough total cryptic genetic homology ("relationship") to warrant further juxtapositioning of families, there are certain exceptional examples of such correlations which are extremely provocative. Mabry has discussed the taxonomic distributions of betacyanins and betaxanthins among certain plant families (see Chapter 14). These compounds have been proposed as a basis for grouping 10 angiosperm families (all of which had been considered by one or more authorities as properly belonging to a single order, Centrospermae). Since betacyanins and betaxanthins are not especially complex or bizarre chemically, there must be other reasons to infer their phyletic value. The distribution of betacyanins and anthocyanins is extremely interesting. Although the phylogenetic inference may still be said to be subjective, it seems plausible that the compounds have a single evolutionary origin. If so, then, as indicated by Mabry (Chapter 14) the Hutchinson system is weakened considerably. Of course there are few objective criteria known to the writer by which one can judge the precise degree of phyletic significance of a specific compound. One can advance certain generalizations to serve as guideposts, but it may require even more than common biosynthetic origins to allow the inference of homology in a given instance. Proof of enzymatic (i.e. genetic) homology may be regarded as a prerequisite, but while the distant prospects of establishing these homologies are not totally negative, the immediate prospects are poor. Since the strongest evidence of homology is probably genetic, and this type of evidence is rarely attainable directly, at least above the generic level, it therefore follows that the most general applicability of secondary chemical data is at the lower taxonomic levels.

D. GENETIC VARIATIONS

Special features of macromolecules render them exceptions to the above generalization. One of the distinctive advantages of serology is the fact that unit interactions are suggestive of protein identity (and possible homology) often at taxonomic levels above those at which genetic evidence of homology is attainable, and other techniques exist for providing, following extensive study, strong circumstantial evidence of macromolecular homologies. Despite the desirability of genetic evidence, it is not essential to have such evidence in order to include chemical data among other criteria of relationship, since the other criteria do not themselves have the support of genetic evidence. Even a good fossil record does not in a strict sense represent genetic evidence as the term is employed above.

Where formal genetic analysis of secondary compounds has been successful, it is obvious that the results might prove to be useful in developing principles for evaluating the systematic meaning of certain distributions. However, such genetic knowledge exists only for the flavonoids and only to a limited extent.

In these compounds, it has been possible to show that not only do minor alterations in the substituent groups of the basic flavonoid nucleus fall under specific genic control, but the genetic mechanism often involves simple Mendelian dominance, or epistasis (which is the interaction of non-allelic genes). Most genetic mechanisms which have been analysed effect what are considered to be superficial changes in the commoner flavonoids (Alston, 1964). Genetic mechanisms governing the relative and absolute amounts of these substances, and their tissue localization, may be much more complex than those having qualitative effects (Alston, 1959). For example, the genetic system which yields blue flowers and green stems in *Baptisia australis* and that which yields white flowers and blue stems in *Baptisia leucantha* may depend upon many more genic differences than those governing the presence of a *group* of flavones in the former species as opposed to a *group* of flavonols in the latter. This real problem involves the necessary introduction of a biological focus upon a chemical feature.

E. ENZYME SPECIFICITY

Another question that is theoretically important in an evolutionary context despite practical difficulties which may stand in the way of an answer, concerns enzyme specificity as it applies to the metabolism of secondary compounds. In an evolutionary context, enzymes may be regarded as the products of a series of changes in a protein which increase the efficiency of that protein to catalyse a reaction. This increased efficiency may be expressed in various ways; i.e. by reduced activation energy, higher turnover rate, or increased specificity. Presumably these changes occur by sequential alterations in the primary structure of the enzyme, reflecting an equivalent change in its genetic template. The intensity of natural selection pressure determines in part the rate of improvement of an enzyme. For example, two independent changes in the primary structure, affecting independently the active site of an enzyme, each change governed by its own allele, would compete with each other, unless rapid replacement of the original allele by the newly evolved one were to occur to minimize competition of this type. It has been said that the selective advantage of secondary compounds must be very slight since it is usually impossible to establish the function of even a whole class of compounds, least of all that of specific molecular subtypes. The fact that there may be a rather large amount of intraspecific variation in flavonoid chemistry (though principally involving relative amounts of specific compounds) would also seem to support the lack of really strong genetic regulation (Brehm and Alston, 1964). Such variation may be the product of an unknown amount of intraspecific genetic heterogeneity, or perhaps may represent the influence of ecological factors upon an unstable feedback-regulating system. The latter explanation is not supported by the study of patterns of flavonoids in genetically identical (clonal) cultures of *Spirodela oligorhiza* grown in over sixty different types of sterile culture conditions (McClure and Alston, 1964). Among these cultures, variation was quite limited and only in special situations (e.g. inhibition of anthocyanin synthesis by benzimidazole) was there any major variation in flavonoid chemistry expressed.

Recently, we have obtained evidence that some of the enzymes governing flavonoid type in *Baptisia* may be highly specific. In *B. leucantha* several

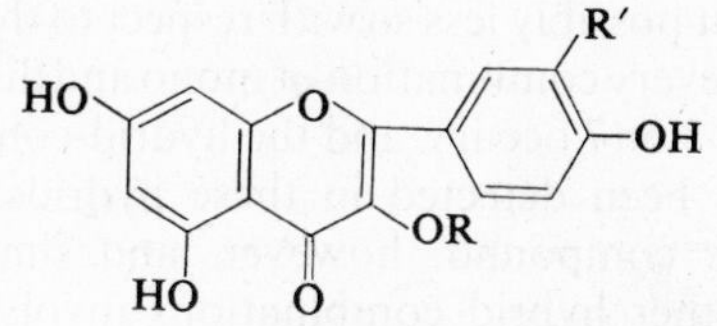

R'=H   R=sugar   Kaempferol 3-glycoside (probably glucoside)
R'=H   R=bioside   Kaempferol 3-diglycoside (probably glucoside)
R'=OH   R=glucose   Quercetin 3-glucoside
R'=OH   R=rhamnosylglucose   Quercetin 3-rhamnoglucoside

Compounds from *B. leucantha*.

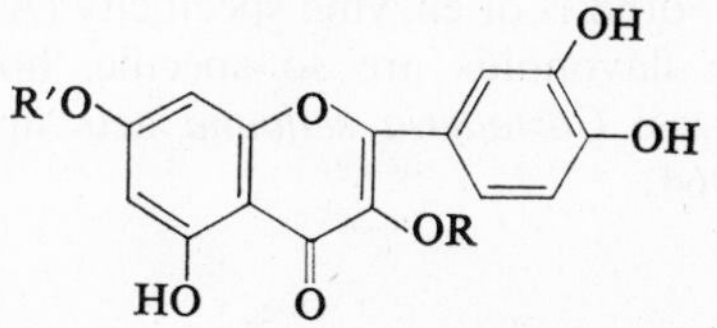

R'=H   R=glucose   Quercetin 3-glucoside
R'=glucose   R=H   Quercetin 7-glucoside
R=R'=glucose   Quercetin 3,7-diglucoside
R'=rhamnosylglucose   R=H   Quercetin 7-rhamnoglucoside
R=glucose R'=rhamnosylglucose   Quercetin 3-glucoside-7-rhamnoglucoside

Compounds from *B. sphaerocarpa*.

Theoretically possible hybrid flavonoids from *Baptisia leucantha* × *B. sphaerocarpa*: quercetin 7-glucoside-3-rhamnoglucoside; kaempferol 7-glucoside kaempferol 7-rhamnoglucoside, kaempferol 3,7-diglucoside; kaempferol 3-rhamnoglucoside-7-glucoside; kaempferol 3,7-dirhamnoglucoside. Analogous possibilities exist for certain flavones, isoflavones, anthocyanidins and flavanones, all of which occur also in these same two species in different specific types.

FIG. 2. Flavonols of *Baptisia leucantha* and *B. sphaerocarpa* and hypothetical hybrid-specific molecules of the hybrid between these two species.

flavonols are produced in the leaves; e.g. kaempferol and quercetin 3-monoglycosides and kaempferol and quercetin 3-rhamnoglucosides. In *B. sphaerocarpa* quercetin 3-glucoside-7-rhamnoglucoside, quercetin 3,7-glucoside, quercetin 7-rhamnoglucoside, quercetin 7-glucoside and quercetin 3-glucoside are produced in the flowers only. In hybrids between *B. leucantha* and *B. sphaerocarpa* several hybrid compounds could be expected on theoretical grounds (Fig. 2). While all of the specific compounds above do appear in the hybrids in the same relative amounts (with minor exceptions among individual plants) none of the theoretically possible kaempferol-type hybrid molecules appears. There is no epistatic suppression of kaempferol, since its 3-glycosides occur in unreduced amount in the hybrids. Similar, but not identical, situations occur in other *Baptisia* hybrids involving *B. alba*. It appears that the enzymes governing specific glycosylation patterns are highly specific with respect to the B-ring substituents but possibly less so with respect to the pre-existing glycosylating pattern. Almost every combination of mono and disaccharide substituents at quercetin positions 3 or 7 occurs, and the hydrid-compound, 7-glucoside-3-rhamnoglucoside has been detected in these hydrids. (Another species, *B. alba*, does make this compound, however, and similar but not identical situations occur in other hybrid combinations involving this species.) It is surprising that such apparent specificity should occur in *B. sphaerocarpa*, which contains 5 other quercetin glycosides. Indeed, this species can also form the kaempferol B-ring pattern since it produces apigenin 7-rhamnoglucoside. It is possible, of course, that some subtle difference in the exact site of synthesis or in the time of synthesis may account for the failure of the new molecular configurations to appear, but proof of this would be difficult to obtain, and we favour the simpler hypothesis of enzyme specificity (Alston *et al.*, 1965). Not all enzymes affecting flavonoids are so specific, however. The *O*-methyl transferase of the grass, *Cortaderia selloana* acts upon various substrates (Finkle and Masri, 1964).

III. DISTRIBUTION OF COMPOUNDS DIFFERING ONLY IN SUBSTITUTION PATTERN

A. FLAVONOIDS

Earlier, it was stated that the functions of the vast majority of secondary compounds in plants are unknown. Comparative studies have contributed indirectly to this enigma through the frequent disclosure of extreme differences in some groups of secondary compounds among species of similar general morphology. In other instances, plants very different in appearance have a nearly identical array of secondary compounds. In the latter situation, one is not able to find any common feature that is peculiar to the two taxa except the chemistry. The existence of a common function based on the presence of one of a series of alternative recurrent complexes of specific compounds is possible, but difficult to establish. Some examples taken from flavonoid chemistry serve to illustrate this pattern resemblance described above. Glycoflavonoids are rather common in the Gramineae (grasses) of the Monocotyledoneae (Harborne and Hall, 1964). In another monocot family, Lemnaceae, glycoflavonoids are common in the genera, *Lemna* and *Spirodela*, but completely absent from *Wolffiella*, and less common in *Wolffia*. Among these extremely simple flowering plants, some of which are less than one millimeter in length, the patterns of flavonoid chemistry are extremely different among even the species of a single genus (Table 1). Their small sizes and structural simplicity render these plants difficult to classify by usual criteria. They are all

TABLE 1. Flavonoid patterns of the family Lemnaceae

Species	Glycoflavones												Antho-cyanins				Flavonols															Flavones									
	1[1]	2	7	15	16	18	20	23	24	25	26	27	8	11	12	19	3	4	5	9	10	22	36	37	38	40	41	42	43	44	45	13	14	17	21	29	32	34	35	46	47
Spirodela																																									
S. intermedia	+	+	+	+	+																											+	+	+							
S. polyrhiza	+	+	+	+	+	+																										+	+								
S. biperforata	+	+	+	+																												+	+								
S. oligorhiza	+	+	+													+						+												+							
Lemna																																									
L. minor	+	+	+	+	+	+							+	+	+																										
L. gibba	+	+	+	+	+	+	+						+	+	+									+	+																
L. obscura			+	+	+		+	+	+														+	+	+	+						+									
L. trisulca		+	+	+	+			+	+	+													+	+	+								+								
L. perpusilla								+	+	+	+	+																													
L. trinervis								+	+																																
L. valdiviana	+											+																													
L. minima	+											+				+						+																			
Wolffiella																																									
W. lingulata																											+	+													
W. oblonga											+																	+	+												
W. gladiata									+	+																		+	+												
W. floridana																+											+	+													
Wolffia																																									
W. punctata																										+	+	+	+	+	+										
W. microscopica																										+	+	+	+	+	+									+	+
W. papulifera																											+	+	+	+	+									+	+
W. columbiana					+	+																			+								+	+							
W. arrhiza				+	+																												+	+						+	+

[1] Numbers in heading column refer to individual compounds detected within the family. Identities of many of these substances have been established within the limits of chromatographic and ultraviolet spectral analyses (McClure, 1964).

aquatic, occupying relatively similar ecological niches. The wide diversity of flavonoid patterns encountered was unexpected (McClure, 1964). In contrast, the legume genus *Psoralea*, which is variable morphologically, produces generally similar flavonoid patterns in all species examined, and the major flavonoid features are the same group of glycoflavonoids as those which occur in the Lemnaceae (Alston and Ockendon, in preparation). Indeed the flavonoid patterns on two-way chromatograms of *Lemna minor* and *Psoralea subacaulis*, despite their relative complexity, are almost indistinguishable. The author has compared the general flavonoid patterns of leaves, flowers, and other parts of a large number of species, and it is likely that most any type of variation which can be imagined can be encountered, including the extent of variability in the genus, degree of complexity of the patterns, and variation within and among individuals. Few generalizations can be made at this time, except that the possibility of predicting the flavonoid types in a particular genus is still not increasing noticeably, at least for the author.

There is a definite possibility that insofar as flavonoid function in the plant is concerned many roads lead to Rome. That is, as noted above, a flavonoid complement might be effective within rather broad limits of specificity, and perhaps a large number of basic flavonoid patterns duplicate each other functionally. For example, if the glycoside type is principally a regulator of solubility of the flavonoid, it is possible to envisage several different patterns based on mono-, di- and tri-glycosides. For example, glycoflavonoids could exist as 6 or 8 and (or) as 6,8 glycosides; flavones could exist as 7-mono- or as 7-diglycosides; and flavonols either as 3-mono- or 3-diglycosides or as 7-mono- or 3,7-diglycosides. It is much more common to find an associated complex representing one of these patterns than to find a mixture of apparently unrelated flavonoids together in a plant. The extent to which a series of flavonoid complexes, each with some functional utility, exists, determines in large measure the number of general flavonoid patterns which may be expected to occur, and accordingly, the systematic significance of a given degree of resemblance.

It is extremely difficult to find any good explanations for the occurrences of the (seemingly) unlikely substitution patterns of some flavonoids. Harborne and Simmonds (1964) have called attention to a number of these types which either show a marked correlation with one or a few plant families, or which otherwise have taxonomically interesting distributions. I have already noted the frequent overlapping of the distributions of isoflavones and rotenones. As observed by Harborne and Simmonds, a 2'-hydroxyl substituent is frequently present in isoflavones, and this substitution may be a requirement for the formation of the second benzopyran ring, characteristic of rotenones. Rotenones also contain isoprenoid substituents at C-8. The rotenones are apparently restricted to the isoflavone-producing groups of Leguminosae.[1] The occurrence

[1] Most rotenones are reported as aglycones but amorphin, a rotenone glycoside, has been reported by Crombie and Peace (1963) from *Amorpha*. According to B. L. Turner, although *Amorpha* is supposedly in the Tribe Psoralinae, not the rotenone producing Tribe Dalbergieae, he has suspected on morphological grounds a relationship of the Psoralineae to certain elements of the Dalbergieae (personal discussion).

of an isoflavone having 8- and 6-carbon isoprenoid substituents, pomiferin, of the osage-orange tree (*Maclura pomifera*—Moraceae) is presumably a coincidence without systematic implications. One wonders, however, what there may be about the isoflavone nucleus which seems to favor the origin of an isoprenoid side chain.

A similar type of coincidence with possibly cryptic significance is the presence of isosaponarin and isolutonarin, the 4'-glucosides of saponaretin and homo-orientin, in a relatively large number of unrelated plants. The 4'-glucoside is uncommon among flavonoids in general but often represented among the glycoflavonoid derivatives. Should one look for purely chemical, or biological, factors which favour such correlations? There are numerous correlations which, indeed, are doubtlessly the result of evolutionary proclivity to advance according to the rules—either the chemical or the biological ones. Thus, the glycoflavonoids, which must have arisen many times, are usually derivatives of common flavones, never the common flavonols whose distributions are also so widespread that statistically one might expect to find glycoflavonols.

Another perplexing situation involves the rather narrowly distributed biflavonyls, which are found in certain gymnosperms and in *Casuarina*. The chief features of the taxonomic distribution of biflavonyls are their absence from the Pinaceae; their generally widespread occurrence in the Cupressaceae and Taxodiaceae and their presence in *Ginkgo biloba* (Ginkgoales). The compounds may occur in other flowering plants, but no extensive search has been made. Chemically, the biflavonyls are of interest in that only the apigenin, or 4'-monohydroxy B-ring type of flavone (plus the 4'-methyl ether) has been reported, and also because of the types of linkages between the flavonoid nuclei (Fig. 3). In the linkage of amentoflavone (III), the apigenin residues are joined by a C—C bond at positions 3' and 8. In *Cryptomeria*, a 4' methyl ether of hinokiflavone (IV), plus another uncharacterized derivative, occur, so that 4' methyl ethers of both types of biflavonyls are now known (Kawano *et al.*, 1964). Although the linkages typical of amentoflavone and hinokiflavone may be regarded as basically different, hence not necessarily catalysed by analogous or homologous enzymes, the distributions of these two types of biflavonyl imply an evolutionary origin involving either: (1) two independent events each with rather low probability, occurring in a restricted taxonomic group, or (2) one event, initially leading to the coupling of flavones to form the basic biflavonyl, followed by a dichotomy or derived state either from the amentoflavone type to hinokiflavone or the reverse. In the second situation, one compound would be the phylogenetic predecessor of the other, and the enzymes, though now functionally different (as judged by the product), would be homologous. It is possible that the enzyme catalysing such a condensation operates to favour the reaction by bringing apigenin monomers into spatial proximity. If so, the actual nature of the oxidative step might depend upon the biochemical-physiological properties of the cell and not specifically the enzyme. It is especially interesting that the biflavonyls, ginkgetin, and isoginkgetin, occur only in yellowed or dying leaves of *Ginkgo biloba* (Harborne

44   R. E. Alston

and Simmonds, 1964). Perhaps, the conditions favoring the oxidative pre-
coupling reaction are governed by the plant's general physiology. An alternative, though less attractive hypothesis, is that the enzyme and its substrate are
physically separated in the living cell, as is the case for some of the mustard
oils and also the cyanogenetic glycosides. The significance of the fact that only
flavones and especially only those flavones of the apigenin B-ring substitution
pattern have been detected does not tempt the author to speculate. It is also
noteworthy that in the Pinaceae, where biflavonyls might be expected but do
not occur, a large number of flavonoids with a non-hydroxylated B-ring are
commonly present, and the significance of this, beyond coincidence, is hidden.

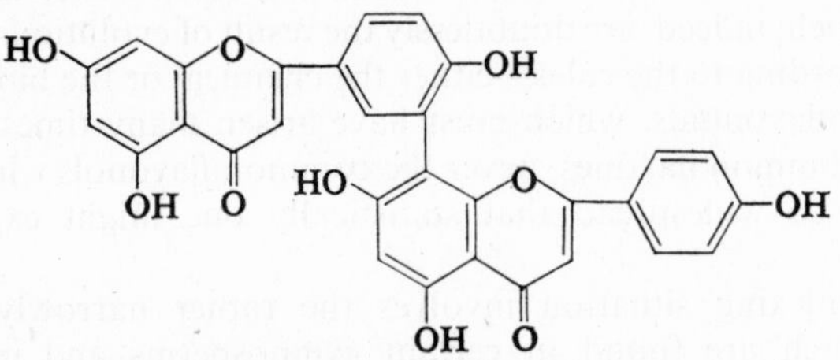

III. Amentoflavone

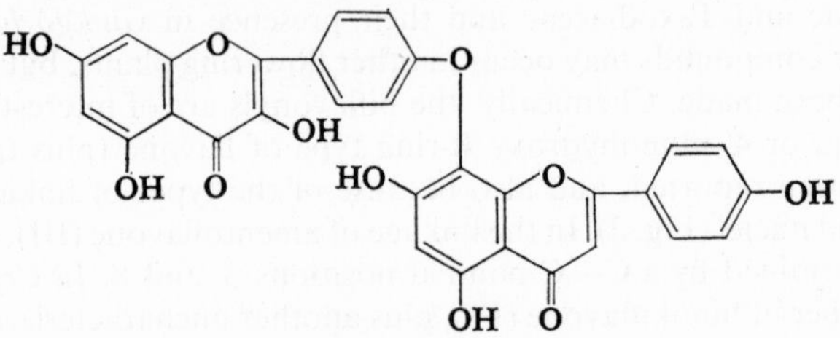

IV. Hinokiflavone

FIG. 3. Linkages typical of biflavonyls.

The recently described biflavonyl, cupressuflavone, is formed by an 8,8 linkage
of apigenin units (Murti *et al.*, 1964). Since there is no obvious involvement of
ring-B in this linkage, there should be no direct requirement for the apigenin-
type ring-B conformation nor could the explanation above be extended as
readily to favour enzymatic homology in the mechanism of the 8,8 linkage.

B. *p*-MENTHANE DERIVATIVES

In contrast, when one is fortunate enough to know something about under-
lying biosynthetic and genetic mechanisms, an evolutionary perspective is not
only available but hardly avoidable. For example, presence of 2-oxygenated-
p-menthane derivatives is governed by a dominant gene, the absence of which is
correlated with the appearance of only 3-oxygenated-*p*-menthane mint oils

(peppermint oils). Another group of mint species, the lemon mints, form only non-cyclical terpenes such as citral and linalool (Reitsema, 1958). One may infer from these distributions that the lemon mints are the most primitive, the spearmint group the most advanced. The problems inherent in inferring a phylogenetic sequence from such distributions have been discussed elsewhere (Alston *et al.*, 1963), but even the knowledge that such data do not represent a sound basis for arriving at a fixed point of view concerning the evolution of mint species cannot wholly suppress the perhaps optimistic view that one's evolutionary perspective is broadened by virtue of having the genetic knowledge. At least the genetic data relate the two series of mint oils in a manner suggestive of *an* evolutionary sequence—the direction of which is uncertain. One other alternative, independent origin of the 2-oxygenated series, is minimized by the dominance relationship. Both groups of species have developed mechanisms to assimilate rather large quantities of their respective oil types, and complete dominance is more readily explained by interconversion (at an earlier precursor stage) than by preferential utilization of a common precursor.

IV. Distribution of Biosynthetically-related Compounds

A. THE ALKALOIDS OF THE LEGUMINOSAE

Examples such as the mint oils, though they may be trivial and merely vehicles for encouraging liberal speculation at present, will be greatly expanded in the future as knowledge of biosynthetic pathways increases. There is no reason to doubt that a natural classification of secondary compounds, integrated appropriately with their dualistic, antecedent basic metabolites, will soon develop. As Mentzer (Chapter 2) has so ably shown, the broad channels of secondary chemistry are already being navigated, and we speak now of the acetate, shikimic acid and the mevalonic acid pathways as firmly established routes to major categories of secondary compounds. Occasionally, as in the rotenones, all of these pathways may contribute a portion of a single compound. A natural system of classification of organic compounds (if correct) will inevitably be based on biosynthetic affinities, and not functional or formal chemical affinities, and such a system will be implicitly phylogenetic. It is interesting to speculate as to how much phylogenetic insight would be provided by total knowledge (macromolecular data excluded) of the steps of biosynthesis of all known secondary compounds of plants. At the present time, progress in acquiring this type of knowledge is slow and, like the fossil record, quite fragmentary. Yet, there is evidence that, unlike the fossil record, progress in elucidating biosynthetic pathways is accelerating. One promising area, which will serve as an example, is that of nitrogen metabolism in the family Leguminosae, with special reference to certain alkaloids, non-protein amino acids and their biogenetic congeners. It has long been recognized that the Leguminosae are a versatile group insofar as secondary products are concerned (though perhaps as Bate-Smith has pointed out, only in an absolute sense because of the size of the family). Aside from numerous peculiar flavo-noids this family produces many unusual non-protein amino acids (see Chapter 12). If we omit those of the sub-family Mimosoideae, interesting enough in their own right, it is possible to group in a general way the amino acids and the alkaloids of the sub-family Papilionoideae on the basis of a broad bio-synthetic derivation of the total group by one of several metabolic excursions from intermediary metabolism of basic amino acids such as lysine, ornithine, arginine, and perhaps others. The following discussion will endeavour to demonstrate that this large metabolic complex, once it is fully understood, may offer clues which are truly relevant to the phylogeny of the plant groups concerned. The first group of substances to be discussed includes alkaloids of

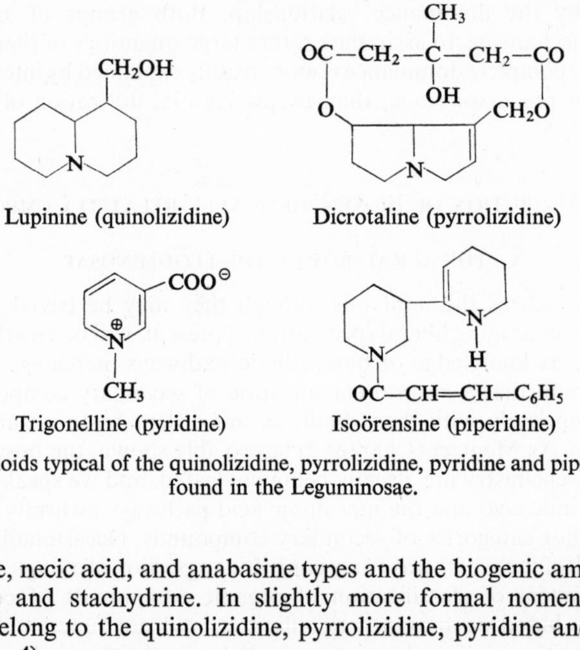

Lupinine (quinolizidine)   Dicrotaline (pyrrolizidine)

Trigonelline (pyridine)   Isoörensine (piperidine)

FIG. 4. Alkaloids typical of the quinolizidine, pyrrolizidine, pyridine and piperidine groups found in the Leguminosae.

the lupinine, necic acid, and anabasine types and the biogenic amines such as trigonelline and stachydrine. In slightly more formal nomenclature, the alkaloids belong to the quinolizidine, pyrrolizidine, pyridine and piperidine groups (Fig. 4).

Certain of the above mentioned nitrogenous substances, viz. trigonelline and the anabasine type alkaloids, contain the pyridine ring. In micro-organisms and higher animals, the pyridine ring is known to be synthesized from nicotinic acid through the established pathway from tryptophan to kynurenine to 3-hydroxyanthranilic acid. This pathway does not apparently represent the source of the pyridine ring of trigonelline. At least, no activity was obtained in trigonelline of soybean (Leguminosae) when radioactive 3-hydroxyanthranilic acid was provided (Aronoff, 1956), nor is nicotine of tobacco labelled when ^{14}C tryptophan is provided despite the fact that tritiated nicotinic acid is incorporated effectively into nicotine of tobacco (Mothes and Schütte, 1963).

Taken together, this evidence suggests that in tobacco (Solanaceae) the pyridine ring is derived from nicotinic acid, but the latter is not synthesized by way of the tryptophan pathway. The latest scheme, offered tentatively by Leete (1965) derives nicotine from aspartic acid and 3-phosphoglyceraldehyde. The evidence also suggests that the pyridine ring of trigonelline is not synthesized via the tryptophan pathway; but there is some evidence that the pyridine rings of trigonelline of soybean and anabasine of tobacco may possibly have a different mode of synthesis. For example, Aronoff (1956), reported a uniform labelling of the pyridine and piperidine rings of anabasine using $^{14}CO_2$, implying their biogenetic equivalence, while no activity was found in trigonelline, using $^{14}CO_2$. However, a recent scheme for the synthesis of the piperidine ring of anabasine proposed by Leete (1965), originates the piperidine ring only, from lysine. Lysine 2-^{14}C was found to yield activity only in the piperidine ring.

Another complication is introduced by the existence of tetrahydroanabasine-type alkaloids such as ammodendrine (V) and adenocarpine (VI) (Fig. 5) in

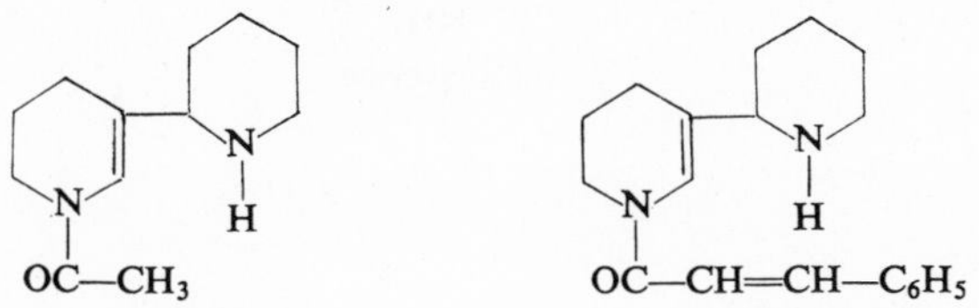

V. Ammodendrine   VI. Adenocarpine

FIG. 5. Tetrahydroanabasine-type alkaloids of the Leguminosae.

several genera of Leguminosae and anabasine itself in *Anabasis* (Chenopodiaceae). The distribution of adenocarpine (VI) and related alkaloids is especially interesting because, in the relatively small genus *Adenocarpus* (Leguminosae), some species reportedly form adenocarpine types while others form the quinolizidine alkaloid, sparteine. A third group of species contain both of these, e.g. *Adenocarpus complicatus* and *A. hispanicus*. Some other genera of Leguminosae produce both types of alkaloids, e.g. *Ammodendron*, *Cytisus* and *Retama*, all in the tribe Genisteae. Additionally, the unrelated *Anabasis aphylla*, produces both of these alkaloid types (Willaman and Schubert, 1961). From this taxonomic distribution, one might speculate that a biosynthetic association exists between the two types of alkaloids in those genera. Lysine can be adduced as a precursor for both groups of alkaloids and indeed has been clearly established as a precursor to the quinolizidine alkaloids. On the basis of known enzymatic conversions of cadaverine to ω-amino-valeraldehyde and spontaneous cyclization, a condensation with piperidine, also from lysine, presents an attractive model for the formation of tetrahydro-anabasine (Fig. 6), but this pathway differs significantly from that proposed by Leete (1965) for the synthesis of anabasine of tobacco. As noted earlier, lysine does not serve as a precursor for the pyridine ring of anabasine in tobacco. Clarification of these problems of comparative aspects of biosynthesis of these alkaloids would obviously be of considerable evolutionary interest.

It may be seen from the previous discussion that the quinolizidine (lupine type) alkaloids are possibly biosynthetically related to the tetrahydroanabasine types and perhaps to the trigonelline-type pyridine derivatives through a pathway emanating from lysine, although possibly a more plausible origin of trigonelline is through *N*-methylation of proline. Before proceeding further with a discussion of the quinolizidine alkaloids, it is worthwhile to point out that the pyrrolizidine group of alkaloids (Fig. 4), also found in the Leguminosae (e.g. *Crotalaria*) may arise from two ornithines by a mechanism which is analogous to the presumed synthesis of the quinolizidine nucleus from lysine. The former alkaloid group thus differs from the latter in consisting of coupled 5-membered heterocyclic rings since ornithine is the 5-carbon homologue of

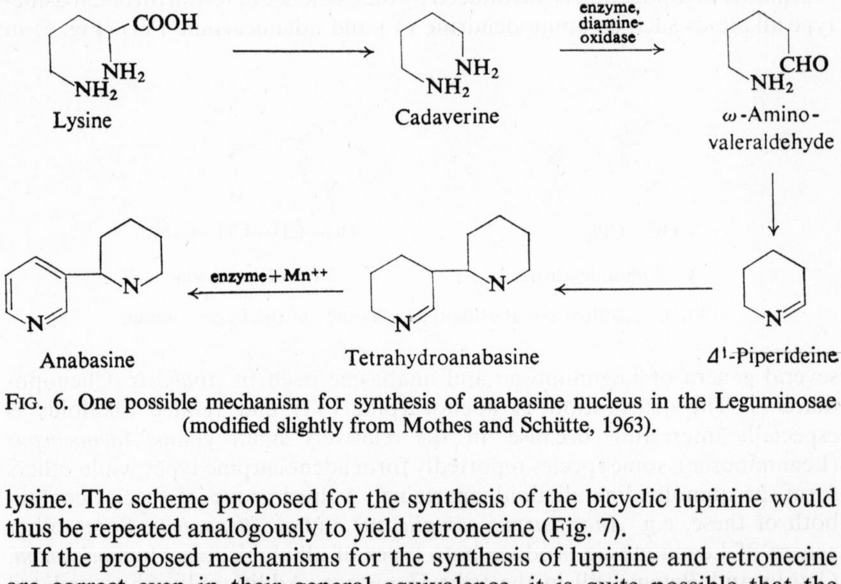

FIG. 6. One possible mechanism for synthesis of anabasine nucleus in the Leguminosae (modified slightly from Mothes and Schütte, 1963).

lysine. The scheme proposed for the synthesis of the bicyclic lupinine would thus be repeated analogously to yield retronecine (Fig. 7).

If the proposed mechanisms for the synthesis of lupinine and retronecine are correct even in their general equivalence, it is quite possible that the mechanisms involve the intervention of homologous enzymes (i.e. enzymes with a single evolutionary origin) which have subsequently undergone a change in specificity. Wagner *et al.* (1958) have disclosed a model of this type of system in *Neurospora crassa*. Two complementary mutants affect equivalent steps involving two different points in the parallel biosyntheses of valine and isoleucine. Presumably, a single enzyme in each instance catalyses the same type of reaction in corresponding precursors (Fig. 8). Nowacki (1963, p. 193) has in fact stated that "the enzyme lysine decarboxylase is probably non-specific in that it partly decarboxylates ornithine into putrescine."

The general rule of parsimony in evolution would seem to favour a channeling of metabolism through either lysine or ornithine, assuming func-

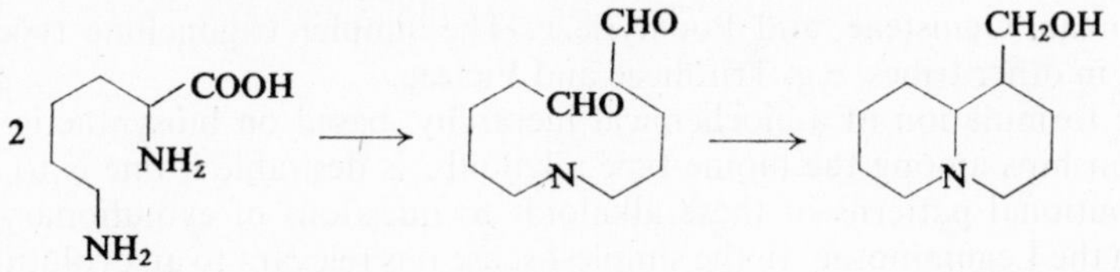

(a) Possible synthesis of lupinine from lysine.

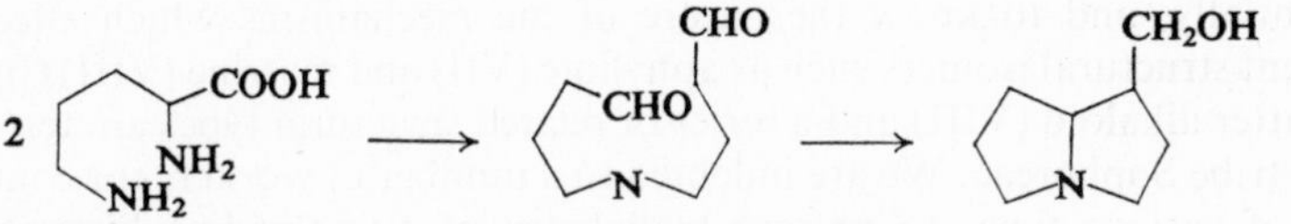

(b) Possible synthesis of retronecine srom ornithine.

FIG. 7. Analogous possible mechanisms of synthesis of (a) quinolizidine and (b) pyrrolizidine alkaloids.

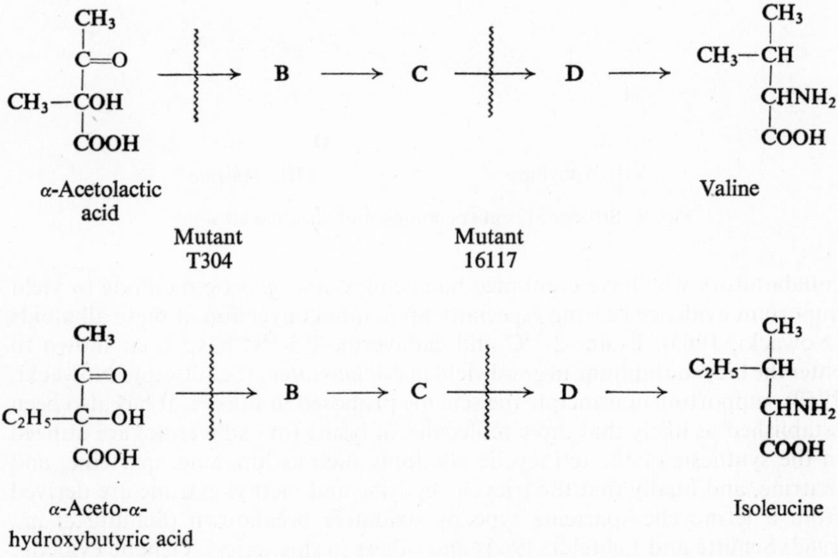

FIG. 8. Effects of two mutations upon valine and isoleucine synthesis in *Neurospora crassa* (from Wagner *et al.* 1958).

tional equivalence of the two groups of alkaloids, either fortuitously or because of greater expendability or availability of one or the other of these two amino acids. Whatever their origins and past distributions, the pyrrolizidines of the Leguminosae are now restricted to the genus *Crotalaria* of the tribe Genisteae while the quinolizidines of the Leguminosae are restricted to the tribes

Sophoreae, Genisteae, and Podalyrieae. The simpler trigonelline types are found in other tribes, e.g. Trifolieae and Vicieae.

The formulation of a biochemical hierarchy, based on biosynthetic inter-relationships among the lupine type alkaloids, is desirable if one is to apply distributional patterns of these alkaloids to questions of evolutionary lines within the Leguminosae. In the simplest sense it is relevant to an evolutionary, though not a taxonomic, perspective to know whether the bicyclic system of lupinine and the tetracyclic system of sparteine are formed alternatively or sequentially, and to know the nature of the mechanisms which effect the different structural isomers such as aphylline (VII) and matrine (VIII) (Fig. 9). The latter alkaloid (VIII) and a series of related structural types are restricted to the tribe Sophoreae. We are indebted to a number of workers for contributions of various types to present knowledge of quinolizidine biosynthesis. Much historical perspective is provided in the succinct review by Leete (1963). Special mention should also be made of the contributions of Schütte and collaborators utilizing ^{14}C tracer techniques and those of Nowacki and

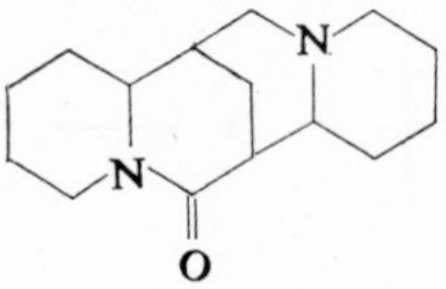

VII. Aphylline 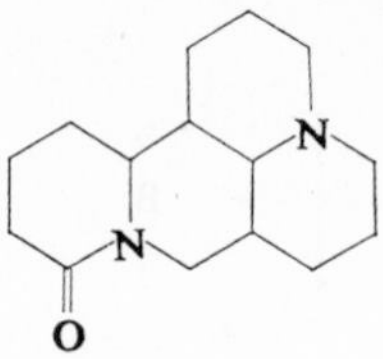   VIII. Matrine

FIG. 9. Structural isomers among quinolizidine alkaloids.

collaborators who have combined biochemical and genetic methods to yield important evidence bearing especially upon interconversion of these alkaloids (Nowacki, 1963). Lysine-2-^{14}C and cadaverine-1,5-^{14}C have been shown to enter the bicyclic lupinine in good yield in *Lupinus luteus* (Schütte and Nowacki, 1959), supporting in principle the scheme proposed in Fig. 7a. It has also been established as likely that three molecules of lysine (or cadaverine) are utilized in the synthesis of the tetracyclic alkaloids such as lupanine, sparteine, and matrine, and finally that the tricyclic cytisine and methyl-cytisine are derived from a tetracyclic sparteine type by oxidative breakdown (Schütte *et al.*, 1962; Schütte and Lehfeldt, 1964; and others in this series). Genetic evidence for a possible two-step enzymatic interconversion of sparteine via lupanine to hydroxylupanine was obtained by Nowacki (1963) by use of the interspecific cross, *Lupinus arboreus* (producing sparteine) × *L. nootkatensis* (producing sparteine, lupanine and hydroxylupanine). All first generation hybrids produced the 3 alkaloids, but in the second (F₂) generation 100% produced sparteine, ±75% produced lupanine and sparteine and ±50% produced hydroxylupanine, lupanine and sparteine. No plants produced hydroxylupanine without both of the other two alkaloids being present. These data fit a genetic model in which the factor governing lupanine synthesis is epistatic to another factor governing hydroxylupanine synthesis. The data do not prove that sparteine is converted to lupanine, however, since no sparteine-deficient plants were available as parents. These genetic techniques offer promising avenues of approaching problems of mechanisms of interconversion, the isolation of specific enzymes affecting quinolizidine synthesis and, eventually, the study of enzyme homologies by means of diverse methods now available. In genera such as *Lupinus* and *Baptisia*, wherein widespread, interspecific hybridization occurs, excellent opportunities for genetic studies exist, although in the latter genus the necessary alkaloid-deficient stocks for such work are lacking. *Baptisia* produces a large array of quinolizidine alkaloids, but at least traces of most alkaloids occur in the majority of species (Cranmer, 1964).

B. THE AMINO ACIDS OF THE LEGUMINOSAE

The tribe Vicieae of the Leguminosae does not produce either pyrrolizidine or quinolizidine alkaloids, but instead certain genera such as *Vicia* and *Lathyrus* accumulate another group of derivatives of the amino acids, asparagine, ornithine, and lysine. These substances, classified as non-protein amino acids, are frequently toxic, and some of them constitute the so-called "lathyrism" principles which produce skeletal or neurological pathology. Early investigations of these toxic principles in *Lathyrus* led to the assumption that a scheme of biosynthesis, originating with asparagine, led by way of a common intermediate, β-cyano-L-alanine, to either β-aminopropionitrile or to α,γ-diaminobutyric acid (Fig. 10). The postulated intermediate was subsequently isolated from two species of *Vicia* though not yet from *Lathyrus* (Ressler, 1962).

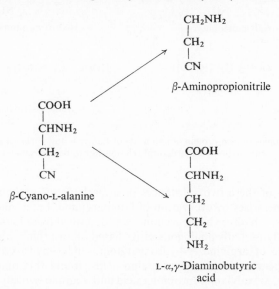

FIG. 10. Possible alternative pathways from β-cyano-L-alanine to two types of lathyrism factors.

The comprehensive analyses of non-protein amino acids by Bell (1964a-c; Bell and Tirimanna, 1964; see also Chapter 12), have led to the disclosure of taxonomically interesting distributions of a series of compounds, some of which may be regarded as derivatives of ornithine or lysine in the two genera, *Vicia* and *Lathyrus* (Fig. 11). The discovery of lathyrine, found only in the single genus *Lathyrus*, was followed by the discovery of homoarginine, a guanidine derivative of lysine. The apparently complementary quantitative

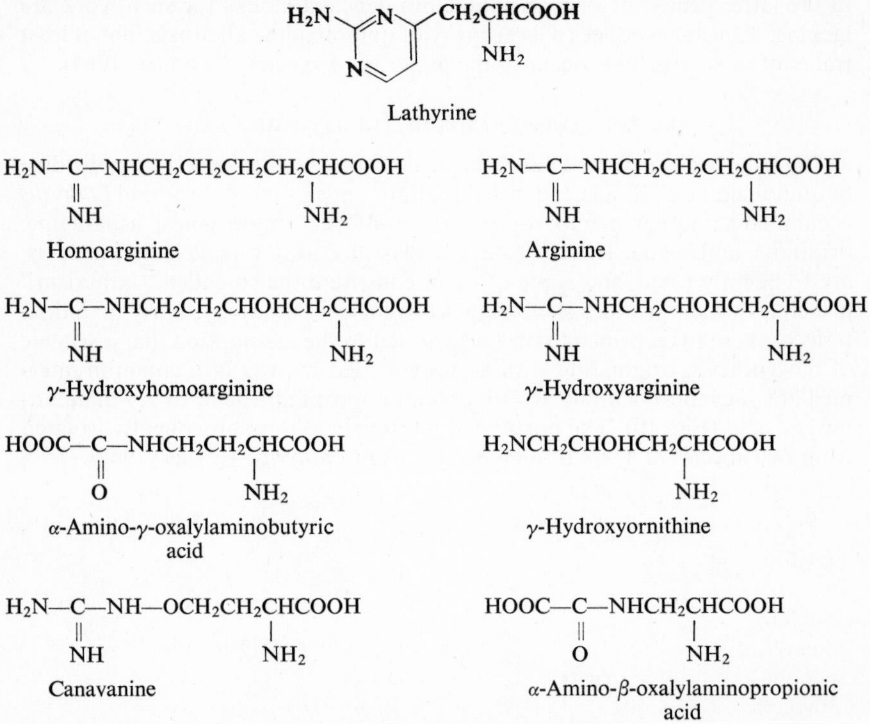

$$H_2N-\underset{\underset{N}{\|}}{N}-CH_2CHCOOH$$
$$\underset{NH_2}{|}$$

Lathyrine

$$H_2N-\underset{\underset{NH}{\|}}{C}-NHCH_2CH_2CH_2CH_2\underset{\underset{NH_2}{|}}{C}HCOOH$$

Homoarginine

$$H_2N-\underset{\underset{NH}{\|}}{C}-NHCH_2CH_2CH_2\underset{\underset{NH_2}{|}}{C}HCOOH$$

Arginine

$$H_2N-\underset{\underset{NH}{\|}}{C}-NHCH_2CH_2CHOHCH_2\underset{\underset{NH_2}{|}}{C}HCOOH$$

γ-Hydroxyhomoarginine

$$H_2N-\underset{\underset{NH}{\|}}{C}-NHCH_2CHOHCH_2\underset{\underset{NH_2}{|}}{C}HCOOH$$

γ-Hydroxyarginine

$$HOOC-\underset{\underset{O}{\|}}{C}-NHCH_2CH_2\underset{\underset{NH_2}{|}}{C}HCOOH$$

α-Amino-γ-oxalylaminobutyric
acid

$$H_2NCH_2CHOHCH_2\underset{\underset{NH_2}{|}}{C}HCOOH$$

γ-Hydroxyornithine

$$H_2N-\underset{\underset{NH}{\|}}{C}-NH-OCH_2CH_2\underset{\underset{NH_2}{|}}{C}HCOOH$$

Canavanine

$$HOOC-\underset{\underset{O}{\|}}{C}-NHCH_2\underset{\underset{NH_2}{|}}{C}HCOOH$$

α-Amino-β-oxalylaminopropionic
acid

FIG. 11. Examples of non-protein amino acids of *Lathyrus* and *Vicia* (arginine is included only to show its structural relationship to homoarginine).

relationship of these two substances in several species of *Lathyrus* suggests that lathyrine arises by cyclization of homoarginine. Later, another member of this series, γ-hydroxyhomoarginine, was described from *Lathyrus*. In *Vicia*, an apparently partially analogous series based on ornithine, was disclosed by the presence of arginine, γ-hydroxyarginine and γ-hydroxyornithine. The latest members in this group of amino acids, found this time in *Lathyrus*, are α-amino, β-oxalylaminopropionic acid and α-amino-γ-oxalylaminobutyric acid. The guanidine derivative, canavanine, found only in the Leguminosae, seems to be allied with these other amino acids, though it has a wider taxonomic distribution within the family. It is not necessary to reiterate the details of the distributions of these compounds in the genera *Lathyrus* and *Vicia*. Bell has demonstrated the existence of definite species-groups among both genera, based on the occurrences of associations of these amino acids. Single amino acids do not serve to delimit the species groups as satisfactorily as do the total patterns. The "holes" in distributional patterns of individual components indicate the possibility of loss mutations, but if parallel evolution of such loss mutations occurs it cannot readily be detected. Bell has stressed the importance of basing any phylogenetic inferences upon intensive biochemical analyses rather than upon patterns of distribution alone and also the need to take into account other lines of evidence, not necessarily chemical in nature. For example, the arbitrary arrangement of *Lathyrus* species into subgroups according to their non-protein amino acid complements is in general agreement with data concerning interspecific hybridization and with cytogenetical data (Bell and Fowden, 1964).

There are a number of still unidentified ninhydrin positive compounds present in *Vicia* and *Lathyrus*, and of course the vast majority of related genera in the Leguminosae have not been investigated thoroughly for such compounds. We may assume that a considerably more elaborate plexus of biosynthetic interrelationships remains to be detected and in due course this will induce intensive study of biosynthetic routes and eventually pertinent enzymology. Collectively, this knowledge should lead to further insight into the evolutionary basis for this elaborate but taxonomically delimited metabolic system.

V. CONCLUSIONS

The purposes of the preceding sections has been to suggest, through examples and allusions, the multiplicity of biological, biochemical and chemical problems that are inherent in comparative phytochemistry. Many challenging biological problems are generated spontaneously, centrifugally, from purely chemical research, but unless the biological perspective is present to sensitize one to such biological implications, an opportunity may be lost. At present, an intellectual bridge has been built between phytochemists and plant systematists through a common interest in the evolution of molecules, of species, or of both. If this interest is to be sustained, nurtured if you will, we must strive continuously to broaden our intellectual perspectives especially perhaps on the biological side. A perspective which views chemotaxonomy as a field in which the only *biological* purpose is acquiring distributional data for more efficient pigeonholing (as an altruistic "service" in the name of science) will eventually alienate the chemists. Conversely, a perspective of taxonomy in which the distributions of plant secondary constituents are viewed as the means by which we can discover whether the Hutchinson, Bessey or some other, taxonomic system is "correct" in a particular instance will have an unfavourable effect not only upon plant systematists but biologists in general. The term biochemical systematics (or chemosystematics) is preferable over chemotaxonomy on the basis of the distinction between systematics and taxonomy as discussed by Simpson (1961). Simpson considers systematics to be the study of the kinds of organisms and their relationships, while taxonomy is the "theoretical study of classification, including its bases, principles, procedures and rules." By biochemical systematics one thus implies a greater latitude in the recognition and pursuit of certain problems spin out of comparative phytochemical investigations, whatever the primary goal of the investigation may be.

We may explore briefly certain implications of the relatively narrow and the relatively broad perspective as applied to comparative phytochemistry in a single example. Recently, two interesting flavonoid compounds were characterized from *Centaurea jacea L.*, subspecies *angustifolia* (Schrank) Gugler, variety *pannonica* (Heuffel) Hayeh, sub-variety *glabrescens* Gugler (Gurniak, 1964). These flavonoids, centaurein (IX) and jacein (X) (Fig. 12), occur respectively in the root and in the shoot of this taxon (Farkas *et al.* 1964). Since, as stated by Gurniak, most varieties do not differ qualitatively in their chemical constituents it is hardly likely that the chemical data will aid in the classification of a taxon which already permits its subvarieties of subspecies to be recognized. With an evolutionary interest, doomed though it may be to

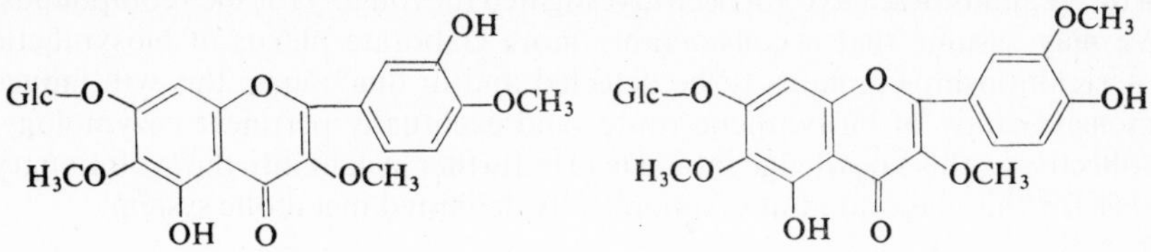

IX. Centaurein   X. Jacein

FIG. 12. Two structurally isomeric flavonoids of the shoot and root of *Centaurea*.

eventual frustration, one might be encouraged to explore the other species of *Centaurea* and allied genera for further distributions of these chemically-interesting highly methylated flavones. If an even broader biological and biochemical perspective is applied, one sees an interesting problem of enzyme specificity in the preferential methylation of the 3' or 4' position (compare the *O*-methyltransferase discussed earlier), but no 3', 4' dimethylation occurs. There is another problem in the mechanism by which the compounds are more or less compartmentalized within the plant, in the root or the shoot and in the possible functional significance of such distributions. The role of general methyl donors such as methionine could be studied with respect to the A and B-ring methylations of these compounds since it has been shown that methionine (methyl-^{14}C) is the methyl donor for some phenolic acids and flavonoids (Hess, 1964). Developmental studies of the sequence and place of occurrence of these compounds in seedlings would be of biological interest. Since the species *Centaurea jacea* hybridizes with several other species, genetic studies could be carried out. What would be the implication, for example, of close linkage of the factors governing the specific occurrences of centaurein (IX) and jacein (X)? Most of the above problems do not fall within the domain of taxonomy, but they do harbor much of interest to evolutionary mechanisms.

REFERENCES

Alston, R. E. (1959). *Genetics* **30**, 261.
Alston, R. E. (1964). *In* "Biochemistry of Phenolic Compounds" (J. B. Harborne, ed.), p. 171. Academic Press, London and New York.
Alston, R. E., Mabry, T. J. and Turner, B. L. (1963). *Science* **142**, 545.
Alston, R. E., Rösler, H., K. Naifeh and Mabry, T. J. (1965). *Proc. nat. Acad. Sci.* (*Wash.*) (In press).
Aronoff, S. (1956). *Plant Physiol.* **31**, 355.
Bell, E. A. (1964a). *Biochem. J.* **91**, 358.
Bell, E. A. (1964b). *Fed. Europ. Biochem. Socs. Abstracts*, **1**, 53.
Bell, E. A. (1964c). *Nature, Lond.* **203**, 378.
Bell, E. A. and Fowden, L. (1964). *In* "Taxonomic Biochemistry and Serology". (C. A. Leone, ed.), p. 203. Ronald Press, New York.
Bell, E. A. and Tirimanna, A. S. L. (1964). *Biochem. J.* **91**, 356.
Brehm, B. G. and Alston, R. E. (1964). *Amer. J. Bot.* **51**, 644.
Constance, L. (1964). *Taxon* 13, 257.
Cranmer, M. (1964). *Amer. J. Bot.* **51**, 687.
Crombie, L. and Peace, L. (1963). *Proc. chem. Soc.* 246.
Dean, F. M. (1963). "Naturally Occurring Oxygen Ring Compounds". Butterworths, London.
Farkas, L., Hörhammer, L., Wagner, H., Rösler, H. und Gurniak, R. (1964). *Chem. Ber.* **97**, 1666.
Finkle, B. J. and Masri, M. S. (1964). *Biochim. biophys. Acta* **85**, 167.
Fraenkel, G. S. (1959). *Science* **129**, 1466.
Grisebach, H. and Ollis, W. D. (1961). *Experientia* **17**, 4.
Gurniak, R. (1964). Dissertation, Universität München.
Harborne, J. B. and Hall, E. (1964). *Phytochemistry* **3**, 421.
Harborne, J. B. and Simmonds, N. S. (1964). *In* "Biochemistry of Phenolic Compounds". (J. B. Harborne, ed.), p. 77. Academic Press, London and New York.
Hendrickson, J. B. (1960). *In* "The Alkaloids" (R. M. F. Manske, ed.), Vol. VI, p. 179. Academic Press, New York and London.
Hess, D. (1964). *Z. Naturf.* **19**, 148.
Kawano, N. H. Miura and Waiss, A. C. (1964). *Chem. & Ind.* **49**, 220.
Leete, E. (1963). *In* "Biogenesis of Natural Compounds" (P. Bernfeld, ed.), p. 739. Macmillan, New York.
Leete, E. (1965). *Science* **147**, 1000.
McClure, J. W. (1964). Ph.D. Dissertation. The University of Texas, Austin, Texas, U.S.A.
McClure, J. W. and Alston, R. E. (1964). *Nature, Lond.* **201**, 311.
Mothes, K. and Schütte, H. R. (1963). *Angew. Chem.* **75**, 265.
Murti, V. V. S., Raman, P. V. and Seshadri, T. R. (1964). *Tetrahedron Letters* **40**, 2995.
Nowacki, E. (1963). *Genetica Polonica* **4**, 161.
Ressler, C. (1962). *J. biol. Chem.* **237**, 733.
Reitsema, R. H. (1959). *J. Amer. Pharm. Ass.* **47**, 267.
Schütte, H. R. and Nowacki, E. (1959). *Naturwissenschaften* **48**, 493.
Schütte, H. R., Nowacki, E. and Schafer, C. (1962). *Arch. Pharm.* **294**, 20.
Schütte, H. R. and Lehfeldt, J. (1964). *J. Prakt. Chem.* **24**, 143.

Simpson, G. G. (1961). "Principles of Animal Taxonomy". Columbia University Press, New York.

Wagner, R. P., Radhakrishman, A. N. and Snell, E. E. (1958). *Proc. nat. Acad. Sci. (Wash.)* **44**, 1047.

Willaman, J. J. and Schubert, B. G. (1961). "Akaloid-bearing plants and their contained alkaloids", p. 287. Agricultural Research Service. U.S.D.A. Technical Bulletin No. 1234, Washington, D.C., U.S.A.

CHAPTER 4

The Distribution of Alkanes

A. G. DOUGLAS and G. EGLINTON

The Chemistry Department, The University, Glasgow, Scotland

I. INTRODUCTION

This topic was last reviewed three years ago (Eglinton and Hamilton, 1963), and the material covered then will receive only brief mention in the present chapter. Little information of taxonomic value relating to the distribution of alkanes has emerged in the meantime, and so the present review will deal mainly with the range of alkanes now known, including those derived from animal sources since these may well act as a guide to the possible occurrence of similar alkanes in plants. Straight and branched alkanes receive the most attention and reference is made to the possible modes of biogenesis of these compounds and of their use as "biological markers" in geological materials and in food cycles. Further, brief mention will be made to olefinic and alicyclic hydrocarbons, for these two classes can be related to the acyclic alkanes. With compounds such as the alkanes, which lack functional groups, separation and identification methods are crucial and must be very efficient to overcome the similarity in physical and chemical properties exhibited by this class of compound.

II. THE N-ALKANES

A. OCCURRENCE OF n-ALKANES

Long-chain n-alkanes are widely distributed in the plant kingdom as components of the cuticular waxes which are common to the surfaces of leaf, stem, flower and pollen: (a few reports refer to alkanes in heartwood, etc.;

Cocker and Shaw, 1963, and Mathis and Ourisson, 1964). The appearance and quantity of the cuticular waxes vary greatly, and the waxy layer has a number of roles to play in the activity of the plant. The coating has a well-defined ultrastructure which although varying greatly from one species to another results in a higher contact angle for water droplets than would a smooth layer of wax (see Leyton and Juniper, 1963). The wax generally consists of a complex mixture of long-chain hydrocarbons, alcohols, alde-hydes, ketones, acids, esters and related compounds.

Table 1 shows the chief types of compound present, the chain lengths being in the region C_{20} to C_{35} and forming what can be regarded as a carbon-number "family". There is evidence that there may be other lipid families of smaller

TABLE 1.	Leaf waxes: long-chain constituents

Dominant carbon number

Even	Odd
$CH_3(CH_2)_nCO_2H$	$CH_3(CH_2)_nCH_3$
$CH_3(CH_2)_nCH_2OH$	$CH_3(CH_2)_nCHOH(CH_2)_mCH_3$
$CH_3(CH_2)_nCHO$	$CH_3(CH_2)_nCO(CH_2)_mCH_3$
$HOCH_2(CH_2)_nCH_2OH$	$CH_3(CH_2)_nCX(CH_2)_4CX(CH_2)_mCH_3$
$HOCH_2(CH_2)_nCO_2H$	$CH_3(CH_2)_nCOCH_2CO(CH_2)_mCH_3$
$HO_2C(CH_2)_nCO_2H$	
	[CX is $>C=O$ or $>CHOH$]

chain length, for example C_{10} to C_{20}, probably related to the normal intra-cellular fatty acids. Recent studies show that some natural sources provide the full range of n-alkanes from around C_{10} to C_{40}, grouped in one or more "families". At the lower end, n-heptane is a prominent constituent of certain pines (Mirov, 1961) and n-nonane and n-undecane have been reported in *Hypericum* species (Mathis and Ourisson, 1964). In Table 1 the various classes of constituents are grouped into *even*-carbon-number and *odd*-carbon-number families. The families of homologues typically show high concentrations of compounds having alternate carbon numbers, with the other isomers—the odd or the even carbon number, as the case may be—only present in small amounts. The components having an oxygen function on the terminal carbon atom are predominantly even-numbered and it seems reasonable to assume that the odd-numbered series generally result by the loss of an oxygenated carbon atom from an even-numbered, acetate-derived, fatty acid or related compound.

B. BIOGENESIS OF n-ALKANES

The characteristic relationship of odd-numbered hydrocarbons and even-numbered acids mentioned above is analogous to that of the naturally occurring polyacetylenes, where the hydrocarbons are nearly always odd-numbered and are believed to be derived from the even-numbered acids by simple decarboxylation at some stage (Bu'Lock, 1964). However, Mazliak (1963, 1964) has claimed that he could not obtain appreciable ^{14}C incorporation into long-chain n-alkanes under conditions (apple cuticle in labelled acetate) which resulted in heavy labelling of the long-chain fatty acids and alcohols of apple wax. He concluded that the alkanes must be formed by an independent route, but this seems inherently unlikely to the authors and further experiments are desirable. In the short chain family of alkanes, Sandermann and Schweers (1960) have shown that n-heptane is derived on an acetate pattern, presumably from n-octanoic acid.

In the even-numbered series, the oxygen functions are typically α-terminal or α,ω-diterminal, in common with the compounds formed by the action of certain bacteria which metabolize n-alkanes to n-alkanoic acids, ω-hydroxy acids and α,ω-dicarboxylic acids of the same carbon number (e.g. Foster, 1962; Baptist *et al.*, 1963; Leadbetter and Foster, 1960; Lukins and Foster, 1963; Kester and Foster, 1963; and Kallio *et al.*, 1963). Certain osmophilic yeasts also ferment n-alkanes to ω-hydroxy acids (Tulloch *et al.*, 1962), and

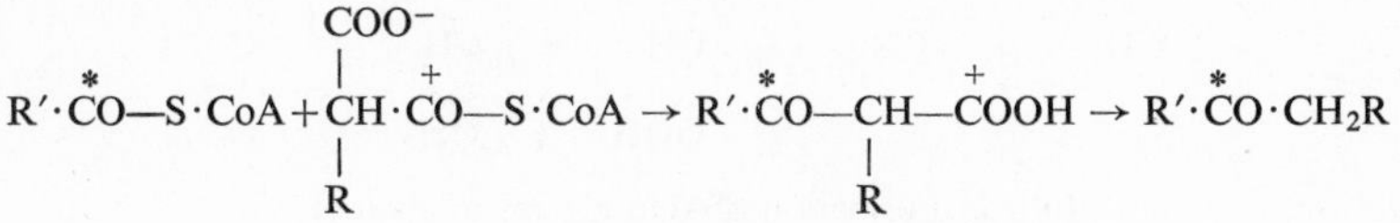

$$R'\cdot\overset{*}{C}O\text{—}S\cdot CoA + \underset{R}{\overset{COO^-}{\underset{|}{CH}}}\cdot\overset{+}{C}O\text{—}S\cdot CoA \rightarrow R'\cdot\overset{*}{C}O\text{—}\underset{R}{\overset{+}{\underset{|}{CH}}}\text{—}COOH \rightarrow R'\cdot\overset{*}{C}O\cdot CH_2R$$

FIG. 1. Mycolic condensation.

mammalian metabolism may also result in terminal oxygenation (McCarthy, 1964; Preiss and Bloch, 1964).

The odd-numbered oxygenated compounds sometimes, but not always, have the expected structure based on the loss of one carbon from an acetate engendered alkanoic acid. Alternative explanations require a starter unit other than acetyl coenzyme A, or subsequent oxygenation as in the conversion of stearic to ricinoleic acid, via oleic acid (Stumpf, 1962; Yamada and Stumpf, 1964). In the case of ketones the mode of biosynthesis may be similar to that of the mycolic acids (Gastambide-Odier and Lederer, 1959). The mid-chain carbonyl group in these compounds has been shown to be derived via the "mycolic condensation" of two palmitic acid units (Fig. 1). The two C_{16} acids unite to form a C_{32} acid which then decarboxylates to give the C_{31} chain (Lederer, 1964). This is believed to be a pathway of general validity.

The occurrence of a C_{29} ketone, which would require two C_{15} acid units, cannot be reasonably explained in this way, since odd-numbered acids are normally present only in trace quantities, formed either by incorporation of a propionate starter or by α-oxidation. Purdy and Truter (1963) have shown that cabbage wax is largely composed of n-C_{29} compounds; the dominant hydrocarbon is n-C_{29} and the other n-C_{29} derivatives are oxygenated either singly or doubly at C-10 and C-15—both as hydroxyl and/or carbonyl groups (e.g. I). This direct carbon number relationship between the most abundant compounds is probably significant and might arise in various ways. For example, the incorporation of a dicarboxylic acid, such as β-ketoadipic acid derived from the Krebs cycle, would generate the 1,6-oxygenation pattern but one or more

$$CH_3(CH_2)_8CO(CH_2)_4CHOH(CH_2)_{13}CH_3$$

I

$$CH_3(CH_2)_{10}COCH_2CO(CH_2)_{14}CH_3$$

II

mycolic acid condensations involving a second dicarboxylic acid followed by multiple decarboxylations, would still be necessary. The simple C_{29} ketone, $CH_3(CH_2)_{13}CO(CH_2)_{13}CH_3$, could similarly be accounted for by union of two C_{16} dicarboxylic acid units. However, highly specific oxygenation of a C_{30} monocarboxylic acid in the fashion already proposed for bacterial attack on carbon chains (Leadbetter and Foster, 1960) is an alternative (Fig. 2).

$$-CH_2- \ \rightarrow \ -\overset{\bullet}{C}H- \ \rightarrow \ \underset{OOH}{-CH-} \ \rightarrow \ \underset{OH}{-CH-} \ \rightarrow \ \underset{O}{-\overset{\parallel}{C}-}$$

FIG. 2. Formation of keto groups in alkanes.

In another type of n-C_{29} compound, the β-diketone (II) reported for *Eucalyptus* species (Horn *et al.*, 1964), the oxygenation is at the 12- and 14-positions, not 10 and 15—as in the cabbage wax compound (I). The inference here is either (a) that there is a mid-chain pairing of monocarboxylic acids, for example n-C_{16} and n-C_{14}, as discussed above, or (b) that the oxygens are relics of a simple polyketide chain derived in a linear C_2 fashion. Horn and his collaborators remark that they could not discern any simple relationships between the chain length distributions of the diketones and of the alkanes occurring with them.

It is interesting to note that the methyl carbinols, are *odd* numbered and of short chain length (in *Eucalyptus*, about C_{15}). They might conceivably arise by hydroxylation of odd-carbon number olefines or by reduction of odd-numbered ketones (formed from β-keto-acids by decarboxylation).

Before leaving the question of biogenesis of the alkanes and related compounds, the site of biosynthesis and the mode of secretion of the leaf waxes deserve mention. The site is believed to be in the subcuticular cells, but definitive studies have yet to be made. It is significant, however, that spinach chloroplasts do not have the ability to synthesize leaf waxes (Zill and Harmon, 1962). Barrera, *et al.* (1964), Mazliak (1963, 1964) and Mazliak and Pommier-Miard (1963) claim that the hydrocarbons and waxy materials present in the deeper cuticular layers are of shorter chain-length than those at the surface and, in *Ruta pinnata*, the iso-alkanes are said to occur only in the upper layers

(Barrera *et al.*, 1964). Again, waxes vary in their general composition from the dorsal to the ventral surface of the leaf (Baker, Batt, *et al.*, 1963), and Oró, Nooner and Wikström (1965) report a lower odd/even ratio for the alkanes in the leaves of *Medicago arabica* as compared with the whole plant.

C. DISTRIBUTION AND TAXONOMIC ASPECTS OF n-ALKANES

Most plants so far studied show a dominance of the odd-carbon numbered n-alkanes, the centre of distribution lying around C_{29} and C_{31}. Recent examples include the wax of the leaves of cabbage, *Brassica oleracea* (Purdy and Truter, 1963); *Eucalyptus* species (Horn *et al.*, 1964); *Aloe* species (Herbin, 1964); *Hypericum* (Mathis and Ourisson, 1964); apple and certain other common fruit trees (Mazliak, 1963 and 1964; Fernandes *et al.*, 1964); clover, *Medicago arabica* (Oró, Nooner and Wikström, 1965); *Trifolium pratense* (Weenink, 1962); sisal (Razafindrazaka and Metzger, 1963); *Vicia faba* (Dixon *et al.*, 1965); and *Chamaecyparis obtusa* (Fukui and Ariyoshi, 1963). Hemming *et al.* (1963) discuss the wax from the spadix of *Arum maculatum* and Hallgren and Larsson (1963) describe that from the pollen of Rye (*Secale cereale*).

However, in the shorter chain length waxes (below C_{25}) there are indications of a much smaller odd/even predominance. Indeed, Stevenson (1961) has reported an odd/even ratio of less than 1 for *Calendula officinalis*, while Kuksis (1964) has claimed that for certain seed oils the odd/even ratio approximates to one below about n-C_{27}. Again, Iwata and Sakurai (1963) have reported that the alkane fraction of *Chlorella* is mainly composed of the n-C_{16} alkane, and certain bacteria, marine plants, algae and phytoplankton are said to have odd/even ratios of about one (Oró, Nooner and Wikström, 1965).

One coral is stated by Ciereszko *et al.* (1963) to contain n-C_{36} alkane as a prominent constituent, while other marine organisms have alkane fractions which are very complex and bear a strong resemblance to petroleum. A similar degree of complexity in the normal, branched and cyclic alkane fractions of wool wax has been reported by Mold *et al.* (1964). In such cases, the problem of contamination either by crude petroleum itself or by the metabolic products of other organisms, such as the bacterial microflora, is ever present: there is scope for using the ^{14}C dating method (Evans *et al.*, 1964), by which contributions from petroleum hydrocarbons can be detected.

Dominantly odd carbon-number distributions of n-alkanes have been reported for the bronze orange bug (Park and Sutherland, 1962) for certain species of Coccidae (Faurot-Bouchet and Michel, 1964), and Dixon *et al.* (1965) have shown that the alkane fraction for one aphid species, *Megoura viciae* differs from that of its host plant, *Vicia faba*. By contrast, Oró, Nooner and Wikström (1965) find that the alkanes extracted from cow dung and from the appropriate pasture plants are very similar.

Taxonomic aspects have received some attention: Aplin *et al.* (1963) have studied the diterpene hydrocarbons of 28 species of Podocarpaceae; Herbin (1964) has examined 16 species of South African *Aloe* (Liliaceae) from various sections and sub-sections of this genus (Fig. 3) and observes the following:

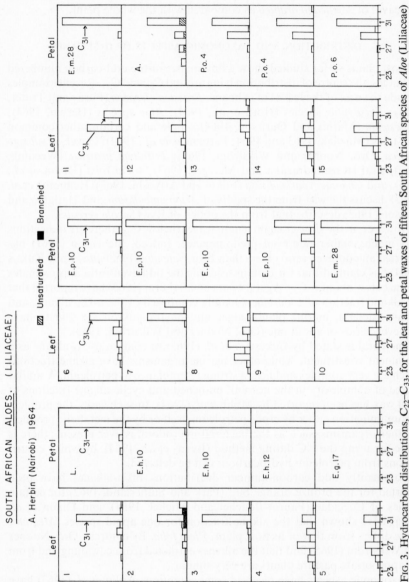

FIG. 3. Hydrocarbon distributions, C$_{22}$–C$_{33}$, for the leaf and petal waxes of fifteen South African species of *Aloe* (Liliaceae) (Herbin, 1964). L. = Anguialoe, E. = Leptoaloe, E. = Eualoe (Subsections: h = Humilies, g = Grandes, p = Prolongatae, m = Magnae). A. = Anguialoe, P. Pachydendron (o = Ortholophae), Numerals = series and groups within subsections.

(i) the petal and leaf waxes differ quite markedly; (ii) the petal wax often contains a single hydrocarbon (n-C_{31}) but the leaves have a range of hydrocarbons [though C_{31} is dominant in most cases; compare Kasprzyk *et al.* (1962) for n-C_{25} in the petal of *Peonia albiflora*]; (iii) chemotaxonomic classifications seem rather unsuccessful since some botanically closely related *Aloes* have very different patterns of n-alkanes, whereas others—widely spaced botanically—may have similar n-alkane patterns. However, the work of Borges *et al.* (In preparation) on a number of New Zealand species related to the genus *Podocarpus* (Podocarpaceae) is more encouraging. Thus, the 21 species of *Podocarpus* examined have mainly n-C_{29} and n-C_{31} compounds whereas two species of *Phyllocladus* (also Podocarpaceae) and seven species of *Cupressus*, *Libocedrus* and *Thuja* (all Cupressaceae) have n-C_{33} and n-C_{35} as the dominant alkanes. As to species constancy, several varieties of tobacco have been examined and found to give similar alkane patterns (Mold *et al.*, 1963; Pyriki and Hofmann, 1963); however, the percentage of alkane was 7% and 40% of the total wax for two varieties of raspberry cane (Baker, Batt *et al.*, 1963).

D. ISOLATION AND CHARACTERIZATION OF ALKANES

This topic has been dealt with at some length in the previous review (Eglinton and Hamilton, 1963) and also by Burchfield and Storrs (1962). Some of the more recent advances in techniques will be mentioned.

There has been some progress in the use of thin layer chromatography for the separation of hydrocarbons from other lipid constituents and for the separation of alkanes from alkenes and aromatic hydrocarbons. (Duncan, 1962; Purdy and Truter, 1963; Haahti *et al.*, 1963; Berg and Lam, 1964; Maier and Mangold, 1964). Trinitrobenzene may be more effective than silver nitrate for selective retention of olefins through complex formation during gas-liquid chromatography (g.l.c.) (Cvetanovic *et al.*, 1964). Molecular sieves (for review, see Barrer, 1964) have been further investigated (Laurent and Bonnetain, 1964) and can be used to cleanly separate n-alkanes by a batch process (Mold *et al.*, 1963; Eglinton *et al.*, 1964) or, with varying success, using them as column materials for direct g.l.c. (Szymanski, 1964; Schenk and Eisma, 1963; Barrall and Baumann, 1964). The difficulty with direct g.l.c. appears to be incomplete entrapment of n-alkanes and/or incomplete exclusion of the branched isomers. The situation seems to vary with the chain length of the alkanes. The modified montmorillonites (Weiss, 1963) also deserve further study for the separation of isomers.

G.l.c. remains the most powerful tool for the separation of homologous alkanes. Programmed-temperature operation is routinely used where a wide range of carbon numbers has to be analysed (Fig. 4). The method has been reviewed by Horning and van den Heuvel (1963) and Levy and Paul (1963). Hydrocarbons of quite long chain length are run routinely (see Jarolimek, Wollrab and Streibl, 1964). Highly polar phases such as tetracyanoethylated pentaerythritol are effective in separating close-boiling point straight and branched alkanes (Eglinton *et al.*, 1964). The use of 7-ring polyphenylether is recommended for selectivity and stability; better separations have been obtained than with SE–30 (West, 1963).

"Fingerprints" obtained by the pyrolysis of organic compounds followed by g.l.c. hold promise as a qualitative tool (Keulemans, and Perry, 1962; Oyama, 1963; Perry, 1963); and dehydrogenation followed by g.l.c. may be useful for cyclic hydrocarbons (Carman, 1963).

Mass spectroscopy is a highly desirable adjunct for the conclusive identification of hydrocarbons on a small scale (e.g. Beynon, 1960; Herlain, 1964; Hood, 1963; Levy *et al.*, 1963). The mass spectra depicted in Fig. 5 exemplify both the structural elucidation of a branched alkane and the detection of small quantities of a structural isomer in an otherwise pure C_{33} alkane (Mold

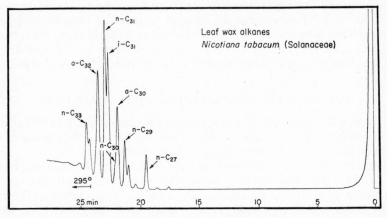

FIG. 4. Gas chromatogram of the leaf wax alkanes of tobacco *Nicotiana tabacum* (Solanaceae) Column conditions; $5' \times \frac{1}{8}''$; 3% SE.30 on 80–100 mesh$_2$; Chromosorb W (DMCS); 30 ml/min N_2 at 50 lb/in^2; temperature programmed at $7.5°$/min; initial column temp 100°; detector 225°; injector 230°; attenuation 10×8 (Eglinton *et al.*, 1965).

et al., 1963; Eglinton *et al.*, 1964, 1965). Thus, the high mass end of the spectra for two g.l.c. cuts taken from branched alkane fractions (excluded from 0·5 nm sieve) are displayed in the lower half of Fig. 5; the left-hand cracking pattern corresponds to that of the anteiso-C_{32} skeleton (3-methylhentriacontane) while the right-hand spectrum, which corresponds in the main to that of the synthetic iso-C_{33} alkane above it, reveals the presence of a small amount (~5%) of the anteiso-C_{33} isomer 2-methyldotriacontane (enhanced peak at m/e 435). The combined gas chromatograph–mass spectrometer (g.c.m.s.) promises important advances (recent references include Banner *et al.*, 1964; Eneroth *et al.*, 1964; Day and Libbey, 1964; Ryhage *et al.*, 1965; Oró, Nooner and Wikström, 1965).

Nuclear magnetic resonance data have been reported for paraffinic chains (Bartz and Chamberlain, 1964). Lack of sensitivity is the main drawback to the use of n.m.r. spectrometers, though special micro-sampling tubes and

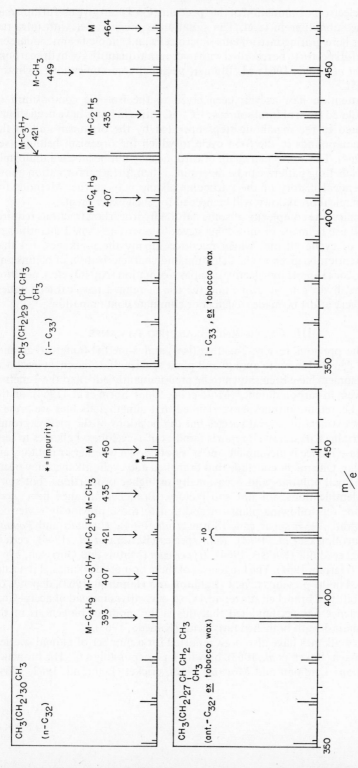

FIG. 5. Mass spectra of synthetic dotriacontane (n-C$_{32}$) and 2-methyldotriacontane (anteiso-C$_{32}$) and of 3-methylhentriacontane (iso-C$_{33}$), and of 3-methylhentriacontane (anteiso-C$_{32}$) and 2-methyldotriacontane (iso-C$_{33}$) from tobacco wax.

repetitive electronic summation of the spectrum (CAT) can give useful information at the submilligram level. The situation is much easier with infra-red spectra for here grating instruments, micro cells and simple beam-condensers permit useful spectra to be recorded with microgram quantities of hydrocarbons (Blumer, *et al.*, 1963, 1964; M. Blumer, personal communication; Eglinton *et al.*, 1965).

Little attention has, as yet, been given to the isotopic composition of alkanes isolated from natural sources: $^{13}C/^{12}C$ ratios in lipids have been shown to be related to the metabolic steps operated by the organism and to the previous compounds in the food cycle to which the organism belongs (see Parker, 1964). The atoms in a chain of carbons may not be labelled uniformly and if the labelling pattern can be determined then further information about the biochemical history of the particular alkane may accrue. Methods for specific chemical breakdown will be needed for such experiments.

In summary, the complexity of some natural hydrocarbon fractions require, (i) that full use be made of molecular sieves to reveal any small quantities of branched or cyclic alkanes which may accompany the n-alkanes. (ii) that careful attention be given to the detection and characterization of olefins and aromatics, for example by selective complex formation ($AgNO_3$ etc.), addition of bromine, hydrogen, or ozone and the use of refined physical techniques. (iii) that every effort be made to detect or eliminate contamination.

III. Cyclic and Branched Alkanes

Since the previous review, 2-methyl-(iso-) and 3-methyl-(anteiso-) alkanes have been detected in more animal and plant species. In tobacco, at least, the 2-methyl isomers have been shown to be predominantly odd, and the 3-methyl isomers even, in carbon number (Mold *et al.*, 1963). Šorm *et al.* (1964), on the basis of g.l.c. retention times, have claimed that dimethyl alkanes are present in the wax extracts of several species but the positions of the methyl groups are not certain. Kuksis (1964) reports families of 1-cyclohexyl alkanes in seed oils but the evidence is incomplete and it is not known whether or not they are extracellular. One might conclude that branched and cyclic alkanes may occur widely in small amounts and occasionally in higher proportions, but some present identifications are not satisfactory. Branched alkanes have been reported for the following plants in addition to those previously reviewed: rose (Wollrab, 1964) sugar cane (Šorm *et al.*, 1964), *Humulus* and *Populus* species (Jarolimek *et al.*, 1964), *Ruta* species (Barrera *et al.*, 1964), certain commercial seed oils (Kuksis, 1964), *Hypericum* (Mathis and Ourisson, 1964) and *Aloes* (Herbin, 1964). The biogenesis of these branched alkanes, if it occurs as indicated in the previous review (Eglinton and Hamilton, 1963), depends on the availability of branched starter acyl CoA derivatives instead of acetyl CoA itself. Evidence has accumulated that this is the mode of biogenesis of the medium chain-length branched fatty acids (Kaneda, 1963).

Branched alkanes have also been reported for a number of animal species, e.g. cockroach (Baker, *et al.*, 1963), blowfly-larvae (possibly a $C_{45}H_{92}$ branched hydrocarbon; Laidman and Morton, 1962), cricket (Leibrand, 1962, 1964), aphid (Dixon *et al.*, 1965), sheep (woolwax, Mold *et al.*, 1964). Cyclic alkanes have been reported for woolwax (Mold *et al.*, 1964) and butter (McCarthy *et al.*, 1964). Woolwax is now found to be an extremely complicated mixture and the various "families" of homologous series of branched and cyclic hydrocarbons reported by Mold and his collaborators will need further study, especially as they differ from previous findings. Tobacco wax is chosen as an example of a wax containing branched alkanes and the programmed g.l.c. is shown (Fig. 4). Improved resolution is easily achieved when the analysis is performed isothermally but the best method of separating the normal from the branched hydrocarbons is certainly by using 0·5 nm molecular sieve. Examples of mass spectra obtained from tobacco wax by such sieving followed by collection of individual g.l.c. peaks, are illustrated in Fig. 5. As already mentioned, these spectra show that a small amount of the anteiso-compound is present in the iso-C_{33} fraction. Such findings underline the need for the preparation of pure reference alkanes in parallel with the study of naturally-occurring hydrocarbons.

TABLE 2. Plant olefins of C_{15}-C_{33} chain length[1]

$CH_3(CH_2)nCH=CH_2$
$CH_3(CH_2)nCH=CHCH_2CH_3$ *cis-*
$CH_3(CH_2)nCH=CH(CH_2)_8CH_3$ *cis-*
$CH_3(CH_2)nCH=CH(CH_2)_mCH_3$ *trans-*
— Dienes —

---

[1] From rose petal, sugar cane (Šorm *et al.*, 1964).

Simple, non-isoprenoid olefins have been reported for a number of plant species in the past (e.g. Warth, 1960, Gerarde and Gerarde, 1962). Recent work (see Table 2) has shown that olefins, like branched alkanes, are frequently present in small amounts and occasionally in large proportions, for example in *Rye* pollen (Hallgren and Larsson, 1963), certain petal waxes— *Aloe* species (Herbin, 1964), *Rosa* species and sugar cane (Wollrab, 1964; Šorm *et al.*, 1964). Hexacos-1-ene has been reported in *Chlorella* (Iwata and Sakurai, 1963). Highly unsaturated systems have been reported, including compounds like aplotaxene (Romanuk *et al.*, 1958) and the numerous poly-acetylenes and related compounds (for a review see Bu'lock, 1964 and Chapter 5). With a few exceptions, the olefins so far reported are dominantly of odd carbon number; this has been shown by hydrogenation to the corresponding n-alkanes. They presumably arise by desaturation of the alkanes, or by dehy-dration of the corresponding secondary alcohols (or some such intermediate). Herbin (1964), however has suggested that the occurrence of alkanes and alkenes in the wax of rapidly developing systems, such as petals, indicates that the olefins may be the precursors of the alkanes.

Simple olefins reported for animal species include a diene from the cockroach

(Baker *et al.*, 1963), but Gilby and Cox (1963) give a different structure for this diene.

IV. Isoprenoid Alkanes and Alkenes

To our knowledge there is only one acyclic, saturated, isoprenoid alkane, farnesane, which occurs in plants; it is found in some *Dipterocarpus* species as about 15% of the resin. Hexahydrofarnesene (containing one double bond) is also present (G. Ourisson, personal communication). Polyunsaturated isoprenoids are of course well known, e.g. farnesene (apple wax—Murray, *et al.*, 1964), squalene, phytoene and the whole range of carotenoid hydrocarbons (Gerarde and Gerarde, 1961, see Chapter 7). The occurrence of new terpene hydrocarbons is usually reviewed annually (e.g. Connolly and Overton, 1964).

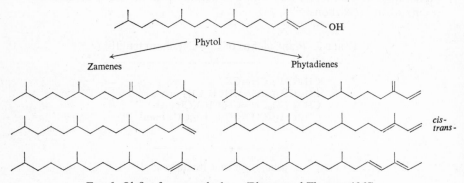

Fig. 6. Olefins from zooplankton (Blumer and Thomas, 1965).

However, certain types of zooplankton (Copepods) and the subsequent members of the food chain (herring, whale, shark, etc.) contain quite large quantities of pristane, the fully saturated norditerpane; together with smaller amounts of the C_{20} phytadienes and the C_{19} mono-olefins—the zamenes (Fig. 6; Blumer *et al.*, 1963, 1964; Blumer and Thomas, 1965a, b). All of these hydrocarbons presumably derive from alteration of the phytyl side-chain of chlorophyll, as there is no evidence for pristane in the phytoplankton. The dienes are claimed to be indigenous to the copepod and are not chromatographic artefacts. There is a recent claim for the presence of squalane and phytane in beef brains (Nicholas and Bombaugh, 1965) but the evidence is not complete.

One saturated, norditerpane, fichtelite (IV), has been known for many years as a constituent of decayed or fossil conifer resin and presumably derives by anaerobic decarboxylation of abietic type resin acids. Its complete stereochemistry has been recently established and there is retention of the C-4 configuration of abietic acid (III) (Burgstahler and Marx, 1964).

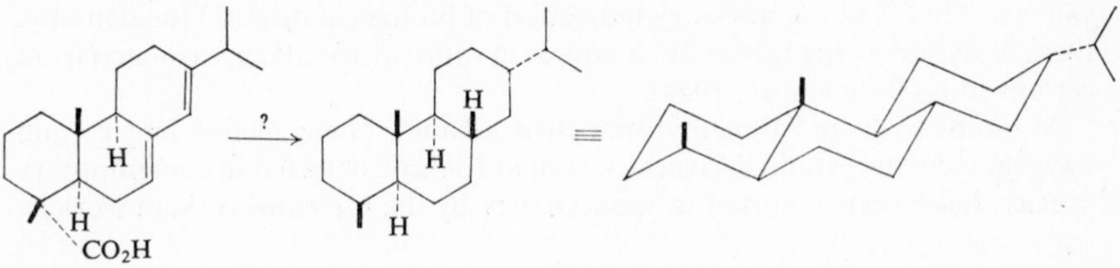

III. Abietic acid   IV. Fichtelite

V. HYDROCARBONS AS BIOLOGICAL MARKERS; AN APPROACH TO ORGANIC GEOCHEMISTRY

Alkanes, as mentioned in the previous review (Eglinton and Hamilton, 1963), occur as constituents of fossil materials, e.g. coal, petroleum, earth-waxes, etc. (see Warth, 1960; Breger, 1963; Colombo and Hobson, 1964). Several workers have proposed that, since alkanes occur widely in living systems and because they are rather insoluble and are resistant to degradation, their occurrence may provide an indication of a formative organism long after its death and interment in a sediment: in fact, Meinschein (1964) maintains that they may be the best preserved and most widely distributed remnants of former life and of particular value on account of their retention of the isotopic, structural and distributional order inherent in their biological origin. Crude petroleum contains a very wide range of hydrocarbon types but there is increasing evidence that a sizeable fraction of the oil is made up of alkanes closely related in their carbon skeletons to lipid molecules common to contemporary organisms (Mair, 1964). Again, many of the aromatic hydrocarbons can be rationalized as dehydrogenation products of these same molecules.

Most contemporary plants show an odd/even carbon number ratio greater than 1, and one might assume that unaltered n-alkane residues from plant debris would show a similar ratio. This is so for relatively unaltered deposits such as lignite or brown coal; montan wax extracts show a clear biological pattern for the straight chain alkanes, alcohols, and acids (Fig. 7, Wollrab *et al.*, 1963).

Crude oils show only a slight (if any) predominance—a general finding, the explanation for which has been the subject of much controversy [for relevant references see the previous Report, Baker (1962); Breger (1963); Cooper and Bray (1963); Ciereszko *et al.* (1963); Colombo and Hobson (1964); Meinschein (1964) and Oró *et al.* (1965)]. Three main arguments have been advanced: (i) oil of non-biological origin makes a substantial contribution (see Sylvester-Bradley and King 1963; and Johnson and Wilson, 1964); (ii) the marked odd/even pattern is destroyed by migration, selective elimination or *de novo* synthesis (see Jurg and Eisma, 1964); (iii) the n-alkanes contributed by the originating organisms already had an odd/even ratio ∼1. Oró *et al.* (1965) have summarized their attitude in regard to the last point: "an odd-to-even carbon-number preference in the distribution of normal alkanes may be a sufficient, but is not a necessary indication of biological origin." Incidentally, there is an interesting report of an odd/even ratio ~ 1 for alkanes isolated from certain soils (Butler *et al.*, 1964).

Monomethyl—and dimethyl-branched alkanes, some olefins and certain oxygenated triterpenoid derivatives, similar to those detected in contemporary species, have been reported in montan wax by the Czechoslovakian workers

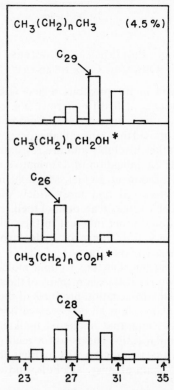

FIG. 7. Distribution of n-alkanes, n-alkanols and n-alkanoic acids from montan wax (Wollrab *et al.*, 1963). *From esters (26% of wax).

(Jarolimek *et al.*, 1964, 1965; Wollrab, 1964). However, the simple, acyclic isoprenoid alkanes have provided the most striking developments in the comparative phytochemistry of contemporary and ancient plant constituents. Evidence is accumulating that these specifically branched hydrocarbons (Fig. 8), especially the C_{20} (phytane) and C_{19} (pristane) compounds, occur quite generally in sediments and crude oils, often as prominent constituents. There is a strong presumption that they are largely derived directly or indirectly from the chlorophyll of green plants and are therefore valid "biological markers". Stereochemical correlations are still lacking for the asymmetric centres of the contemporary and the fossil alkanes, but the gross structures have been assured by the full range of physical methods (see Eglinton *et al.*, 1965). Interestingly, Cason and Graham (1965) have isolated the acyclic C_{14}, C_{15}, C_{19} and C_{20} isoprenoid acids from a relatively "young "petroleum, thereby providing a parallel for the alkane distribution noted above.

Pristane has been identified in brown coal distillates (Kochloefl *et al.*, 1963) and in a number of crude oils (see Mair, 1964 for references). However, the first positive identification in a sedimentary rock would appear to be due to Cummings and Robinson (1964) and Robinson *et al.* (1965).

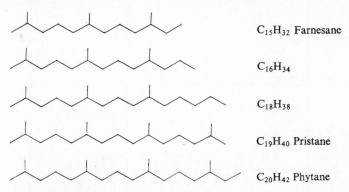

$C_{15}H_{32}$ Farnesane

$C_{16}H_{34}$

$C_{18}H_{38}$

$C_{19}H_{40}$ Pristane

$C_{20}H_{42}$ Phytane

FIG. 8. Branched alkanes of isoprenoid type isolated from ancient sediments and crude oils.

Programmed temperature g.l.c. records for the alkane fractions of the shale examined by Eglinton *et al.* (1964, 1965), the Green River shale (Colorado), are shown in Fig. 9. Three charts are given; the total alkane fraction after separation from the rock extract; the branched-cyclic alkanes; and the n-alkanes resulting from treatment of the total alkanes with 0·5 nm molecular sieve. The shale is of Eocene age ($\sim 60 \times 10^6$ years old) but the preponderance of odd-numbered n-alkanes typical of contemporary plant waxes and of the C_{19} and C_{20} isoprenoid alkanes is striking and accords well with the morphological evidence for plant debris in the shale. Even more impressive, the higher branched-cyclics (Fig. 9, retention time *ca.* 23 min) have now been shown to be steranes and triterpanes (Calvin, 1965).

Pristane, phytane and other acyclic isoprenoids have recently been isolated from, and identified in, much older sedimentary rocks of the Precambrian, for example, the dark Nonesuch shale ($\sim 1 \times 10^9$ yr old) of upper Michigan (Eglinton *et al.*, 1964, 1965; Meinschein *et al.*, 1964; Barghoorn *et al.*, 1965) and the Gunflint chert ($\sim 2 \times 10^9$ yr old) of Ontario (Oró *et al.*, 1965). The presence of these alkanes of presumed biological origin are paralleled by micropalæontological observations made on these rocks which indicate the presence in them of microfossils with the morphology of algae and simple fungi. This chemical characterization of Precambrian life has now been taken one step further by Calvin's group at Berkeley (Belsky *et al.*, 1965), who have examined the alkane fractions (Fig. 10) derived from an even older Precambrian shale (Soudan iron formation of Minnesota), the age of which ($\sim 2.5 \times 10^9$ yr) extends beyond that of rocks for which morphological remains have been accepted as identified. The normal hydrocarbons have an unusually narrow

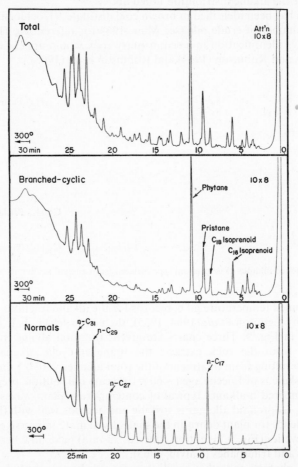

FIG. 9. Gas-liquid chromatograms of alkane fractions from the Green River shale, Colorado. (A) Total, (B) branched-cyclic and (C) normal alkanes. Column conditions as for Fig. 4. (Eglinton *et al.*, 1965.)

distribution around n-C_{17}, without any odd/even carbon-number predominance. The isoprenoids, however, are well-marked and 2- and 3-methyl alkanes are also prominent. Although the isoprenoids are highly suggestive of a chlorophyll-based life as early as 2.5×10^9 years ago the authors point out that there remains the possibility of infiltration of the sediment by a younger oil. This point is under active consideration.

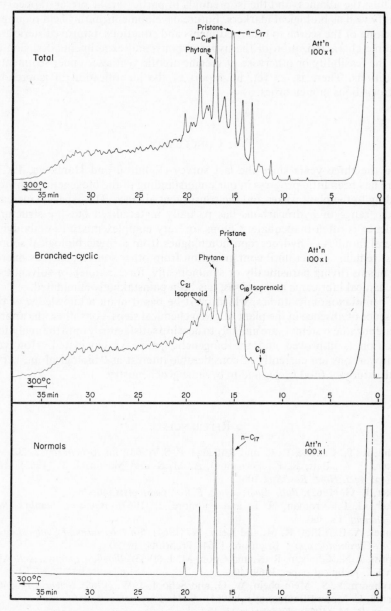

Fig. 10. Gas-liquid chromatograms of alkane fractions from the Soudan, iron formation Minnesota. Total, branched-cyclic and normal alkanes. Column conditions as for Fig. 4, but 10′ and $\frac{1}{16}$″ of 1% SE-30.

Thus, the alkanes, and the isoprenoids in particular, at present appear to qualify well as biological markers. Future developments in this field require a widening of the search to other sediments and situations, improved steric and isotopic characterization of the hydrocarbons, and experiments designed to test the feasibility or otherwise of specific abiotic syntheses under sedimental conditions. There is, as yet, no report of the identification of isoprenoid hydrocarbons in meteorites.

VI. CONCLUSION

In the three years since the last survey (Eglinton and Hamilton, 1963), there has been little progress in our understanding of the biogenetic pathways leading to the plant waxes and the constituent alkanes. The expected increase in the range of hydrocarbons has partially materialized but the structural evidence is often inadequate. Claims for very complex mixtures embodying several families of hydrocarbon homologues from a single biological source must remain suspect until contamination from other sources, such as micro-organisms (living parasitically or symbiotically, for example), or solvents, or mechanical lubricants and greases, has been painstakingly eliminated.

Chemotaxonomic studies, which must be based upon a knowledge of the presence of absence in the plant of given chemical steps—or rather, the appropriate enzyme systems—are unlikely to develop satisfactorily until the analytical situation is improved and the biogenetic pathways elucidated. However, hydrocarbons are currently of considerable interest as "biological markers" in the study of food cycles and in organic geochemistry.

REFERENCES

Aplin, R. T., Cambie, R. C. and Rutledge, P. S. (1963). *Phytochemistry* 2, 205.
Baker, E. A., Batt, R. F., Fernandes, A. M. S. and Martin, J. T. (1963). *Ann. Rep. Agri. Hort. Res. Sta.* 106.
Baker, E. G. (1962). *Bull. Amer. Assoc. Petrol. Geologists,* 46, 76.
Baker, G. L., Vroman, H. E. and Padmore, J. (1963). *Biochem. Biophys. Res. Commun.* 13, 360.
Banner, A. E., Elliot, R. M. and Kelly, W. (1964). *5th International Symposium on Gas-Chromatography*, Brighton, 1964. Preprints, p. 20.
Baptist, J. N., Gholson, R. K. and Coon, M. J. (1963). *Biochim. biophys. Acta* Also A.E.I. technical paper 3. 69, 40.
Barghoorn, E. S., Meinschein, W. G. and Schopf, J. W. (1965). *Science,* 148, 461.
Barrall, E. M. and Baumann, F. (1964). *J. Gas Chromatog.* 2, 256.
Barrer, R. M. (1964). *Endeavour* 23, 122.
Barrera, J. B., Reyes, R. E. and Gonzales, A. G. (1964). *Anal. Fis. Quim.* 608, 601.
Bartz, K. W. and Chamberlain, N. F. (1964). *Anal. Chem.* 36, 2151.
Belsky, T., Johns, R. B., McCarthy, E. D., Burlingame, A. L., Richter, W. and Calvin, M. (1965). *Nature, Lond.* 206, 446.

Berg, A. and Lam, J. (1964). *J. Chromatog.* **16**, 157.

Beynon, J. H. (1960). "Mass Spectrometry and its Application to Organic Chemistry", p. 325. Elsevier Publishing Co., Holland.

Blumer, M., Mullins, M. M. and Thomas, D. W. (1963). *Science* **140**, 974.

Blumer, M., Mullins, M. M. and Thomas, D. W. (1964). *Heligolaender Wiss. Meeresunters.* **10**, 187.

Blumer, M. and Thomas, D. W. (1965a). *Science* **147**, 1148.

Blumer, M. and Thomas, D. W. (1965b). *Science* **148**, 370.

Borges, J., Brooks, C. J. W., Cambie, R. C., Eglinton, G., Hamilton, R. J. and Pellitt, P. (In preparation).

Breger, I. A. (ed.) (1963). "Organic Geochemistry", Pergamon Press, Oxford.

Bu'Lock, J. D. (1964). *In* "Progress in Organic Chemistry" (J. W. Cook and W. Carruthers, eds.), p. 86, Butterworth, London.

Burchfield, H. P. and Storrs, E. E. (1962). "Biochemical Applications of Gas Chromatography", Academic Press, New York and London.

Burgstahler A. W. and Marx J. N. (1964). *Tetrahedron Letters* **45**, 3333.

Butler, J. H. A., Downing, D. T. and Swaby, R. J. (1964). *Aust. J. Chem.* **17**, 817.

Calvin, M. (1965) *Graham Young Lecture*, Glasgow.

Carman, R. M. (1963). *Aust. J. Chem.* **16**, 225.

Cason, J. and Graham, D. W. (1965). *Tetrahed. Letters* **21**, 471.

Ciereszko, L. S., Attaway, D. H. and Wolf, M. A. (1963). *Amer. Chem. Soc. Petrol. Res. Fund. 8th Ann. Rep.* p. 33.

Cocker, W. and Shaw, S. J. (1963). *J. chem. Soc.* 677.

Colombo, U. and Hobson, G. D. (eds.) (1964). "Advances in Organic Geochemistry", Pergamon Press, Oxford.

Connolly, J. D. and Overton, K. H. (1964). "Alicyclic Compounds". *Ann. Rep. Chem. Soc.*

Cooper, J. E. and Bray, E. E. (1963). *Geochim. Cosmochim. Acta* **27**, 1113.

Cummings, J. J. and Robinson, W. E. (1964). *J. Chem. Engng. Data* **9**, 304.

Cvetanovic, R. J., Duncan, F. J. and Falconer, W. E. (1964). *Canad. J. Chem.* **42**, 2410.

Day, E. A. and Libbey, L. M. (1964). *J. Fd. Sci.* **29**, 583.

Dixon, A. F. G., Martin-Smith, M. and Subramanian, G. (1965). *J. chem. Soc.* 1562.

Duncan, G. R. (1962). *J. Chromatog.* **8**, 37.

Eglinton, G. and Hamilton, R. J. (1963). *In* "Chemical Plant Taxonomy", (T. Swain, ed.) p. 187. Academic Press, London and New York.

Eglinton, G., Scott, P. M., Belsky, T., Burlingame, A. L. and Calvin, M. (1964). *Science* **145**, 263.

Eglinton, G., Scott, P. M., Belsky, T., Burlingame, A. L., Richter, W. and Calvin, M. (1965). "Advances in Organic Geochemistry", Vol. 2, Pergamon Press, London.

Eneroth, P., Hellstrom, K. and Ryhage, R. (1964). *J. Lipid Res.* **5**, 245.

Evans, C. D., Oswald, J. and Cowan, J. C. (1964). *J. Amer. Oil Chem. Soc.* **41**, 406.

Faurot-Bouchet, E. and Michel, G. (1964). *J. Amer. Oil Chem. Soc.* **41**, 406, 418.

Fernandes, A. M. S., Baker, E. A. and Martin, J. T. (1964). *Ann. appl. Biol.* **53**, 43.

Foster, J. W. (1962). *J. Microbiol. Serol.* **28**, 241.

Fukui, Y. and Ariyoshi, H. (1963). *Yakugaku Zasshi. J. Pharm. Soc. Japan* **83**, 1106.

Gastambide-Odier, M. and Lederer, E. (1959). *Nature, Lond.* **184**, 1563.

Gerarde, H. W. and Gerarde, D. F. (1961). Preprint from *Association of Food and Drug Officials of the U.S.* Vols. XXV and XXVI, 1961, 1962. Presented at 65*th*

Annual Conference of the Association of Food and Drug Officials of the U.S., Washington D.C. June, 1961.

Gilby, A. R. and Cox, M. E. (1963). *J. Inst. Physiol.* **9**, 671.

Haahti, E., Nikkari, T. and Juva, K. (1963). *Acta chem. scand.* **17**, 538.

Hallgren, B. and Larsson, S. (1963). *Acta chem. scand.* **17**, 1822.

Hemming, F. W., Morton, R. A. and Pennock, J. F. (1963). *Biochem. J.* **85** B., 291.

Herbin, G. A. (1964). *E. African Acad. Sci.* 2nd Symposium.

Herlain, A. (1964). *Brennstoff-Chem.* **45**, 244.

Hood, A. (1963). *In* "Mass Spectrometry of Organic Ions" (F. W. McLafferty, ed.), p. 597. Academic Press, New York and London.

Horn, D. H. S., Kranz, Z. H. and Lamberton, J. A. (1964). *Aust. J. Chem.* **17**, 464.

Horning, E. C. and Van den Heuvel, W. J. A. (1963). *Ann. Rev. Biochem.* **32**, 709.

Horning, E. C. and Van den Heuvel, W. J. A. (1964). *J. Amer. Oil Chem. Soc.* **41**, 707.

Iwata, I. and Sakurai, Y. (1963). *Agr. Biol. Chem.* **27**, 253.

Jarolimek, P., Wollrab, V. and Streibl, M. (1964). *Coll. Czech. Chem. Commun.* **29**, 2528.

Jarolimek, P., Wollrab, V., Streibl, M. and Šorm, F. (1964). *Chem. & Ind.* 237.

Jarolimek, P., Wollrab, V., Streibl, M. and Šorm, F. (1965). *Coll. Czech. Chem. Commun.* **30**, 880.

Johnson, C. B. and Wilson, A. T. (1964). *Nature, Lond.* **204**, 111.

Jurg, J. W. and Eisma, E. (1964) *Science* **144**, 1451.

Kallio, R. E., Finnerty, W. R., Wawzonek, S. and Klimstra, P. D. (1963). *Symp. Marine Microbiol.* Ch. **42**, 453.

Kaneda, T. (1963). *Biochem. Biophys. Res. Commun.* **10**, 283.

Kasprzyk, Z., Kochman, K. and Pass, L. (1962). *Bull. Acad. Polon. Ser. Sci. Biol.* **10**, 457.

Kester, A. S. and Foster, J. W. (1963). *J. Bacteriol.* **85**, 859.

Keulemans, A. I. M. and Perry, S. G. (1962). *Gas Chromatog.* Butterworth, London.

Kochloefl, K., Schneider, P., Rericha, R., Horák, M. and Bazant, V. (1963). *Chem. & Ind.* 692.

Kuksis, A. (1964). *Biochem. J.* **3**, 1086.

Laidman, D. L. and Morton, R. A. (1962). *Biochem. J.* **84**, 386.

Laurent, A. and Bonnetain, L. (1964). *C. R. Acad. Sci., Paris* **258**, 180.

Leadbetter, E. R. and Foster, J. W. (1960). *Arch. Mikrobiol.* **35**, 92.

Lederer, E. (1964). *6th Congress Biochem. N.Y. Proceedings of Plenary Sessions,* p. 63; *Angew. Chem.* **3**, 393.

Leibrand, R. J. (1962). M.Sc. Thesis. Bozeman, Montana.

Leibrand, R. J. (1964). Ph.D. Thesis, Bozeman, Montana.

Levy, E. J., Galbraith, F. J. and Melpolder, F. W. (1963). *In* "Advances in Mass Spectrometry", Vol. II (R. M. Elliott, ed.), p. 395, Pergamon Press Oxford.

Levy, E. J. and Paul, D. G. (1963). *Facts and Methods for Sci. Res.* **4**, no. 1, 10.

Leyton, L. and Juniper, B. E. (1963). *Nature, Lond.* **198**, 770.

Lukins, H. B. and Foster, J. W. (1963). *J. Bact.* **85**, 1074.

Lukins, H. B. and Foster, J. W. (1963). *Z. Allg. Mikrob.* **3**, 251.

McCarthy, M. J., Kuksis, A. and Beveridge, J. M. R. (1964). *J. Lipid Res.* **5**, 609.

McCarthy, R. D. (1964). *Biochim. biophys. Acta* **84**, 74.

Maier, R. and Mangold, H. K. (1964). *Arch. Anal. Chem. Inst.* **3**, 369.

Mair, B. J. (1964). *Geochim. Cosmochim. Acta* **28**, 1303.

Mathis, C. and Ourisson, G. (1964). *Phytochemistry* **3**, 115.

Mazliak, P. (1963). *Ph.D. Thesis*, Paris.
Mazliak, P. (1964). *6th International Congress of Biochemistry*. New York (Abstracts).
Mazliak, P. and Pommier-Miard, J. (1963). *Fruits* **18**, 177.
Meinschein, W. G. (1964). *Space Science Reviews* **5**, 1.
Meinschein, W. G., Barghoorn, E. S. and Schopf, J. W. (1964). *Science* **145**, 261.
Mirov, N. T. (1961). *U.S. Dept. Agric. Bull.* 1239, 1.
Mold, J. D., Means, R. E., Stevens, R. K. and Ruth, J. M. (1964). *Biochemistry* **3**, 1293.
Mold, J. D., Stevens, R. K., Means, R. E. and Ruth, J. M. (1963). *Biochemistry* **2**, 605.
Murray, K. E., Huelin, F. E. and Davenport, J. B. (1964). *Nature, Lond.* **204**, 80.
Nicholas, H. J. and Bombaugh, K. C. (1965). *Biochim. biophys. Acta.* **98**, 372.
Orò, J., Nooner, D. W. and Wikström, S. A. (1965). *Science* **147**, 870; *J. Gas Chromatog.* **3**, 105.
Oró, J., Nooner, D. W., Zlatkis, A., Wikström, S. A. and Barghoorn, E. S. (1965). *Science* **148**, 77.
Oyama, V. I. (1963). *Nature, Lond.* **200**, 1058.
Park, R. J. and Sutherland, M. D. (1962). *Aust. J. Chem.* **15**, 172.
Parker, P. L. (1964). *Geochim. Cosmochim. Acta* **28**, 1155.
Perry, S. G. (1963). *J. Gas Chromatog.* **2**, 54.
Preiss, B. and Bloch, K. (1964). *J. biol. Chem.* **239**, 85.
Purdy, S. J. and Truter, E. V. (1963). *Proc. roy. Soc.* B **158**, 536, 544, 553.
Pyriki, C. and Hofmann, F. (1963). *Ber. Inst. Tabakforsch. Dresden* **10**, 69.
Razafindrazaka, J. and Metzger, J. (1963). *Bull. Soc. chim. Fr.* 1630.
Robinson, W. E., Cummings, J. J. and Dinneen, G. U. (1965). *Geochim. Cosmochim. Acta* **29**, 249.
Romanuk, M., Herout, V. and Šorm, F. (1958). *Chem. listy* **52**, 1965.
Ryhage, R., Wikström, S. and Waller, G. R. (1965). *Anal. Chem.* **37**, 435.
Sandermann, W. and Schweers, W. (1960). *Chem. Ber.* **93**, 2266.
Schenk, P. A. and Eisma, E. (1963). *Nature, Lond.* **199**, 170.
Šorm, F., Wollrab, V., Jarolimek, P. and Streibl, M. (1964). *Chem. & Ind.* 1833.
Stevenson, R. (1961). *J. Org. Chem.* **26**, 5228.
Stumpf, P. K. (1962). *Nature, Lond.* **194**, 1158.
Sylvester-Bradley, P. C. and King, R. J. (1963). *Nature, Lond.* **198**, 720.
Szymanski, H. (1964). *J. Gas Chromatog.* **2**, 154.
Tulloch, A. P., Spencer, J. F. T. and Gorin, P. A. J. (1962). *Canad. J. Chem.* **40**, 1326.
Warth, A. H. (1960). "Chemistry and Technology of Waxes", 2nd Ed. Reinhold, New York.
Weenink, R. O. (1962). *Biochemistry* Vol. 1, 523.
Weiss, A. (1963). *Angew. Chem. (Int.)* **2**, 134.
West, W. W. (1963). *Am. Chem. Soc. meeting, New York*, 1963. Abstract 12B.
Wollrab, V. (1964). *Riechstoffe Aromen Koerperpflegemittel* **14**, 321.
Wollrab, V., Streibl, M. and Šorm, F. (1963). *Coll. Czech. Chem. Commun.* **28**, 1904.
Yamada, M. and Stumpf, P. K. (1964). *Biochem. Biophys. Res. Commun.* **14**, 165.
Zill, L. P. and Harmon, E. A. (1962). *Biochim. biophys. Acta* **57**, 573.

CHAPTER 5

The Biogenesis of Natural Acetylenes

J. D. BU'LOCK

Department of Chemistry, The University, Manchester

I. INTRODUCTION

When in 1962 Professor Sørensen surveyed the distribution of the natural acetylenes (Sørensen, 1962), he observed that "our present knowledge of acetylenic compounds is probably sufficient to examine taxonomic problems at the level of genus, section, or species, but is far too incomplete for problems concerning higher groups in the classification." At that time one could discern amongst the natural acetylenes sets of compounds which were obviously inter-related in various ways, but since the biogenetic explanations of these inter-relationships were still largely unknown, the natural acetylenes had perforce to be treated mainly as individual compounds. It is with advances and with prospects in these biogenetic aspects that this chapter is mainly concerned, in the belief that an understanding of biosynthetic mechanisms is crucial to our understanding of natural products in general and of chemotaxonomy in particular.

All naturally occurring compounds are the products of sequences of reaction steps which are sometimes simple, sometimes complex. The chemical structures of the products which result are cryptic and ambiguous summaries of the biogenetic processes; if we have nothing better, then these chemical structures can be used as taxonomic markers, but only with serious reservations—as outlined by Professor Hegnauer in his contribution (Chapter 13). However, the actual biogenetic reaction steps are obviously better markers. Indeed, if we were to accept the equation, one reaction = one enzyme = one gene, then each unit reaction step would directly represent one unit of inherited character. In fact, in the realm of natural products we are commonly dealing with enzymes of relatively low specificity, which effect similar transformations on a variety of substrates. Thus each gene or its corresponding enzyme is associated with a range of possible reactions, and will be manifested in a variety of final products in ways which depend upon the type of substrate made available to the enzyme by other processes. Moreover the activity of the enzymes of secondary metabolism is not invariably apparent, but is usually a function of physiological history and conditions. Ultimately, therefore, our examination of natural products must attempt to disentangle the modes of expression, both phenotypic and genotypic, which the operative biosynthetic pathways display.

ACETYLENES

Order	*Families*
Santalales[2]	Santalaceae, Opiliaceae, Olacaceae, Loranthaceae
Ranales	Lauraceae
Rosales	Leguminosae
Geraniales	Simarubaceae, Euphorbiaceae (*allene*)
Malvales (*cyclopropenes*)	Malvaceae, Sterculiaceae
Umbelliflorae[2]	Araliaceae, Umbelliferae
Tubiflorae	Labiatae (*allene*)
Campanulatae	Compositae[2]

FATTY ACID EPOXIDES

Rhoeadales	Cruciferae
Rosales	Leguminosae
Geraniales	Euphorbiaceae
Malvales	Malvaceae
Myrtifolae	Onagraceae
Umbelliflorae	Umbelliferae
Campanulatae	Compositae

FIG. 1. Plant families[1] from which acetylenes (above) and fatty acid epoxides (below) have been reliably recorded.

[1] Arranged after Engler. [2] Major source.

For example, let us compare two of the "unit processes" displayed by natural acetylenes, namely thiophene formation and epoxidation. Both transformations are experimentally attested though their enzymic mechanism is, of course, unknown. In thiophene formation a conjugated diyne unit affords the four carbons of the thiophene ring which is effectively completed by the addition of H_2S (see Fig. 3) although it more probably involves a multi-stage process involving cysteine. The reaction occurs widely in Compositae and it is probable that when sufficient data are available the distribution of the thiophenes will be shown to have taxonomic significance. The enzymes responsible cannot be wholly non-specific, and indeed their precise specificity may also show generic variations, since isomeric thiophenes formed from the same precursor by variant steps have been isolated from different sources. However, the whole of this taxonomic range must be confined within that of polyacetylene-producing species, since without the precursor diynes thiophene formation obviously cannot occur. We, therefore, cannot ask whether the two characters are independent since the expression of one is dependent upon the operation of the other.

The conversion of double bonds to the corresponding epoxides forms an interesting comparison. This reaction plays a considerable part in leading to the final structures of natural acetylenes; as with the thiophenes, isomeric epoxides formed from common precursors occur in different species. However, this type of reaction is not necessarily confined to the acetylenic olefines, and in view of the lower specificity of enzymes in secondary metabolism we must allow that the epoxidation "character" can be manifested independently in non-acetylene-producing species. The epoxidation reaction has, potentially, its own chemotaxonomy. For example, it is interesting to consider its occurrence in co-operation with normal fatty acid synthesis. Hilditch and Williams (1964) list fatty acid epoxides as being rather frequent in Compositae and present also in Cruciferae, Euphorbiaceae, Leguminosae, Malvaceae, and Umbelliferae (Fig. 1). Acetylenes are of course characteristic of Compositae but they also occur in two more of the above families, Umbelliferae and Leguminosae; the occurrence of cyclopropene acids also suggests that acetylene formation may occur in Malvaceae. There appears, therefore, to be some parallelism between acetylene formation and epoxidation despite their potential independence.

II. PRECURSORS OF POLYACETYLENES

With very minor exceptions the natural acetylenes all contain unbranched chains of carbon atoms (even when cyclized) and can be derived from suitably unsaturated carboxylic acids with even numbers of carbon atoms. The broad application of this "n-C_{2n} rule" has been given in detail elsewhere (Bu'Lock, 1964), and there is no need to give extended examples here. However, we should note that there is good experimental evidence bearing upon two aspects of this rule, viz. (*a*) the initial assembly of the n-C_{2n} molecule by a standard process and (*b*) the transformation reactions by which n-C_{2n} precursors are transformed into the full range of natural acetylenes.

A priori one might expect any n-C_{2n} acids to arise either by the standard mechanism of fatty acid synthesis or by some variant thereof, such as is seen in the biosynthesis of aromatic polyketides. The experimental confirmation is exemplified in Fig. 2, which illustrates a range of natural acetylenes which incorporate labelling from ^{14}C-acetate in the manner which this hypothesis requires. Moreover, in the particular case of dehydromatricarianol (IV) the C_{10} chain has been shown to comprise one C_2 unit derived directly from acetyl-CoA while the remaining four units are derived by way of malonyl-CoA (Bu'Lock and Smalley, 1962), precisely as in fatty acid and polyketide synthesis by the malonyl-CoA type of pathway. Biosynthesis by such a pathway is of course ubiquitous and of itself has no taxonomic significance. In passing, we should note that experimental data of this kind do not establish the stage at which unsaturation is introduced into the n-C_{2n} acids, nor do they preclude the formation of such acids by way of longer carbon chains. However, they

$$\overset{*}{H}C{\equiv}C\cdot\overset{*}{C}{\equiv}C\cdot\overset{*}{C}H{=}CH\cdot\overset{*}{C}HOH\cdot\overset{*}{C}H_2\cdot CH_2\cdot\overset{*}{C}OOH$$

I. Nemotinic acid

$$CH_3\cdot\overset{*}{C}H_2\cdot CH{=}\overset{*}{C}H\cdot(C{\equiv}\overset{*}{C})_3\cdot(CH_2\cdot\overset{*}{C}H_2)_3\cdot\overset{*}{C}H_2COOH$$

II.

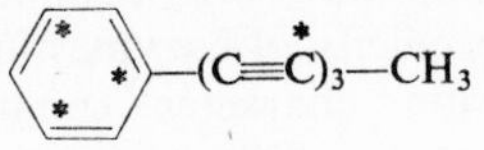

III.

$$CH_3\cdot(C{\equiv}C)_3\cdot CH{=}CH\cdot CH_2OH \quad \text{IV Dehydromatricarianol}$$

FIG. 2. Some of the natural acetylenes which incorporate labelling from 1-^{14}C-acetate ($\overset{*}{C}$) in the manner predicted by the n-C$_{2n}$ rule. (I), from a Basidiomycete, contains six C$_2$ units and has lost C$_{(12)}$ by terminal oxidation and decarboxylation (Bu'Lock and Gregory, 1959). (II) contains nine C$_2$ units and is a typical fatty acid from Santalaceae (Bu'Lock and Smith, 1963). Compound (III) is one of the aromatic polyacetylenes from Compositae and contains seven C$_2$ units, the original carboxyl C$_{(1)}$ having been lost at some stage (Bohlmann and Kleine, 1965; Jones, 1961).

Transformations involving triple bonds

$$-C{\equiv}C- \;\rightarrow\; -CH{=}\underset{\displaystyle S\cdot CH_3}{\overset{|}{C}}-$$

$$-C{\equiv}C-C{\equiv}C- \;\rightarrow\; -C{=}CH-CH{=}C- \overset{\hspace{2em}}{\underset{\rule{3em}{0.4pt}\; S \;\rule{2em}{0.4pt}}{}}$$

$$-C{\equiv}C-CH_2- \;\rightarrow\; -CH{=}C{=}CH-$$

$$-C{\equiv}C-CH{=}CH-CH_2OH \;\rightarrow\; -CH_2-C{=}CH-CH{=}CH \;\underset{\rule{4em}{0.4pt}\; O \;\rule{2em}{0.4pt}}{}$$

Other transformations

$$-CH{=}CH- \;\rightarrow\; -CH{-}CH- \atop \diagdown O \diagup$$

and derivatives

$$-CH{=}CH- \;\rightarrow\; CH_2{-}CHOH$$
(via epoxide?)

$$R\cdot CH_3 \;\rightarrow\; R\cdot CH_2OH \;\rightleftarrows\; R\cdot CHO \;\rightleftarrows\; R\cdot COOH \;\rightarrow\; R\cdot H$$
(R=alkenyl, alkynyl, aryl or thienyl)

FIG. 3. Experimentally-attested unit steps in the biogenesis of some natural acetylenes.

establish quite adequately the general context in which acetylene synthesis occurs.

The processes which transform the n-C_{2n} acids into the final range of natural acetylenes can largely be defined by structure comparisons (Bu'Lock, 1964), but here also there is welcome support from experiments in which labelled acetylenes have been shown to undergo the expected transformations. Most of this work has come from Jones and co-workers in Oxford (Jones, 1965) and from Bohlmann *et al.* in Germany (Bohlmann and Kleine, 1965) and a substantial part is not yet published.[1]

Naturally all the evidence of this kind pertains to individual cases, but in Fig. 3 I have presented in a general manner some of the "unit steps" which are thus substantiated. The potentialities of this kind of breakdown for chemotaxonomic applications have already been discussed, and perhaps in the not-too-distant future an analysis on this basis will be attempted. Clearly it is the chemotaxonomy of such unit steps which underlies the observations at genus level described by Sørensen (1962). The wider questions concern the biogenesis of the acetylenic n-C_{2n} acids, a problem to which we must now turn.

III. FORMATION OF ACETYLENES

A. FROM ENOLS

As noted above, the observed formation of natural acetylenes from assemblies of C_2 units carries no indication as to the stage at which unsaturation is introduced. One attractive hypothesis is that the triple and double bonds arise by elimination reactions from enols and alcohols respectively, and that such reactions take place at the keto-acid and hydroxy-acid levels which are known to occur in the normal process of fatty acid synthesis. The formation of olefinic fatty acids in such a way is a widely-recognized possibility; for the analogous formation of acetylenes, elimination from suitably activated enol derivatives can be considered (Jones, 1961). Good laboratory analogies for this type of reaction have now been established (Harley-Mason and Fleming, 1961; Craig and Moyle, 1962). Such a mechanism has other attractions, since the formation of acetylenes would take its place alongside the formation of aromatic polyketides as a non-reductive variant of fatty acid synthesis; indeed, surprisingly similar mechanisms can be suggested for these two processes, superficially very different.

The hypothesis, that acetylenes are formed by an elimination process from enolic intermediates in a type of fatty acid assembly process, is adequate to explain the available data, and it is attractive; aside from this, however, there is little to be said for it. As a mechanism, it does not lead to any particularly distinctive structural features which we might hope to find as confirmatory evidence in the products, nor does it lend itself to any very convenient experimental test within the limits of our present techniques. For these reasons, it is

[1] I am grateful to Professor Sir Ewart Jones and to Professor F. Bohlmann for many of these results, which were presented to the Chemical Society at their Glasgow symposium in April 1965.

4

advisable to examine any alternative hypothesis for the biogenesis of triple bonds.

B. BY DEHYDROGENATION

The alternative hypothesis is that triple bonds arise by removal of hydrogen from double bonds, just as olefins are formed by dehydrogenation of saturated compounds in many biological systems. Thermodynamically the second dehydrogenation would be very little more exacting than the first; as model biological reactions, therefore, the dehydrogenations of saturated compounds may be instructive. In a variety of organisms the double bond of oleic acid (V) is introduced in this way, and in accompanying acids of different chain length the unsaturation remains at Δ^9. Less commonly, isomers of oleic acid (V) are produced with similar specificity. Here hydrogen is being selectively removed from positions in the chain which are not chemically activated but which are situated at specific distances from the carboxyl group. In the further conversions

oleic acid (V) → linoleic acid → linolenic acid

and similar reactions, hydrogen is removed at specific distances from the first double bond and in the distal part of the chain, i.e. that part more remote from the carboxyl group. Some micro-organisms can attack n-alkanes and here the oxidation is terminal (as in the ω-oxidation of fatty acids in animals). Though none of these reactions has a known mechanism, these instances serve to show the terms in which their specificity is determined. From our present knowledge of plant lipids it is also clear that their characteristic fatty acid compositions result from the interplay of two sets of factors, one set determining the distribution of chain lengths and the other determining the distribution of unsaturation or other modifications.

$$CH_3 \cdot (CH_2)_7 \cdot CH{=}CH \cdot (CH_2)_7 \cdot COOH$$

V. Oleic acid

$$CH_3 \cdot (CH_2)_7 \cdot C{\equiv}C \cdot (CH_2)_7 \cdot COOH$$

VI. Stearolic acid

$$CH_3 \cdot (CH_2)_5 \cdot CH{=}CH \cdot C{\equiv}C \cdot (CH_2)_7 \cdot COOH$$

VII. Ximenynic acid

Into such a context the acetylenic fatty acids from plant lipids fit most satisfactorily. Consider first the main series of such acids from plants of the Santalaceae, shown in Fig. 4, of which the best-known is ximenynic acid, octadec-9-yn-11-enoic (VII). The simplest of the series is stearolic acid (VI), recently found in *Pyrularia pubera* (Hopkins and Chisholm, 1964); Dr. G. N. Smith confirms that this acid is also present in *Santalum acuminatum*. The acids of this series have a well-defined taxonomic distribution (cf. Sørensen, 1962) and frequently occur mixed together: thus in our earlier study of *Santalum acuminatum* we found at least six of the series, differently distributed between various parts of the plant (Bu'Lock and Smith, 1963). By far the most plausible biogenetic explanation for this series of acids is that all are derived from oleic acid (V), first by further desaturation at the double bond to give stearolic acid (VI), and then by extending conjugated unsaturation in the distal part of the chain. The continued desaturation seems to follow a regular pattern, viz:

$$
\begin{matrix}
& a & & b & & b & & c \\
\text{-ene} & \rightarrow & \text{-yne} & \rightarrow & \text{-enyne} & \rightarrow & \text{-dienyne} & \rightarrow & \text{enediyne}
\end{matrix}
$$

Co-occurring with these acids are a series of 8-hydroxy-acids, which correspond simply to oxidation of the main series at the reactive (propargylic) methylen group (Fig. 4). The idea that these acids are formed sequentially from oleic is at least consistent with some of our isotope-incorporation data (Bu'Lock and Smith, 1963) and further studies—somewhat delayed by the reluctance of *S. acuminatum* to germinate—are in hand. An interesting feature of the seed-fat of typical Santalaceae is that linoleic acid appears to be quite absent. Thus in our *S. acuminatum* the principal C_{18} acids are oleic (V) and ximenynic (VII), with small amounts of stearolic (VI) and (possibly) stearic, and of the C_{16}

$$CH_3 \cdot (CH_2)_3 \cdot (CH=CH)_2 \cdot C\equiv C \cdot (CH_2)_7 \cdot COOH$$

$$CH_3 \cdot (CH_2)_3 \cdot CH=CH \cdot (C\equiv C)_2 \cdot (CH_2)_7 \cdot COOH$$

$$CH_3CH_2 \cdot (CH=CH)_2 \cdot (C\equiv C)_2 \cdot (CH_2)_7 \cdot COOH$$

$$CH_3CH_2 \cdot CH=CH \cdot (C\equiv C)_3 \cdot (CH_2)_7 \cdot COOH*$$

$$CH_2=CH \cdot CH=CH \cdot (C\equiv C)_3 \cdot (CH_2)_7 \cdot COOH$$

FIG. 4. The main series of acetylenic acids from Santalales, arranged biogenetically. *8-Hydroxy-derivatives of these acids are also known.

acids, presumably palmitic. In such species, then, the conversion of oleic acid into acetylenes literally replaces the usual conversion into linoleic acid.

Other genera in related families of the Santalales also produce these acids, but there are interesting variations. In the case of *Ongokea gore* (Olacaceae) (Gunstone and Sealy, 1963) the variant series shown in Fig. 5 is found, again accompanied by the corresponding 8-hydroxy-acids. This series seems to result by a variation in the sequence of continued desaturation together with an additional reaction which is responsible for introducing the terminal unsaturation, Δ^{17}. Another interesting acid is that recently found in *Acanthosyris spinescens* (Santalaceae) (Powell and Smith, 1965a), which is a C_{17} acid of structure (VIII).

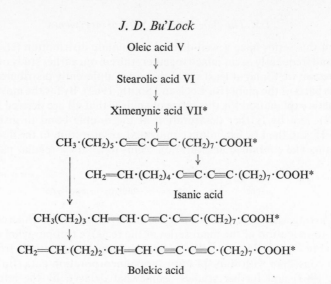

Oleic acid V

Stearolic acid VI

Ximenynic acid VII*

$$CH_3 \cdot (CH_2)_5 \cdot C{\equiv}C \cdot C{\equiv}C \cdot (CH_2)_7 \cdot COOH*$$

$$CH_2{=}CH \cdot (CH_2)_4 \cdot C{\equiv}C \cdot C{\equiv}C \cdot (CH_2)_7 \cdot COOH*$$

Isanic acid

$$CH_3(CH_2)_3 \cdot CH{=}CH \cdot C{\equiv}C \cdot C{\equiv}C \cdot (CH_2)_7 \cdot COOH*$$

$$CH_2{=}CH \cdot (CH_2)_2 \cdot CH{=}CH \cdot C{\equiv}C \cdot C{\equiv}C \cdot (CH_2)_7 \cdot COOH*$$

Bolekic acid

FIG. 5. The variant series of acetylenic acids in *Ongokea gore* arranged biogenetically. *8-Hydroxy-derivatives of these acids also occur.

We can immediately recognize (VIII) as an 8-hydroxyximenynic acid variant which has undergone two further modifications; one is the terminal desaturation reaction already noted in the *Ongokea* series, and the second is chain-shortening by α-oxidation, a process which has been found to occur rather widely in plants (Hitchcock and James, 1964) and which is also involved in the biogenesis of malvalic acid (Bu'Lock and Smith, 1964).

$$CH_2{=}CH \cdot (CH_2)_4 \cdot CH{=}CH \cdot C{\equiv}C \cdot CHOH \cdot (CH_2)_5 \cdot COOH$$

VIII

$$CH_3 \cdot (CH_2)_{10} \cdot CH{=}CH \cdot (CH_2)_4 \cdot COOH$$

IX. Petroselenic acid

$$CH_3 \cdot (CH_2)_{10} \cdot C{\equiv}C \cdot (CH_2)_4 \cdot COOH$$

X. Tariric acid

If in the Santalales stearolic acid (VI) and its derivatives are formed from oleic acid (V), then similarly tariric acid (X) should be formed from octadec-6-enoic, petroselenic acid (IX). So far, tariric acid is recorded only from species of the genus *Picramnia*, family Simarubaceae, whilst petroselenic acid is characteristically found in the Umbellales. However, the latter (IX) is also recorded from one genus, *Picrasma*, in Simarubaceae. The properties of the monoethynoid acids such as stearolic (VI) or tariric (X) are such that they would very easily escape detection in the ordinary procedures of fat analysis, and I am confident that with more refined techniques such acids will turn out to be much more frequent than our present data imply. For example, I am informed that laballenic acid, recently isolated from one of the Labiatae, *Leonotis nepetaefolia* (Bagby, Smith and Wolff, 1964) is octadeca-5,6-dienoic acid (XI), i.e. the allene corresponding to tariric acid and very probably derived from it.

$$CH_3 \cdot (CH_2)_{10}CH\!=\!C\!=\!CH \cdot (CH_2)_3 \cdot COOH$$

XI. Laballenic acid

A most interesting recent discovery is a non-conjugated isomer of ximenynic acid, namely crepenynic acid (XIII), for which Mikolajczak *et al.* (1964)

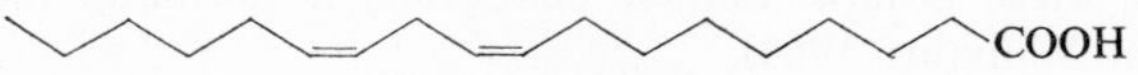

XII. Linoleic acid

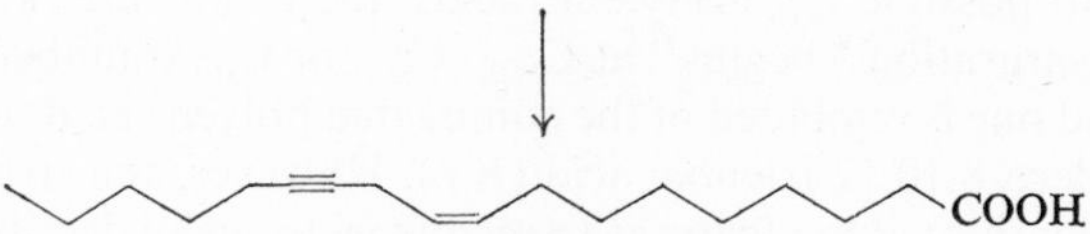

XIII. Crepenynic acid

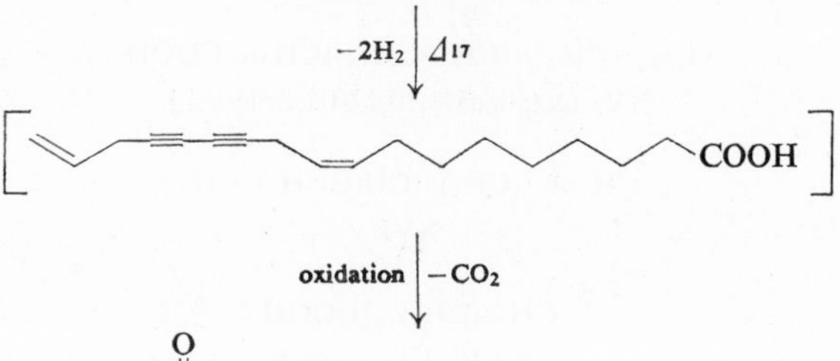

$-2H_2$ | Δ_{17}

COOH

oxidation | $-CO_2$

XIV. Falcarinone

Fig. 6. Biogenesis of falcarinone from crepenynic acid. Oxygenation also occurs at the position marked*.

established the structure *cis*-octadec-9-en-12-ynoic acid. This was isolated from the seed-fat of *Crepis foetida*, belonging to the Cichorieae or Liguliflorae, the section of Compositae in which the more easily-detected polyacetylenes seem to be quite absent (cf. Sørensen, 1962). Biogenetically we can at once see (Fig. 6) that crepenynic acid (XIII) bears the same relationship to linoleic acid (XII) as does stearolic (VI) to oleic (V), or tariric (X) to petroselenic (IX). Since linoleic acid itself (XII) is in any case formed by a dehydrogenation reaction (from oleic acid, (V)), the plausibility of our suggestion of a second such step is enhanced.

Moreover, from crepenynic acid (XIII) we can plausibly derive a surprising variety of further acetylenes. For example, the C_{17} ketone falcarinone (XIV) (Fig. 6) and its derivatives occur widely in Compositae, Umbelliferae, and

Araliaceae (Bohlmann *et al.*, 1962). The C_{17} chain of these compounds obviously implies a C_{18} acid, and on inspection we see that the required precursor can be derived from crepenynic acid by further dehydrogenation in the distal part of the chain, possibly with insertion of the $\varDelta^{17}$ double bond as a distinct step, and oxygenation at allylic centres—all being reactions which we have already encountered in the "stearolic" series, e.g. in *Ongokea* (Fig. 5). Further compounds which make up a "crepenynic series" are considered subsequently, but at this stage we might note that compounds which can fairly clearly be assigned to the linoleic/crepenynic biogenetic series occur in Compositae, Umbellales, and (amongst fungi) Hymenomycetes; in the lipids of all these groups, linoleic acid (XII) is normally very prominent (Hilditch and Williams, 1964).

If we examine the other C_{17} acetylenes which are now known, for their relationship to possible C_{18} acetylenic acids, the results are rather less clearcut. Their unsaturation "begins" at $C_{(10)}$, $C_{(8)}$, or $C_{(6)}$ (numbering the *parent* C_{18} chain) and one is reminded of the conjugated polyene acids of Compositae such as octadeca-8,10,12-trienoic acid (XV). However, the structures do not display sufficiently striking features to enable us to make decisive use of them in elaborating a biogenetic hypothesis.

$$CH_3 \cdot (CH_2)_4 \cdot (CH{=}CH)_3 \cdot (CH_2)_6 \cdot COOH$$
XV. Octadeca-8,10,12-trienoic acid

$$CH{\equiv}C \cdot (CH_2)_7 \cdot CH(OCH_3) \cdot CH_3$$
XVI

$$CH_3 \cdot (CH_2)_{10} \cdot COOH$$
XVII. Lauric acid

$$CH_3 \cdot (CH_2)_2 \cdot CHOH \cdot (CH_2)_9 \cdot COOH$$
XVIII

If for a moment we turn to some of the minor natural acetylenes, however, further agreement with the desaturation hypothesis is found. The compound 2-methoxyundec-10-yne (XVI) occurs in a number of Lauraceae (Matthews, *et al.*, 1963); if we derive the terminal acetylene by dehydrogenation and decarboxylation from an α,β-unsaturated acid, then this compound is seen as a derivative of lauric acid (XVII)- the characteristic lipid acid of this plant family. Analogies for the oxygen atom at $C_{(11)}$ can be found, e.g. convolvulinic acid (11-hydroxymyristic acid (XVIII)) from jalap. Again, the unusual cyclic ether (XX) shown in Fig. 7 has been isolated by Irie *et al.* (1965) from a seaweed, *Laurencia glandulifera*. This curious structure is best derived from a bis-epoxide, as shown, and if we then allow the terminal ethynyl group to arise by the same sequence as postulated for the Lauraceae, viz:

$$-CH_2 \cdot CH_2 \cdot COOH \rightarrow -CH{=}CH \cdot COOH \rightarrow -C{\equiv}C \cdot COOH \rightarrow -C{\equiv}CH$$

then the parent acid of the *Laurencia* compound will be hexadeca-4,7,10,13-tetraenoic acid (XIX). Such "interrupted" C_{16} polyene acids are particularly characteristic of the marine Algae (Hilditch and Williams, 1964) and it is precisely this acid which has been most fully characterized from *Scenedesmus* (Klenk and Knipprath, 1959).

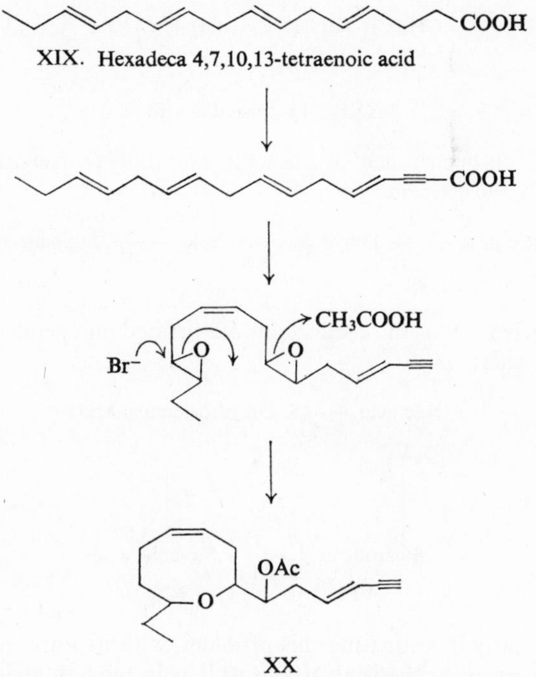

XIX. Hexadeca 4,7,10,13-tetraenoic acid

Fig. 7. Probable biogenesis of a C_{15} acetylene in Algae.

Finally we should note the cyclopropene sterculic acid (XXI), and its α-oxidation product malvalic acid (XXII), which occur in seed-fats of Malvaceae and Sterculiaceae together with the corresponding cyclopropanes. We have shown that the cyclopropene system here arises by the addition of

$$CH_3 \cdot (CH_2)_7 \cdot C = C \cdot (CH_2)_7 \cdot COOH$$
$$\diagdown \diagup$$
$$CH_2$$

XXI. Sterculic acid

$$CH_3 \cdot (CH_2)_7 \cdot C = C \cdot (CH_2)_6 \cdot COOH$$
$$\diagdown \diagup$$
$$CH_2$$

XXII. Malvalic acid

one carbon atom at $C_{(9)}$-$C_{(10)}$ of a C_{18} precursor (Bu'Lock and Smith, 1964), the extra carbon coming from methionine. [In Lactobacilli, an analogous alkylation converts vaccenic acid (octadec-11-enoic, XXIII) into lactobacillic

$$CH_3 \cdot (CH_2)_5 \cdot CH = CH \cdot (CH_2)_9 \cdot COOH$$

XXIII. Vaccenic acid

$$CH_3(CH_2)_5 \cdot CH\!-\!CH \cdot (CH_2)_9 \cdot COOH$$
$$\diagdown \diagup$$
$$CH_2$$

XXIV. Lactobacillic acid

(XXIV)]. The co-occurrence of sterculic and dihydrosterculic acid could therefore imply the sequence

$$\text{Oleic acid} \xrightarrow{+C_1} \text{Dihydrosterculic acid} \xrightarrow{-2H} \text{Sterculic acid}$$
$$\text{V} \qquad\qquad\qquad\qquad\qquad\qquad\qquad \text{XXI}$$

but an alternative is that the cyclic acids are formed independently from oleic and stearolic acids:

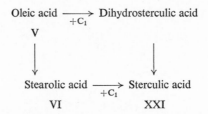

We are currently investigating this problem, with its implication that stearolic acid may occur in the Malvales as well as in the Santalales. Should this be the case, incidentally, it would provide a further case of taxonomic parallelism between the occurrences of acetylene formation and of epoxidation, frequent in the fats of Malvales*.

Thus far, therefore, we can be reasonably satisfied with the hypothesis that the acetylenes are formed from normal fatty acids by a selection of processes including successive dehydrogenations leading to double and triple bonds. Taxonomically, the remarks made earlier about such "unit processes" will apply; it is in this sense satisfactory that the choice of fatty acid used as starting-material in any particular case seems fully concordant with what is already known concerning the chemotaxonomy of fatty acids. The examples considered up to now, however, have all been either fatty acids themselves or compounds very closely related to the fatty acids. Can our hypothesis clarify the structures of less closely-related polyacetylenes, and in particular those of shorter chain-lengths?

* *Added in proof.* The presence of stearolic acid and its α-oxidation product is now confirmed.

IV. Chain Shortening

A. PLANT ACETYLENES

If we classify the structures, not of the natural acetylenes themselves, but of their presumed parents, the correspondingly unsaturated nC_{2n} acids, several points emerge. For instance, as regards chain-lengths; in fungi the chains range from C_8 to C_{14} with C_{10} predominating; in Compositae the range is C_{10} to C_{18}, with C_{14} greatly predominating; elsewhere the range is C_{12} to C_{18}, and here C_{18} predominates. We also find that on average the shorter-chain compounds contain the highest proportion of unsaturated centres, and the longer-chain compounds contain the largest proportion of fully-saturated (CH_2) groups. Such a situation is probably significant and requires explanation. One possibility is that there are say, C_{18} precursors, with the unsaturation largely concentrated at one end of the chain, which could afford the shorter chains by removal of carbon from the other, more saturated, portion. Removal could be effected by the ordinary process of β-oxidation. Now we have already seen, in the "stearolic" and "crepenynic" series, C_{18} acids with the requisite distribution of unsaturation, viz. in the distal moiety; can we find more specific and circumstantial evidence for the chain-shortening process?

In interpreting chemical structures biogenetically it is essential to have "markers", features which will identify units despite metabolic vicissitudes—such as the isoprene skeleton, or the alternate oxygenation of aromatic polyketides. For the compounds so far considered, the chain-length has itself been one such marker and the position of unsaturation is another. With shorter-chain compounds the first of these markers is of course lost, but other features can also be used.

Recent studies of *Dahlia* sp. by Bohlmann and Kleine (1965) and by Jones, (1965) have provided a most interesting series of compounds, including in particular four C_{16} alcohols (XXV–XXVIII).

$$CH_3 \cdot (CH{=}CH)_2 \cdot C{\equiv}C \cdot CH_2 \cdot \overset{cis}{CH{=}CH} \cdot CH_2 \cdot (CH_2)_4 \cdot CH_2OH$$

XXV

$$CH_3 \cdot (CH{=}CH)_2 \cdot C{\equiv}C \cdot CH{=}CH \cdot CH{=}CH \cdot (CH_2)_4 \cdot CH_2OH$$

XXVI

$$CH_3 \cdot CH{=}CH \cdot (C{\equiv}C)_2 \cdot CH_2 \cdot \overset{cis}{CH{=}CH} \cdot CH_2 \cdot (CH_2)_4 \cdot CH_2OH$$

XXVII

$$CH_3 \cdot CH{=}CH \cdot (C{\equiv}C)_2 \cdot CH{=}CH \cdot CH{=}CH \cdot (CH_2)_4 \cdot CH_2OH$$

XXVIII

These compounds presumably represent reduction products of the corresponding C_{16} acids. Now two of these compounds (XXV and XXVII) have the unusual group *cis*-$CH{=}CH.CH_2.C{\equiv}C$—, which we have already seen in crepenynic acid (XIII) and the falcarinones (e.g. XIV), here occupying $C_{(7)}$–$C_{(11)}$ of the C_{16} chain and thus corresponding to the required location, $C_{(9)}$–$C_{(13)}$, in any C_{18} precursor. These two compounds would therefore be formed by extending conjugated unsaturation into the distal moiety of crepenynic acid as before and shortening the chain by β-oxidation at the carboxyl end.

The two other C_{16} alcohols (XXVI and XXVIII) are fully-conjugated, and they could arise from the former pair by the conversion

$$-CH_2-CH\overset{cis}{=}CH-CH_2- \xrightarrow[-2H]{} -CH=CH-CH=CH-$$

A similar overall reaction might also be required to account for the conjugated polyene fatty acids which, as already noted, occur frequently in Compositae; the reaction could be realized indirectly by a rearrangement involving adjacent acetylenic groups. Figure 8 shows the four C_{16} compounds

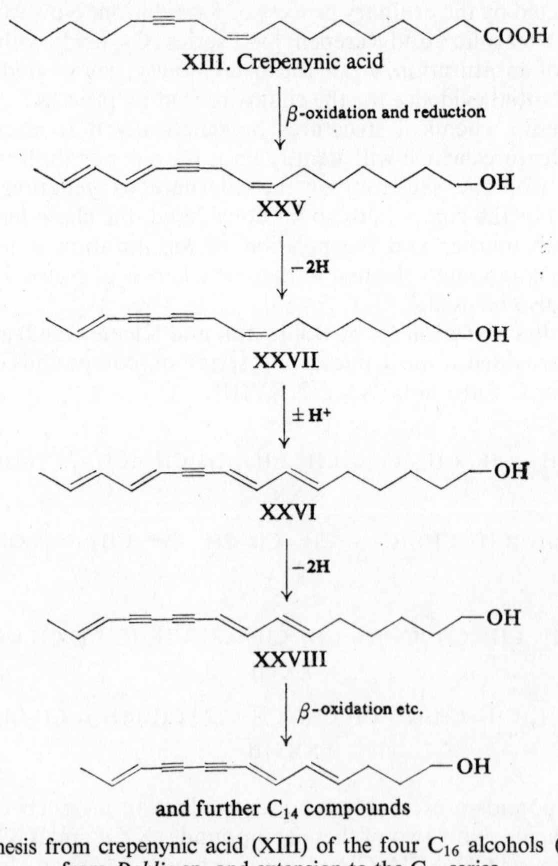

XIII. Crepenynic acid

β-oxidation and reduction

XXV

−2H

XXVII

±H⁺

XXVI

−2H

XXVIII

β-oxidation etc.

and further C_{14} compounds

FIG. 8. Biogenesis from crepenynic acid (XIII) of the four C_{16} alcohols (XXV–XXVIII) from *Dahlia* sp. and extension to the C_{14} series.

in such a sequence, and also shows how a further β-oxidation step would then lead to compounds of the C_{14} series which occur in the same species.

Other evidence that this type of rearrangement can indeed occur is provided by the isolation by Powell and Smith (1965b) from another of the Compositae, *Helichrysum* sp., of the acid (XXIX) which is presumably

$$CH_3 \cdot (CH_2)_4 \cdot C \equiv C \cdot CH = CH \cdot CHOH \cdot (CH_2)_7 COOH$$

XXIX

derived from a hydroxylated crepenynic acid by anionotropic rearrangement (note that the double bond is now *trans*, as we should expect). Unfortunately by such rearrangements to fully-conjugated unsaturation the "crepenynic marker" is lost. Consequently it is not particularly useful to explore further ramifications of such a scheme; one can merely state that by processes of the general type outlined one can plausibly derive the full range of natural acetylenes from Compositae, though whether crepenynic acid is always required as the precursor it is impossible to say. By the time we reach the C_{14} derivatives in Compositae, the more distinctive structural features have been obscured to all but the eye of faith.

B. FUNGAL ACETYLENES

With the typical polyacetylenes from higher fungi (Hymenomycetes) we encounter a similar situation: the long series of C_{10} derivatives shows no features which *necessarily* relate to longer-chain precursors though such an origin remains perfectly possible. However, there is a small group of longer-chain products which do preserve more distinctive features, and which in fact can be related to crepenynic acid (XIII). Figure 9 shows how the distal introduction of triple bonds, combined with β-oxidation, would lead to C_{14} and C_{12} acids which are closely related to the C_{11}–C_{14} fungal metabolites. The longest-chain compound, from *Poria sinuosa*, preserves the "crepenynic marker" system intact, whilst in the remainder we can see that it has been converted by Favorski rearrangements into the distinctive allene group, etc. In mycomycin (XXX), from *Odontia bicolor*, the *cis*-double bond remains, now adjacent to

$$CH \equiv C \cdot C \equiv C \cdot CH = C = CH \cdot (CH = CH)_2 \cdot CH_2 \cdot COOH$$

XXX. Mycomycin

the allene system. The isolated methylene group of the "crepenynic" system, at once propargylic and allylic, is mechanistically quite an apt precursor for the fungal allenes, and a biogenesis from crepenynic acid would moreover explain one curious feature of these compounds; all are diyne allenes and no other arrangement is yet known. This now appears as a simple consequence of the limited space available in the distal part of the molecule for conjugated triple bonds in the non-allenic precursors.

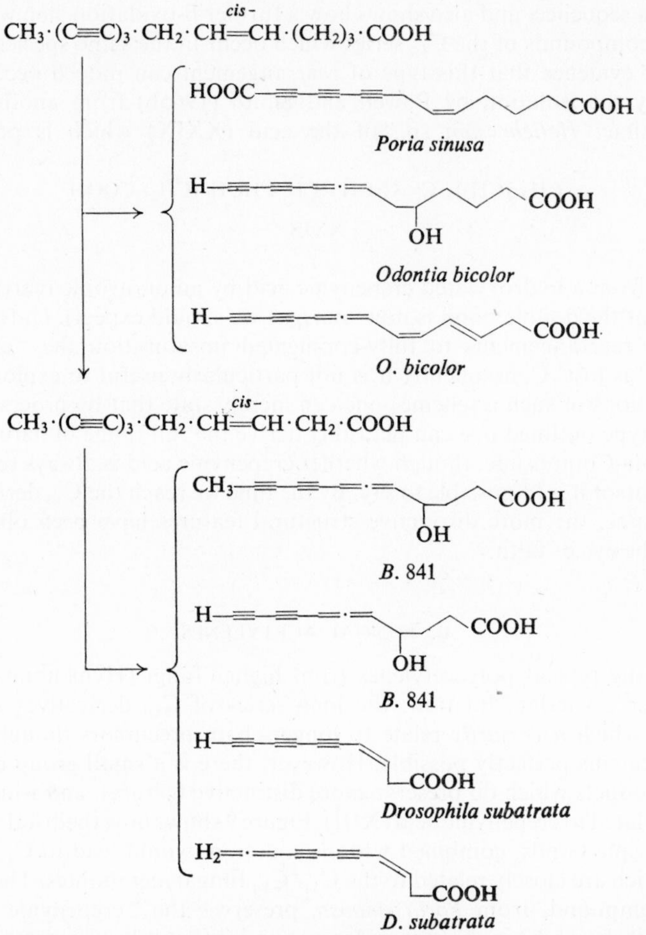

cis-

$$CH_3 \cdot (C \equiv C)_3 \cdot CH_2 \cdot CH = CH \cdot (CH_2)_3 \cdot COOH$$

Poria sinusa

Odontia bicolor

O. bicolor

cis-

$$CH_3 \cdot (C \equiv C)_3 \cdot CH_2 \cdot CH = CH \cdot CH_2 \cdot COOH$$

B. 841

B. 841

Drosophila subatrata

D. subatrata

FIG. 9. Relationship of fungal allenes to C_{14} and C_{12} acids derived from crepenynic by distal desaturation and β-oxidation.

V. CONCLUSIONS

Even in the absence of experimental evidence, biogenetic speculation should preserve decent limits, and in the foregoing passages I hope that these have not been transcended. If I have dealt more with biogenesis than with chemotaxonomy, this is because, in my view, the one is an essential prerequisite for the other, and in the present case biogenetic certainties are lacking. In default, then, I have tried to delimit the biogenetic possibilities, in a way which may prove susceptible to experimental analysis. The dehydrogenation hypothesis for the biogenesis of acetylenes from fatty acids seems at least worthy of experimental investigation. Should it be proved correct,* then the relationship of acetylene formation to the wider topic of lipid composition will provide ample scope for chemotaxonomy.

* *Added in proof.* Dr. G. N. Smith has now demonstrated that in *Tricholoma grammo-podium*, added 10-^{14}C-oleic is converted via 10-^{14}C-linoleic, -crepenynic, and 14,15-dehydro-crepenynic into 2-^{14}C-dehydromatricarianol, thus confirming the key reactions of dehydro-genation and chain shortening.

REFERENCES

Bagby, M. O., Smith, C. R. and Wolff, I. A. (1964). *Chem. & Ind.* 1861.

Bohlmann, F., Arndt, C., Bornowski, H., Jastrow, H. and Kleine, K.-M. (1962). *Ber. dtsch. chem. Ges.* **95**, 1320.

Bohlmann, F. and Kleine, K. M. (1965). *Ber. dtsch. chem. Ges.* **98**, 872.

Bu'Lock, J. D. (1964). *Progress in Organic Chemistry* **6**, 86.

Bu'Lock, J. D. and Gregory, H. (1959). *Biochem. J.* **74**, 322.

Bu'Lock, J. D. and Smalley, H. M. (1962). *J. chem. Soc.* 4662.

Bu'Lock, J. D. and Smith, G. N. (1963). *Phytochemistry* **2**, 289.

Bu'Lock, J. D. and Smith, G. N. (1964). *Biochem. Biophys. Res. Commun.* **17**, 433.

Craig, J. C. and Moyle, M. (1962). *Proc. chem. Soc.* 149 *and later papers.*

Gunstone, F. D. and Sealy, A. J. (1963). *J. chem. Soc.* 5772.

Harley-Mason, J. and Fleming, I. (1961). *Proc. chem. Soc.* 245 *and later papers.*

Hilditch, T. P. and Williams, P. N. (1964). "The Chemical Constitution of Natural Fats", 4th Ed. Chapman and Hall, London.

Hitchcock, C. and James, A. T. (1964). *Biochem. J.* **93**, 22P.

Hopkins, C. Y. and Chisholm, M. J. (1964). *Tetrahedron Letters*, 3011.

Irie, T., Suzuki, M. and Masamune, T. (1965). *Tetrahedron Letters* 1091.

Jones, E. R. H. (1965). *Chem. Engng. News* **39**, 46.

Klenk, W. and Knipprath, W. (1959). *Z. phys. Chem.* **317**, 243.

Matthews, W. S., Pickering, G. B. and Umoh, A. T. (1963). *Chem. & Ind.* 122.

Mikolajczak, K. L., Smith, C. R., Bagby, M. O. and Wolff, I. A. (1964). *J. org. Chem.* **29**, 318.

Powell, R. G. and Smith, C. R. (1965a). *Chem. & Ind.* 470.

Powell, R. G. and Smith, C. R. (1965b). *J. org. Chem.* **30**, 610.

Sørensen, N. A. (1962). *In* "Chemical Plant Taxonomy" (T. Swain, ed.), p. 219. Academic Press, London.

CHAPTER 6

The Distribution of Terpenoids

G. WEISSMANN

*Federal Research Organization for Forestry and Forest Products,
Hamburg, Germany*

I. INTRODUCTION

The terpenoids are a group of substances which are widely distributed in the plant and animal kingdoms. They are compounds whose structures are all based on the linking together of isoprene (C_5H_8) units and the majority contain 2, 3, 4, 6, or 8 multiples of this basic structure. The various groups of terpenes are distinguished according to the number of such units as shown in Table 1.

TABLE 1. Types of terpenoid compound

Type	Number of isoprene units	Occurrence
Hemiterpenes	1	In coumarins, quinones. etc.
Monoterpenes	2	Turpentine oil
Sesquiterpenes	3	Ethereal oils
Diterpenes	4	Resin acids
		Phytol
		Vitamin A
Triterpenes	6	Squalene
		Steroids
		Bile acids
		Sex hormones
Tetraterpenes	8	Carotenoids
Polyterpenes	n	*cis*-Rubber
		trans-Gutta, balata

A. THE BIOSYNTHESIS OF TERPENES

Our fundamental knowledge of the chemistry of the terpenoids was laid down by the work of Wallach (1887) and Ruzicka (1938), and led to the proposal of the "isoprene hypothesis". This original idea was not, however, quite correct, for it is not free isoprene itself which is the ultimate precursor of the terpenes. The search for the true isoprenoid precursor occupied a number of different research groups for many years, and several C_5 compounds were proposed and subsequently discarded (Sandermann, 1962). The problem was finally solved by the groups associated with Bloch (Shaykin *et al.*, 1958) and Lynen (1960) who elucidated the structure of the precursor and its mode of biosynthesis. According to them, the precursor, mevalonic acid, (I) is formed

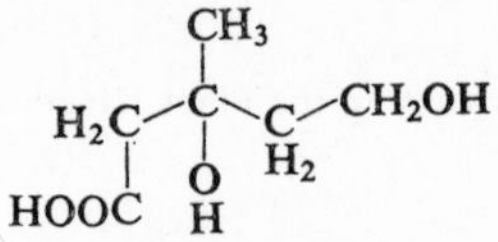

I. Mevalonic acid

by reduction of β-hydroxy-β-methylglutaric acid which itself is produced by the addition of acetic acid to acetoacetic acid after activation by coenzyme A. As has been shown by various experiments, mevalonic acid is better than any previously suggested compound as a precursor of a wide variety of terpenoids. Mevalonic acid (MVA) can also be formed from leucine via isovaleryl-CoA, β-methylcrotonyl-CoA, and β-methylglutaconyl-CoA. The subsequent reaction steps in the biosynthesis of the terpenes are enzymatically catalysed phosphorylation reactions. Mevalonic acid-5-phosphate and 5-pyrophosphate (MVA-5-PP) are formed successively under the action of ATP and MVA-kinase and phospho-MVA-kinase respectively. Subsequently, MVA-5-PP is phosphorylated at the tertiary hydroxyl group, and the product is stabilized by elimination of phosphoric acid and CO_2 to form isopentenyl pyrophosphate (IPP (II)), which is the so-called "active isoprene" and the true structural unit of all terpenoids.

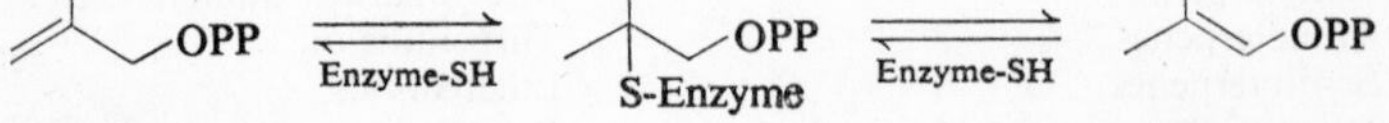

II. Isopentenyl pyrophosphate

III. Dimethylallyl pyrophosphate

FIG. 1. Interconversion of isopentenyl- and dimethylallyl-pyrophosphates.

The transformation of (II) to dimethylallyl-pyrophosphate (III) takes place under the influence of isopentenyl-pyrophosphate-isomerase (Fig. 1). This equilibrium reaction is essential for the further interlinkage of the terpene units. The IPP-isomerase is totally inhibited by compounds such as iodoacetamide or p-chloromercuribenzoate. Lynen therefore assumes that the active group of the enzyme contains a sulfhydryl group possibly in the form of a cysteine residue with a peptide-like bond, and that the isomerization takes place via the addition of the SH-enzyme to the double bond in II or III (Lynen *et al.*, 1959).

The linking of II and III to give geranyl pyrophosphate (V) is the starting point for the formation of the majority of plant terpenes (Fig. 2). The multiplicity of the natural terpenoids is formed by variation in the subsequent mode

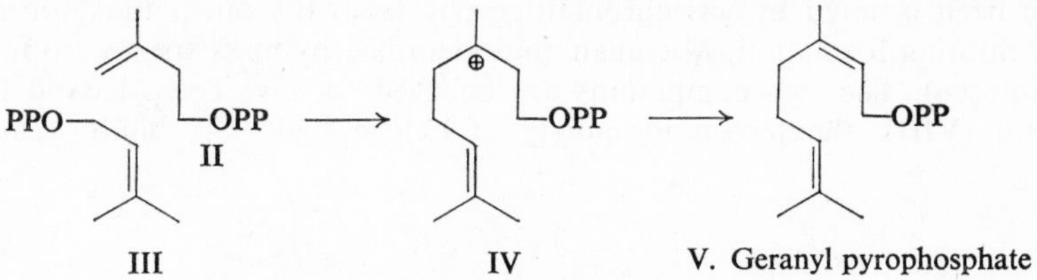

PPO— —OPP ⟶ —OPP ⟶ —OPP

II

III   IV   V. Geranyl pyrophosphate

Fig. 2. The formation of geranyl pyrophosphate.

of condensation. The monoterpenes arise from geranylpyrophosphate (GPP) by cyclization, rearrangement, or oxidation. The addition of another IPP unit gives farnesyl pyrophosphate which leads to the sesquiterpenes. The diterpenes are derived from geranylgeranyl pyrophosphate which is produced by the condensation of two GPP units. The presence of the various K_2 vitamins and ubiquinones in micro-organisms and animals which contain isoprenoid side chains of up to 50 C-atoms shows that the condensation of polyterpene chains does not stop at the stage of farnesyl pyrophosphate. The biosynthesis of the polyprenes, rubber and gutta-percha, can be envisaged as proceeding in a similar manner. Archer *et al.* (1961) and Lynen and Henning (1961) have found a rapid incorporation of [1-^{14}C]-isopentenyl-PP into rubber in *Hevea brasiliensis*.

In addition to the "head-to-tail" condensation of the isoprenoid units which leads to the above compounds, nature possesses a further system of building at the stage of farnesyl-PP and geranylgeranyl-PP. Tri- and tetraterpenes can be formed by the "tail to tail" dimerization of the C_{15} and C_{20} units. Thus, the dimerization of farnesyl-PP forms squalene (C_{30}), from which the cyclic triterpenes and steroids are derived. Tetraterpenes (C_{40}), the dimerization products of geranylgeranyl-PP, are also widely distributed in the plant kingdom, for example the carotenoids (see Table 1 and Chapter 7).

B. THE DISTRIBUTION OF TERPENES IN THE PLANT KINGDOM

Organisms can, in principle, be divided into two groups. The first are those which can utilize absorbed energy and synthesize all the substances which they require for their existence from simple inorganic substances. The second group, which includes animals, are those organisms which need to ingest complicated organic materials. Among the autotrophic organisms, the green plants are the most important. They are capable of utilizing the sun's energy for the synthesis of hundreds of complicated energy-rich compounds. The starting materials are CO_2, water, and inorganic salts, with ammonia or nitrate as the source of nitrogen. The key substance in photosynthesis is chlorophyll, and thus a terpenoid compound is at the centre of plant life.

Proof of the existence of isoprenoid compounds millions of years ago has recently been given (Anon., 1964). The hydrocarbons phytane (VI 2,6,10,14-tetramethylhexadecane) and pristane (VII 2,6,10,14-tetramethylpentadecane) have been isolated by gas chromatography from the oil of the Nonesuch Precambrian formation, Michigan, and identified by mass spectroscopy (see Chapter 4). The two compounds are believed to have been derived from phytol (VIII), the terpenoid moiety of chlorophyll. The billion-year-old

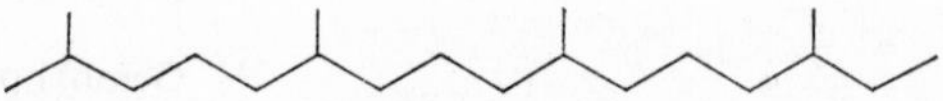

VI. Phytane

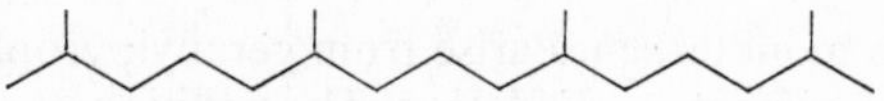

VII. Pristane

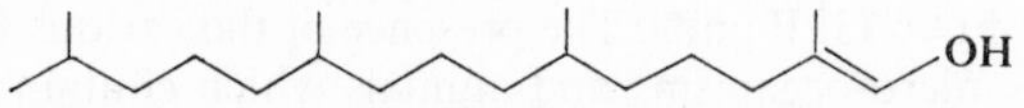

VIII. Phytol

formation also contains microfossils and porphyrins, and it may be concluded that photosynthetically active organisms existed at this period.

In addition to phytol, other terpenoids also occur in the primitive forms of plant life (Fig. 3). However, because of their large number, and to some extent of the amounts in which they occur, we can regard the mono-, sesqui-, di- and poly-terpenes as being typical constituents of the higher land plants. The steroids and the carotenoids, on the other hand, are distributed generally, and some of them have a vital function. Moritz (1958) calls these latter compounds the "true aims" in terpene metabolism. Terpenes in Pteridophyta, which include only a few living species, have scarcely been investigated.

II. The Monoterpenes

As stated above, the first stage in the formation of monoterpenes is the linking of one molecule of isopentenyl-pyrophosphate (II) and one molecule of dimethylallyl-pyrophosphate (III) under the catalytic action of farnesyl-pyrophosphate-synthetase, to give geranyl-pyrophosphate (V) via the cation

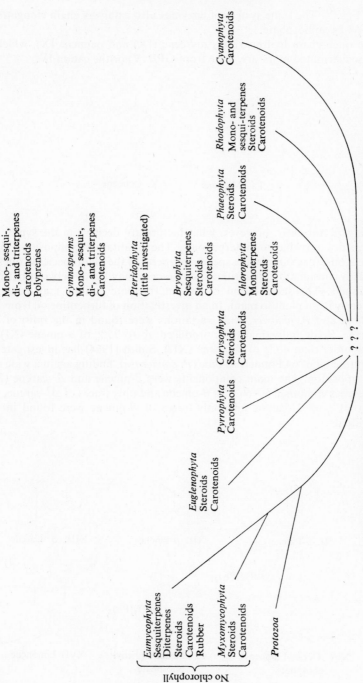

Fig. 3. Evolution of the plant kingdom (after Smith, 1955) and occurrence of terpenoids.

(IV) (Fig. 2). The same system of enzymes also catalyses chain elongation to farnesyl-pyrophosphate.

The open-chain hydrocarbons myrcene (IX) and ocimene (X), which are widely distributed in nature, arise from GPP (V) or the cation IV.

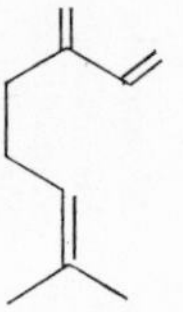

IX. Myrcene 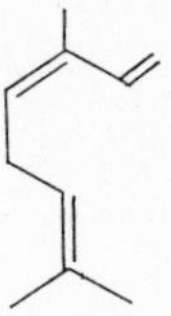   X. Ocimene

It is still not clear, however, what factors are decisive in the synthesis of the various possible monoterpenoids. The qualitative composition of the ethereal oils, for example, is a specific property of the plant species in question, although considerable quantitative differences may occur between particular individuals. The stabilization of the intermediate cations may take place differently from plant to plant. In an investigation of turpentine oil from *Pinus maritima* and *P. pinaster*, wide variations were found in the ratio of α- to β-pinene. Some individual trees produced up to 95% of α-pinene (XI), and others more than 40% of β-pinene (XII). Smith (1964) has investigated the turpentine oil of 64 Ponderosa pine (*P. ponderosa*), finding again a wide range in composition. The main components were β-pinene and Δ^3-carene (XIII). Three trees contained 50–60% of β-pinene and only traces of Δ^3-carene, while up to 80% of Δ^3-carene and only traces of β-pinene were found in other individuals.

XI. α-Pinene 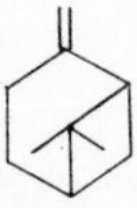   XII. β-Pinene 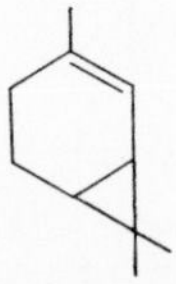   XIII. Δ^3-Carene

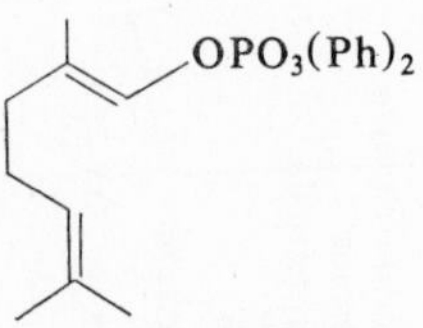

XIV. Geranyl diphenyl phosphate 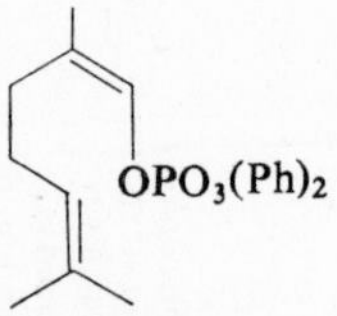   XV. Nerol diphenyl phosphate 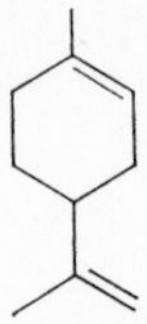   XVI. Limonene

An interesting contribution to the formation of the monoterpenes has been provided by Miller and Wood (1964). When a solution of geranyl diphenyl phosphate (XIV) in ether was allowed to stand for a long time, the open-chain terpenes myrcene (IX) and ocimene (X) could be isolated from the reaction product. Similar treatment of the diphenyl phosphate of nerol (XV, *cis*-configuration at the allylic double bond) gave limonene (XVI) in 45% yield. The reactions correspond to the formation of C—C bonds from pyrophosphates *in vivo*, and indicate that the stereochemistry at the allylic double bond in the precursor of the monoterpene hydrocarbons determines which terpene is formed.

In a study on the biogenesis of limonene (XVI) from (2-^{14}C)MVA in *Pinus pinea* L., Sandermann and Bruns (1962a,b) investigated the question of whether the methyl groups of the isopropyl grouping are equivalent. To decide between the reaction paths formulated (a, b or c, Fig. 4), the terminal methylene groups

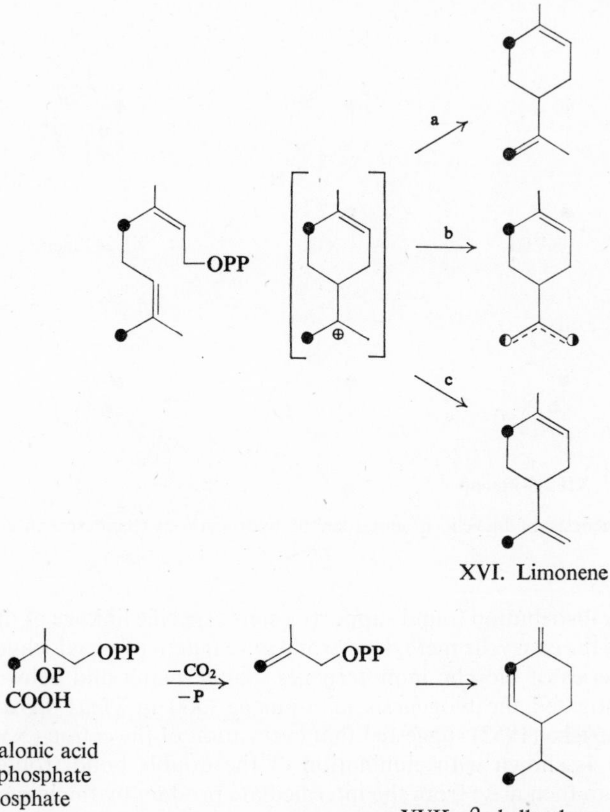

XVI. Limonene

I. Mevalonic acid
5-pyrophosphate
3-phosphate

XVII. β-phellandrene

FIG. 4. Biogenesis of limonene and β-phellandrene (Sandermann and Bruns, 1962; Bruns, 1964).

of the isolated radioactive limonene were split off by oxidation, and the formaldehyde produced was isolated as the dimedone derivative. No activity could be detected in this compound. The stabilization of the cation in the form of limonene therefore takes place specifically in accordance with pathway c (Fig. 4).

Likewise, Bruns (1964) has been able to establish a specific incorporation of [2-^{14}C]–MVA into β-phellandrene (XVII) from *Pinus contorta*. Here again,

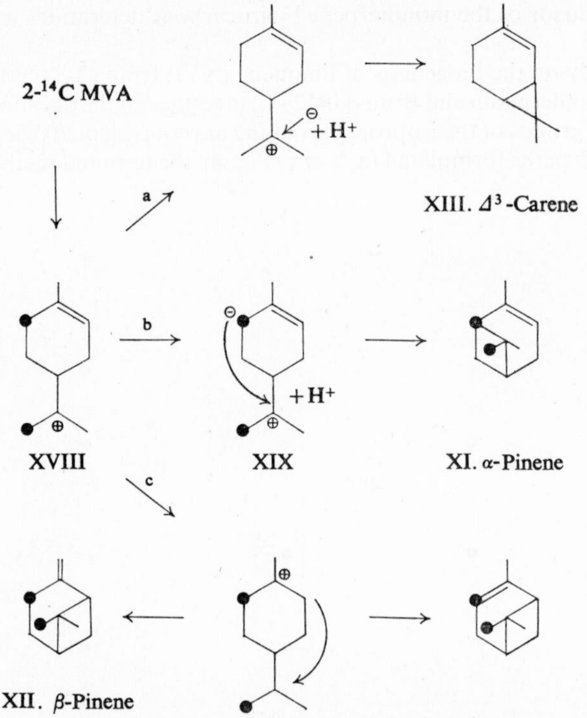

FIG. 5. Biogenesis of bicyclic monoterpenoid hydrocarbons (Sandermann and Schweers, 1962).

the activity distribution found supports a stereospecific linkage of the C_5 units. In this case the exocyclic methylene group gave inactive formaldehyde (Fig. 4).

In the series of bicyclic monoterpenes, Sandermann and Schweers (1962) have investigated the biogenesis of α-pinene (XI) in *Pinus nigra* Austriaca (Fig. 5). Ruzicka (1953) suggested that cyclization of the cation (XVIII) might take place as shown with elimination of the double bond (route c); α- or β-pinene can then arise from the intermediate product by the loss of a proton (Fig. 5). By the use of [2-^{14}C]–MVA, it was possible to show that XVIII actually reacts by the *direct* elimination of a proton (route b) in the way which was formulated by Ruzicka for the formation of carene (XIII) (route a). However, while the formation of β-pinene (XII) can also be explained by route c, it is not possible to do so by the intermediate (XIX) postulated by Sandermann and Schweers (route b). As a further possibility for the synthesis of β-pinene, Ruzicka discussed a radical mechanism and considers myrcene (IX) as a possible precursor. In agreement with this theory, Crowley (1962) found that β-pinene was formed when an ethereal solution of myrcene (IX) was irradiated with U.V. light.

Of the tricyclic monoterpene hydrocarbons, so far only tricyclene (XX) has been found in some coniferous oils. Porsch and Farnow (1962b) isolated it

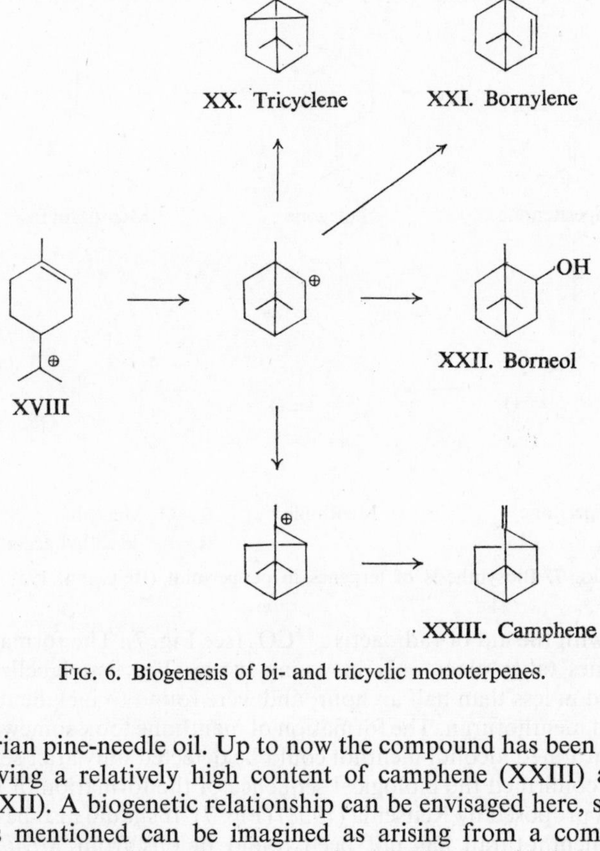

FIG. 6. Biogenesis of bi- and tricyclic monoterpenes.

from Siberian pine-needle oil. Up to now the compound has been found only in oils having a relatively high content of camphene (XXIII) and bornyl acetate (XXII). A biogenetic relationship can be envisaged here, since all the substances mentioned can be imagined as arising from a common inter-mediate (Fig. 6). The formation of bornylene (XXI) can also be explained by this mechanism.

Oxygen-containing compounds of the monoterpenoid series—alcohols, aldehydes, and ketones are found to an overwhelming extent in the ethereal oils of the angiosperms. The formation of the alcohols is regarded by Mayer (1961) as a competing reaction with that leading to the hydrocarbons, and not as a secondary one. The alcohols arise by the elimination of protons from hydrated cations. Alcohols can also be formed in secondary reactions but only to a lesser extent. When a plant is capable of synthesizing alcohols at all, the alcohols and the corresponding hydrocarbons are frequently found together, particularly in the pairs limonene-α-terpineol and pinene-borneol.

As yet, no light has been shed on the biosynthesis of the terpene ketones. Sandermann (1938) regards limonene as the precursor for the formation of carvone in caraway (*Carum carvi*), while Kremers (1922) assumes that the synthesis of the oxygen-containing terpenes which are found in *Mentha* species takes place via citral. Battaile and Loomis (1961) have investigated the biosynthesis of terpenes in young peppermint plants (*Mentha piperita* L., var.

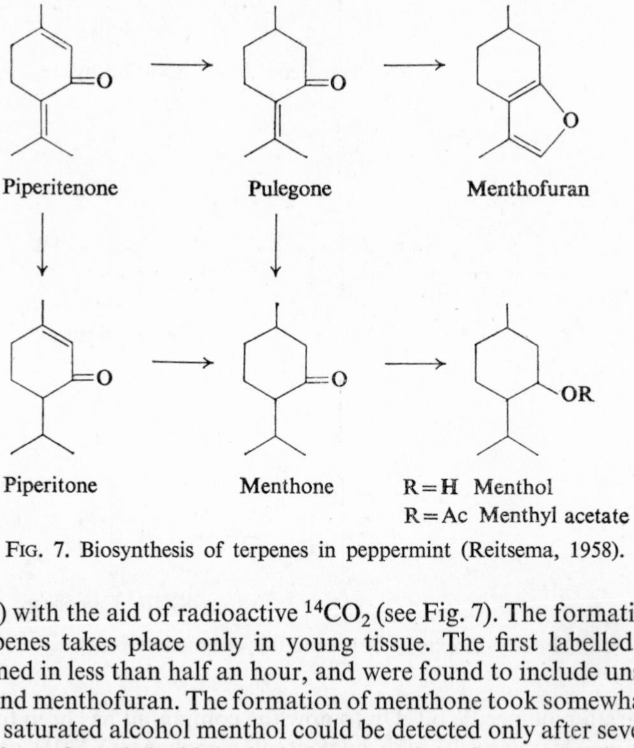

Piperitenone   Pulegone   Menthofuran

Piperitone   Menthone   R = H Menthol
                                R = Ac Menthyl acetate

FIG. 7. Biosynthesis of terpenes in peppermint (Reitsema, 1958).

Mitcham) with the aid of radioactive $^{14}CO_2$ (see Fig. 7). The formation of the monoterpenes takes place only in young tissue. The first labelled terpenes were formed in less than half an hour, and were found to include unsaturated ketones and menthofuran. The formation of menthone took somewhat longer, while the saturated alcohol menthol could be detected only after several days. This result confirmed the biological sequence of the formation of terpenes in peppermint proposed by Reitsema (1958) (Fig. 7). It should also be mentioned here that menthofuran has not been found in oils from *Mentha arvensis* (Porsch and Farnow, 1962a).

III. THE SESQUITERPENES

The structural unit for the formation of sesquiterpenes is assumed to be *cis*- or *trans*-farnesol (XXIV and XXV) (Ruzicka, 1953). According to

Hendrickson (1959), the biogenesis of this class of compound proceeds, after the elimination of the allylic hydroxyl group, via an unstable cation which forms the different classes of ring compounds by linkage to one or other of the isolated double bonds (Fig. 8). The resulting compounds can then be stabilized either by reaction with the solvent or by the liberation of a proton.

The involvement of the central double bond in cyclization is possible only

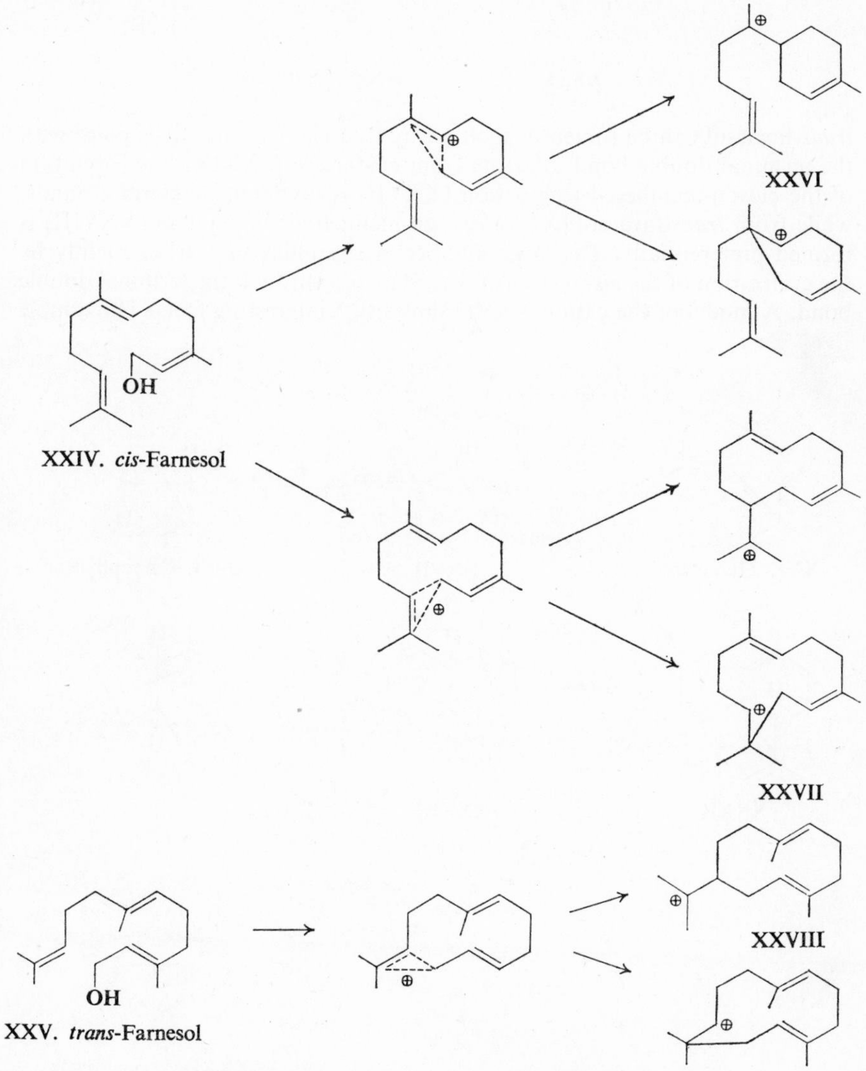

FIG. 8. Biogenesis of sesquiterpenes (Hendrickson, 1959).

with *cis*-farnesol (XXIV). Almost all monocyclic sesquiterpenes with six-membered rings have the skeleton of bisabolene (XXIX), which arises from XXVI by the simple splitting out of a proton. On the other hand, *cis*- and

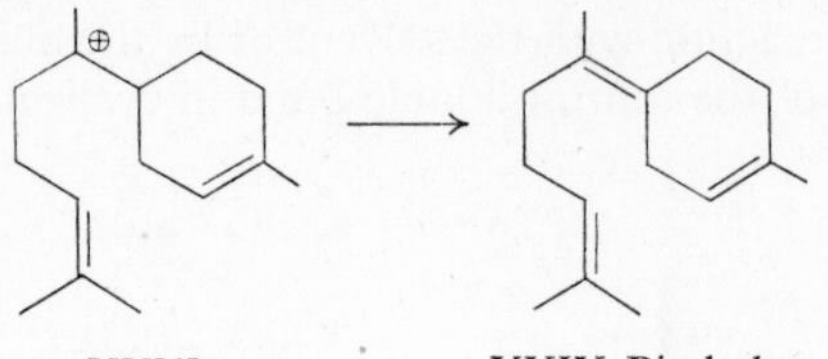

XXVI   XXIX. Bisabolene

trans-farnesol can be folded in such a way that ring closure takes place with the terminal double bond. Starting from *cis*-farnesol (XXIV), the formation of the eleven-membered ring cation (XXVII) is favoured on steric grounds, while from *trans*-farnesol (XXV) the ten-membered ring system (XXVIII) is formed preferentially. The large number of sesquiterpenes arises mainly by the cyclization of the *cis*- or *trans*-farnesyl cation through the terminal double bond. A model of the cation XXVII shows two interesting facts. The double

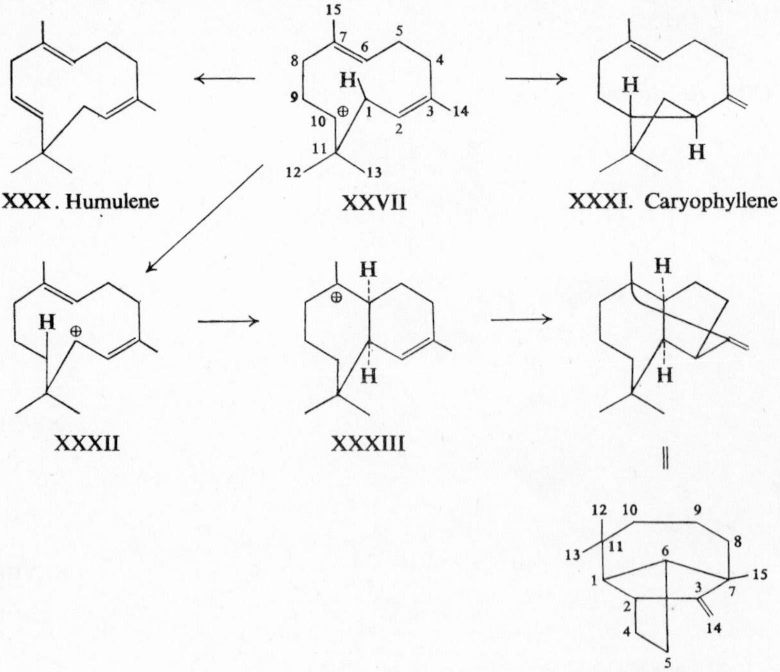

XXX. Humulene   XXVII   XXXI. Caryophyllene

XXXII   XXXIII

XXXIV. Longifolene

Fig. 9. Biogenesis of sesquiterpene hydrocarbons.

bonds are not near enough together for internal cyclization. Furthermore, the H atom extending inwards on C-1 blocks the double bond at C-6–7. The simplest method for the neutralization of the cation is the elimination of a proton, which leads to the hydrocarbon humulene (XXX) (Fig. 9).

The reaction of the C-10 cation with the C-2–3 double bond is not sterically hindered. Here, probably, a proton is simultaneously split off at C-3 and the hydrocarbon caryophyllene (XXXI) is formed (Fig. 9).

Hendrickson assumes a 1,3-hydride migration in XXVII to be primarily responsible for the formation of longifolene (XXXIV). The cation (XXXII) resulting from this migration is transformed into the ion (XXXIII) by ring closure, and this is stabilized as longifolene by rearrangement and the loss of a proton (Fig. 9). Support for Hendrickson's proposed synthetic route for the sesquiterpenes was given by Sandermann and Bruns (1962) with their work on the biosynthesis of longifolene in *Pinus longifolia*. By the incorporation of [1-^{14}C]-acetate in young plants and chemical degradation of the resulting active longifolene, these authors were able to show that the formally regular structure is produced by rearrangements. The two possibilities available for consideration give different distributions of the activity (Fig. 9). In the case of a regular structure, the formaldehyde produced by the ozonization of the exocyclic double bond must be active, while if the rearrangement discussed above takes place an inactive methylene group must be formed. The dimedone derivative isolated showed practically no activity, and this is, therefore, in favour of the synthetic route proposed by Hendrickson (Fig. 9).

Their common occurrence in certain types of plants is also in favour of similar synthetic routes for these sesquiterpenes and the related juniperol (XXXV) Table 2. Thus longifolene frequently occurs in pines together with either juniperol or caryophyllene. In juniper, it is present with juniperol, while

TABLE 2. Distribution of sesquiterpenes (Akiyoshi *et al.*, 1960; Hendrickson, 1959).

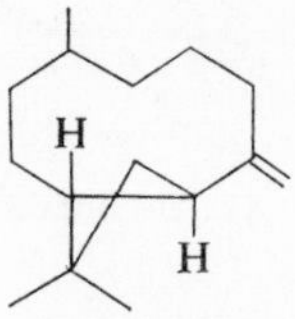

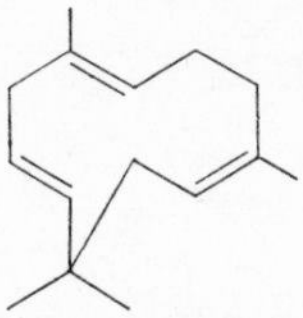

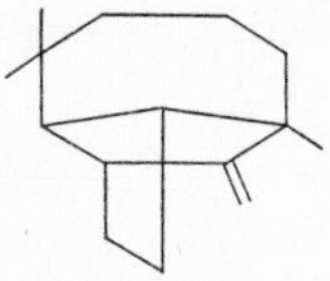

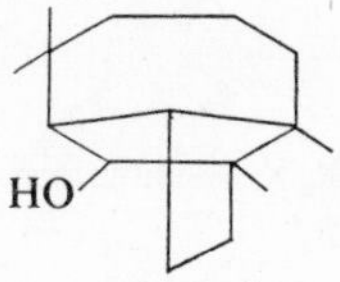

	Caryophyllene XXXI	Humulene XXX	Longifolene XXXIV	Juniperol XXXV
Eugenia caryophyllata	+	+		
Juniperus communis			+	+
Pinus densiflora			+	+
P. thunbergii			+	+
P. longifolia	+		+	
P. maritima	+		+	

in clove oil (*Eugenia caryophyllata*) caryophyllene occurs together with humulene (Table 2).

From the turpentine of *Pinus longifolia* Roxb., Nayak and Dev (1963) isolated, in addition to longifolene (XXXIV), another fully saturated sesquiterpene, longicyclene (XXXVII). This comprises a tetracyclic system with a three-membered ring. The treatment of longicyclene with cupric acetate in boiling glacial acetic acid leads to its partial isomerization to longifolene. If

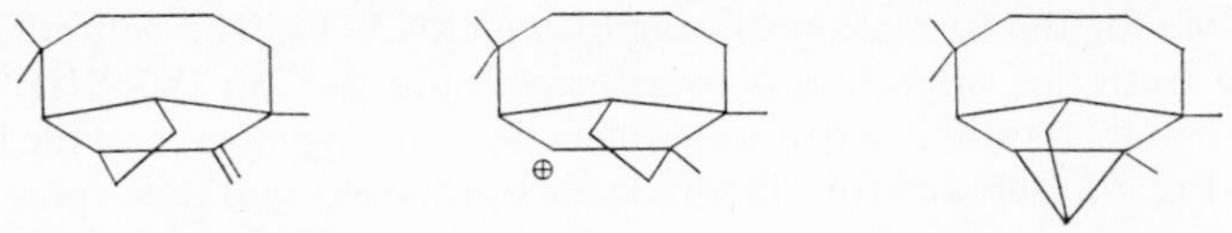

XXXIV. Longifolene   XXXVI.   XXXVII. Longicyclene

we assume that the cation XXXVI formulated by Ourisson (1955) is the precursor, the biogenetic relationship between longifolene and longicyclene can easily be demonstrated. Furthermore, this gives an interesting parallel to the formation of camphene and tricyclene from a common precursor (see p. 105).

The biosynthetic route for the bicyclic sesquiterpene hydrocarbons α- and β-himalachene (XXXIX, XL) (Brendenburg and Erdtman, 1961) isolated from *Cedrus* spp., is again assumed to go via the cation (XXXIII) (Rao *et al.*, 1952; Bredenberg and Erdtman, 1961).

A common biological precursor can also be assumed for the sesquiterpene alcohols elemol, eudesmol, and hinesol (Fig. 10) isolated from "atractylis concret", an ethereal oil from China (Yosioka *et al.*, 1959b). Hinesol is a tertiary alcohol of the vetivane type (Yosioka *et al.*, 1959a), for which Motl

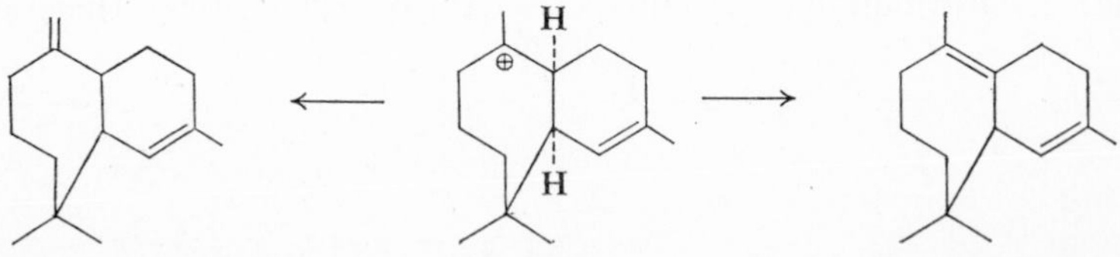

XXXIX. α-Himalachene   XXXIII.   XL. β-Himalachene

et al. (1961) have proposed the structure (XLIV). It is possible to derive the three alcohols from the *trans*-cation (XXVIII, see Fig. 8), from which a tertiary alcohol (XLI) can arise directly by simple hydration. The isomeric eudesmols (XLIII) can be formed by subsequent elimination of water (Bates and Hendrickson, 1962). After previous isomerization of the double bonds, hinesol (XLIV) can be formed in a similar reaction, and elemol (XLII) is produced by a simple cyclization to a six-membered ring system (Fig. 10).

Various sesquiterpene alcohols have been isolated from the heartwood of the Eastern White Cedar (*Thuja occidentalis* L.). The main products are

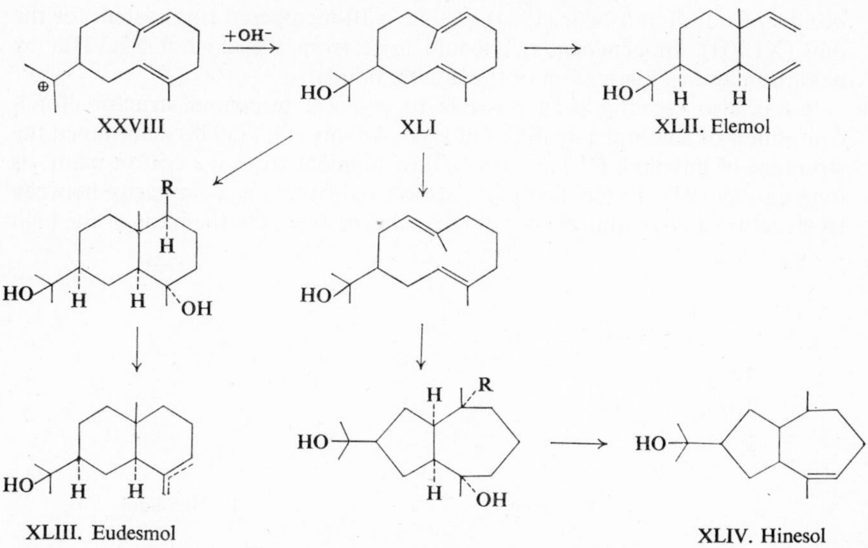

FIG. 10. Biogenetic relationships of some sesquiterpene alcohols.

occidentalol (XLV) and occidol (XLVI) (Nakatsuka and Hirose, 1956; Hirose and Nakatsuka, 1959).

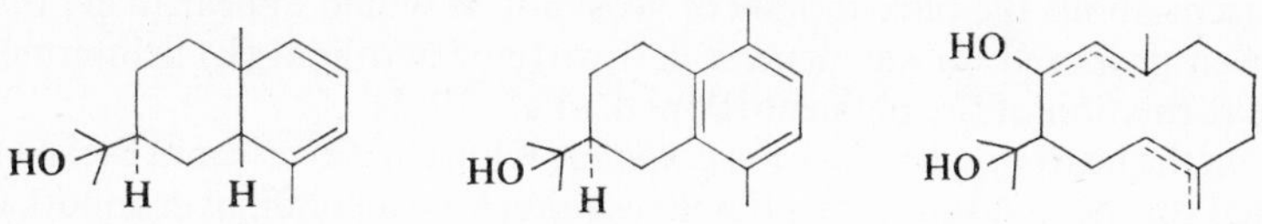

XLV. Occidentalol XLVI. Occidol XLVII. Occidiol

In addition, traces of eudesmol and a new sesquiterpene diol have been discovered (Rudloff and Nair, 1964). According to Rudloff and Erdtman (1962), the rings in occidentalol (XLV) are *cis*-coupled and the compound should therefore be biogenetically related to occidol. It is assumed that the precursor is a tertiary alcohol (XXVIIIa) produced from the ion XXVIII (see Fig. 8), as in the case of eudesmol. If, however, the methyl groups are located on opposite sides, a similar reaction mechanism would lead first to the *cis*-coupled intermediate (XLVIII), which would then lead to occidentalol and

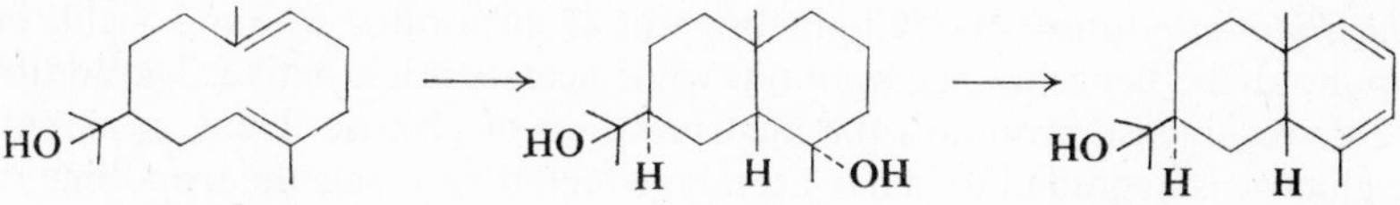

XXVIIIa XLVIII XLV. Occidentalol

occidol. Rudloff and Nair (1964) assume a 10-membered ring system for the diol (XLVII). Biogenetically, it could arise from the alcohol XXVIIIa by oxidation and isomerization of the double bonds.

It has also recently been possible to prove a terpenoid structure for a compound of a completely different type. Adams *et al.* (1938) determined the structure of gossypol (IL), a toxic yellow pigment from the cotton plant, as long ago as 1938. Purely formally, at first sight there is a similarity between its structure and sesquiterpenes of the cadalene type. On the basis of the high

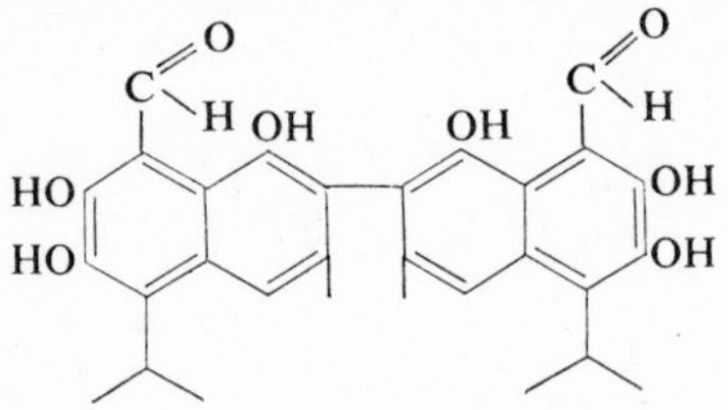

IL. Gossypol 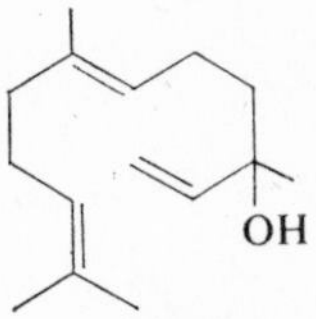   L. Nerolidol

rate of incorporation of [1-^{14}C]-acetate, Heinstein *et al.* (1962) were able to show that acetate is a direct precursor of gossypol. Furthermore, the specific activity of isolated degradation products was in agreement with that expected on theoretical grounds. When [2-^{14}C]-mevalonic acid was used, it was possible to achieve an incorporation rate of 22%. Only hypotheses can be put forward at present about the biosynthesis of gossypol. It would appear likely that, as has been proposed for squalene, a derivative of nerolidol (L) is intermediate in the formation of the pigment (Popjak *et al.*, 1961).

From the heartwood of *Cupressus lindleyi* Klotsch, Zavarin and Bicho (1963) isolated for the first time a tropolone sesquiterpene alcohol, hydronootkatinol (LI), and elucidated its structure. In addition to the alcohol, nootkatin (LII), β-thujaplicin (LIII), and the terpene phenol carvacrol (LIV) were also found. A biogenetic relationship between the tropolone derivatives is obvious. It is remarkable that the same mechanism of cyclization appears to take place at the stage of thujaplicin and nootkatin.

The biosynthesis of the tropolones is still obscure. Although they do not have the normal terpenoid structure, Erdtman (1952) regards them as terpenes. This is shown by their frequent occurrence together with terpenoid phenols, and by the fact that the tropolones can readily be converted into isopropyl benzoic acid derivatives by a benzylic acid rearrangement.

On the other hand, Tanenbaum *et al.* (1959, 1962) deny that the tropolones have a true isoprenoid structure. In the synthesis of stipitatic acid (LV) by *Penicillium stipitatum*, [6-^{14}C]-glucose was incorporated in good yield, into troponoid and benzenoid compounds while acetate is less active. The addition of 2-deoxyglucose promotes the incorporation of glucose. Here, apparently, the glucose is degraded to active acetate, which then forms the tropolone ring with either deoxyglucose or trioses formed from it.

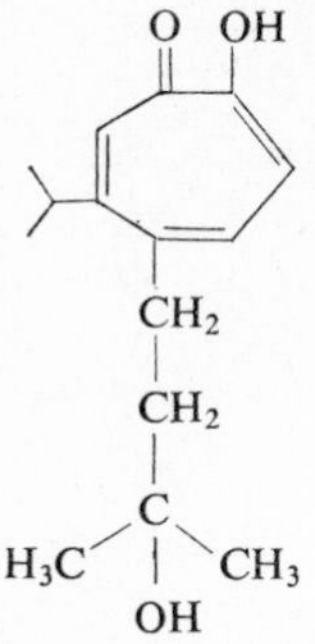

LI. Hydronootkatinol 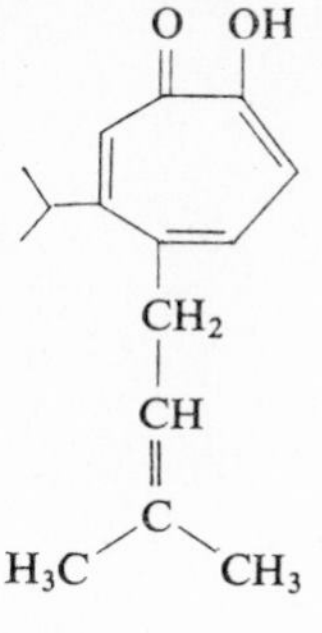   LII. Nootkatin 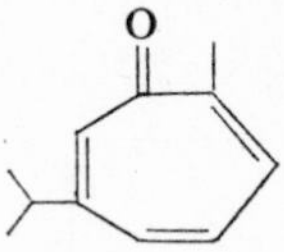   LIII. β-Thujaplicin

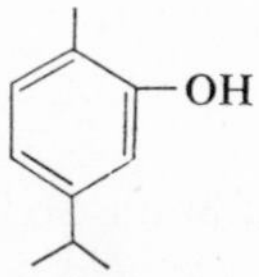

LIV. Carvacrol 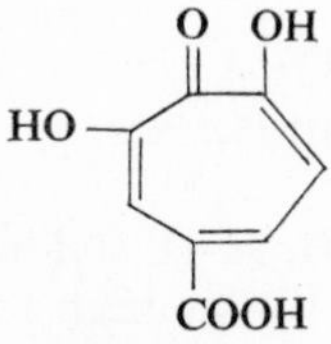   LV. Stipitatic acid

IV. THE DITERPENES

The basic structural unit of the diterpenes is geranylgeraniol, in which four isoprene units are linked in a head-to-tail sequence. With a few exceptions, this normal structure is retained in the diterpenes. Formally, the diterpenes can be derived most simply from geranyl-linalool (LVI) (Ruzicka, 1963). While the conversion of all-*trans*-squalene in a stereospecific non-stop cyclization leads directly to tetra- or pentacyclic cations, the cyclization of geranyl-linalool (LVI) stops at the stage of the bicyclic cation (LVII). In nature, the cation is often stabilized to give a bicyclic end-product. The addition of a hydroxyl group gives sclareol (LVIII), while the elimination of a proton gives manool (LIX). Up to this stage, as in the formation of the A and B rings of the triterpenes, the cyclization leads to an anti-*trans*-configuration of the substituents at carbon atoms 5, 10, and 9. Further cyclizations and rearrangements lead to the tri- and tetracyclic diterpene compounds.

Diterpenes are widely distributed in the conifers where they are present primarily as carboxylic acids. In the last few years, however, various alcohols and aldehydes have also been found, which have long been assumed to be biological precursors of these resin acids. In an investigation of kauri gum from *Agathis australis* (New Zealand), Enzell and Thomas (1964) were able to isolate from the neutral constituents a series of hydroxy-ketones for which they proposed the following structures (Fig. 11). The main products are araucarolone (LX) and araucarone (LXI), while araucarol (LXII) and araucarenolone (LXIII) are present in only small amounts. All four compounds have

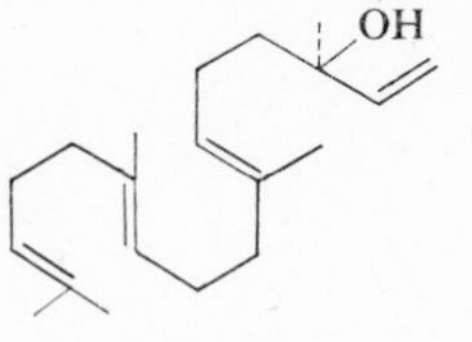

LVI. Geranyl-linalool

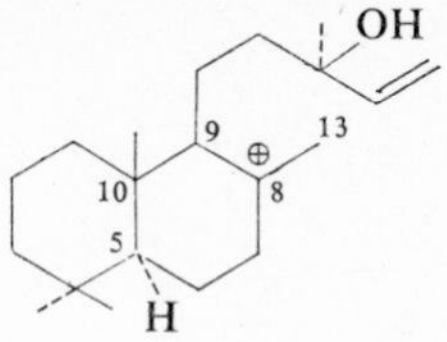

LVII

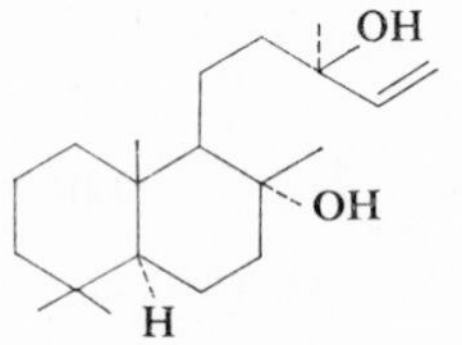

LVIII. Sclareol

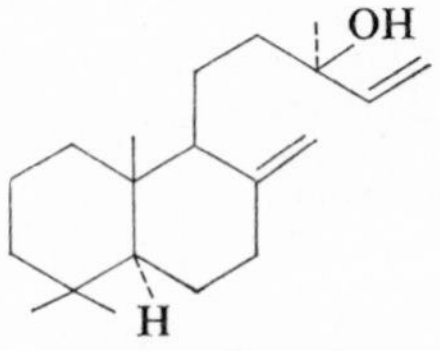

LIX. Manool

the "normal" 5α-H, 9α-H, 10β-methyl configuration. So far, only the bicyclic agathic acid (LXIV) has been found among the acid components from *A. australis* and also those from *A. alba* and *A. microstachya* which is a native of North Queensland (Carman, 1964). On the other hand, agathic acid has not been detected in *A. robusta* from South Queensland. From the gum of this tree Carman and Dennis (1964) isolated only laevopimaric acid (LXV) and communic acid (LXVI). The associated occurrence of these two acids is interesting for two reasons. It is unusual for bi- and tricyclic diterpene acids to occur together. In fact such a combined occurrence was only shown for the first time two years earlier by Enzell and Theander (1962) who isolated pinifolic acid (LXVII) from the needles of *Pinus silvestris*. Furthermore, this is in fact the first time that diterpene acids with α- and β-C(4)-carboxyl groups have been found together in a plant.

Chandra *et al.* (1964) have recently isolated some new diterpenes from *Araucaria imbricata* Pavon (= *A. araucana*), a hitherto little studied species of the Araucariaceae, and have elucidated their structure. By extracting the bark, they obtained a diol (LXVIII) and the corresponding hydroxycarboxylic acid (LXIX). The intermediate stage, the hydroxy-aldehyde (LXXI) was also detected. On the basis of chemical and physical investigations it was shown that the stereochemical configuration corresponded to that of labdanolic acid (LXXII). In addition to the compounds mentioned, the acetate of the acid (LXX) was also found in the balsam of *A. imbricata* (Weissmann and Bruns, 1965). (See also Weissmann *et al.* 1965.)

Communic acid (LXVI) found in *Agathis robusta* (Carman and Dennis, 1964) was first isolated from the bark of *Juniperus communis* L., and its structure was established by Arya *et al.*(1961). Magoni and Belardini (1964a,b) succeeded in isolating, in addition to communic acid, two other bicyclic diterpene acids, cupressic acid (LXXIII) and isocupressic acid (LXXIV) from the balsam of *Cupressus sempervirens*, which also belongs to the family of Cupressaceae.

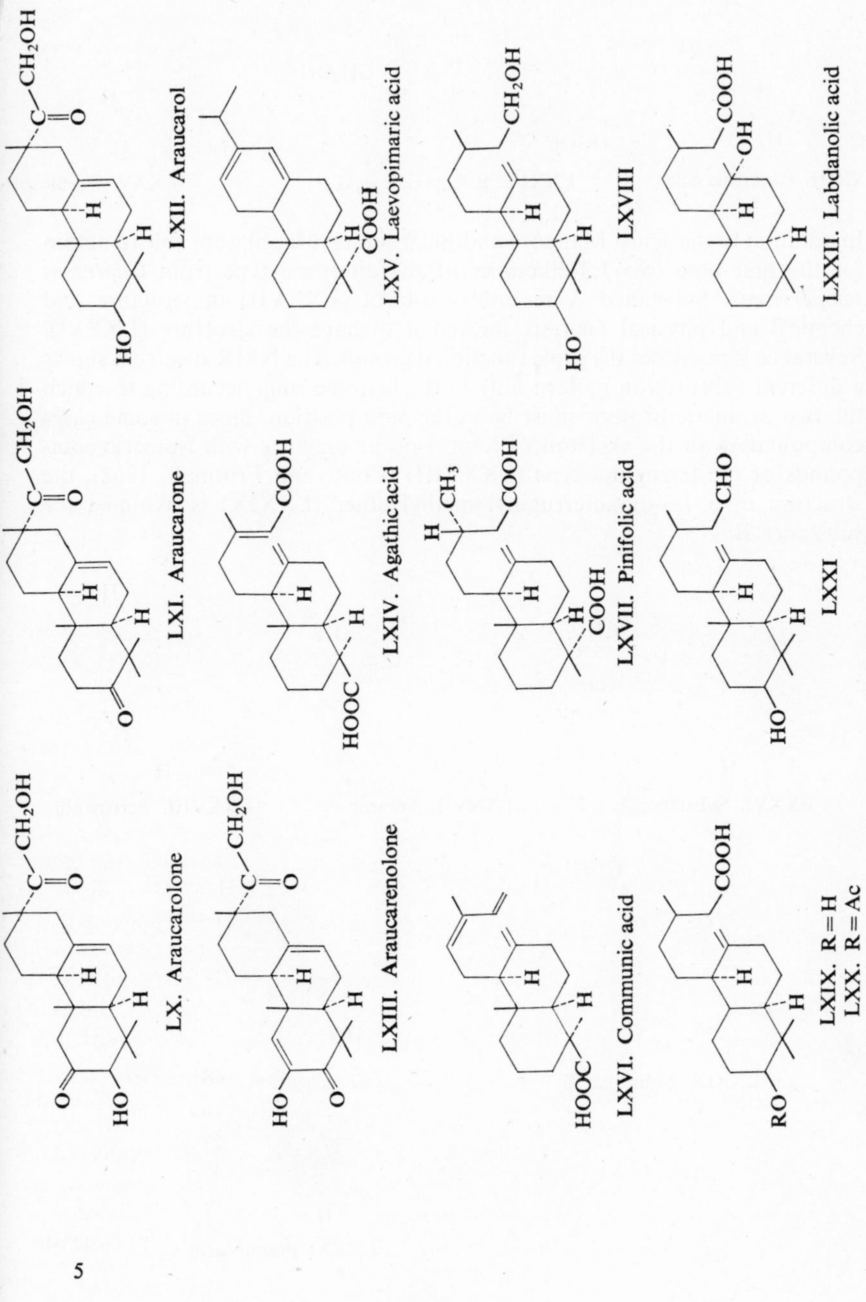

FIG. 11. Diterpenes from the Araucariaceae.

LXII. Araucarol

LXV. Laevopimaric acid

LXVIII

LXXII. Labdanolic acid

LXI. Araucarone

LXIV. Agathic acid

LXVII. Pinifolic acid

LXXI

LX. Araucarolone

LXIII. Araucarenolone

LXVI. Communic acid

LXIX. R = H
LXX. R = Ac

The structure of cupressic acid points to a biogenetic relationship with the torulosol (LXXV) described by Enzell (1960).

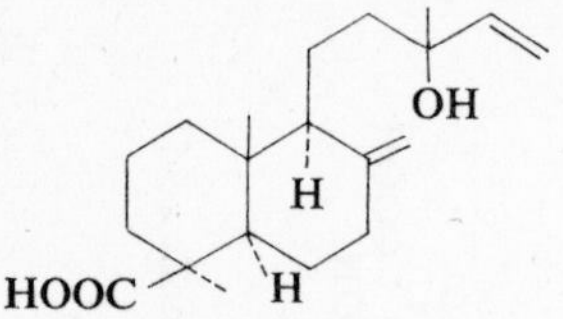

LXXIII. Cupressic acid 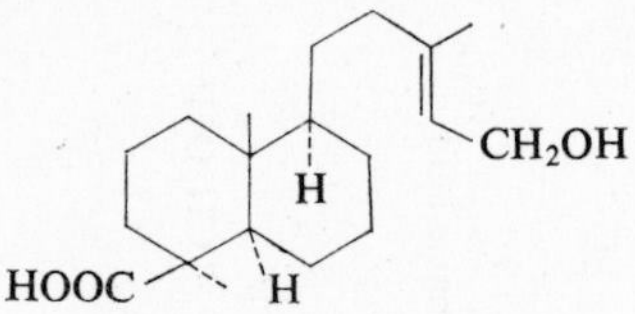   LXXIV. Isocupressic acid 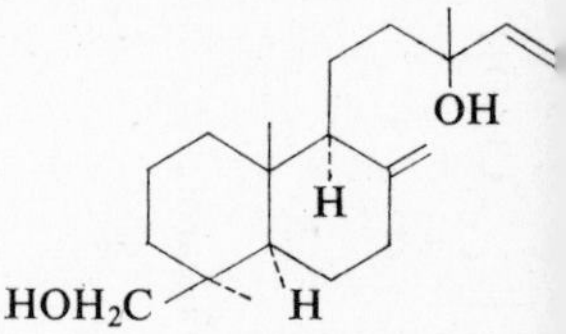   LXXV. Torulosol

In addition to the acids, Mangoni and Belardini (1964a, b) were able to obtain for the first time two 1,3-diketones of the diterpene type from *Cupressus sempervirens*. Substance A resembles totarol (LXXVII) in structure and chemical and physical analysis showed it to have the structure (LXXVI). Substance B possesses the same functional groups. The NMR spectrum shows a different substitution pattern only in the benzene ring, according to which the two aromatic protons must be in the *para*-position. Since in some cases compounds with the skeleton of totarol occur together with isomeric compounds of the ferruginol type (LXXVIII) (Chow and Erdtman, 1962), the structure of a 1,3-dioxoferruginyl methyl ether (LXXIX) is assumed for substance B.

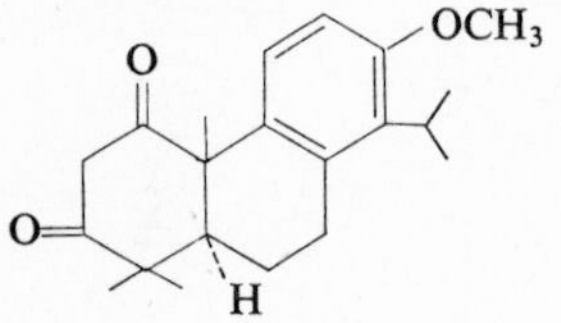

LXXVI. Substance A 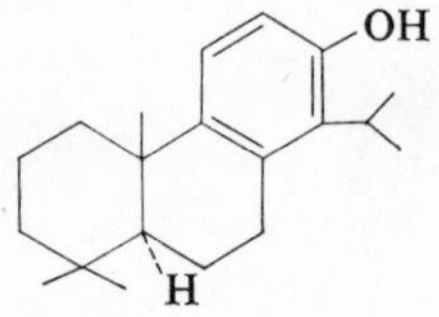   LXXVII. Totarol 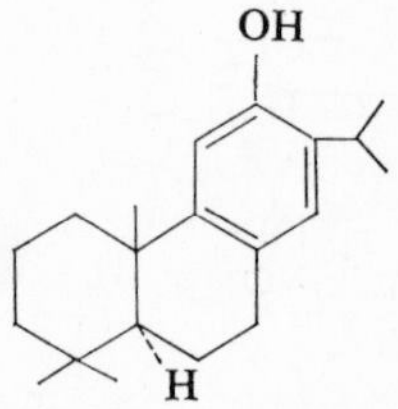   LXXVIII. Ferruginol

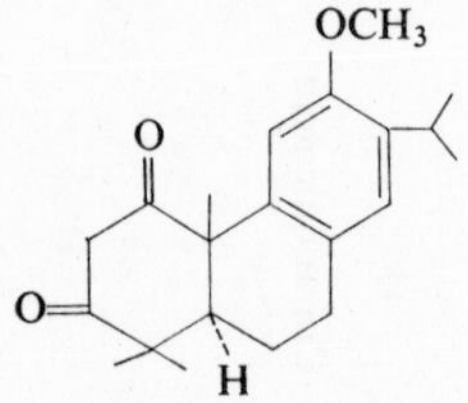

LXXIX. Substance B

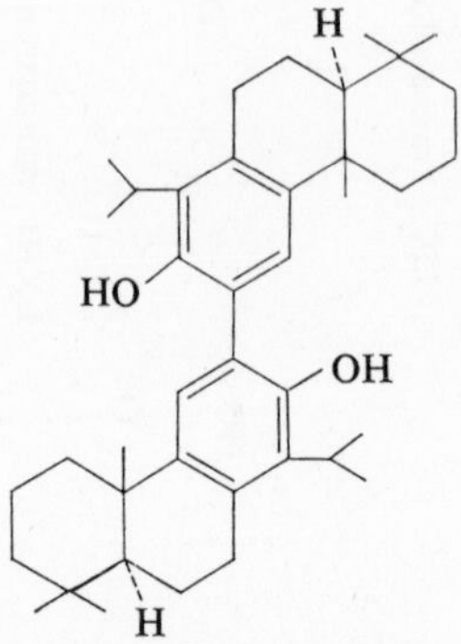

LXXX. Podototarin

There is a very close genetic relationship between totarol and podototarin (LXXX) from *Podocarpus totara* G. Benn. This substance, obtained from the heartwood, is a bisditerpenoid whose structure has been elucidated by Cambie and Simpson (1962) and has been confirmed by synthesis from totarol. The linkage of such bisditerpenes is formed as in gossypol, between the C-atoms in the *ortho*-position to the phenolic hydroxyl group. The chemistry of the dimerization can be assumed to involve an oxidation of the phenolic group by the action of a phenol-dehydrogenating enzyme, as has been proposed for the formation of lignin from coniferyl alcohol. The oxygen radical produced can then become stabilized by the union of the mesomeric *ortho*-radicals.

A series of diterpene alcohols has recently been obtained from the bark of the Lodgepole pine (*Pinus contorta* Dougl.) (Rowe and Scroggins, 1964). The bark of *P. contorta* is characterized by the fact that benzene extraction gives almost 30% of soluble components nearly one-third of which consists of diterpene alcohols. The main constituent is 1, 3-epimanool (LXXXI). It was possible to establish the structure of this compound by comparison with manool (LIX) from which it differs only by the position of the hydroxyl group on C_{13}. The epimeric diol (LXXXII) corresponding to torulosol could also be obtained in small amounts. In addition, contortadiol (LXXXIV) and the corresponding monoaldehyde contortolal (LXXXIII) were found.

In addition to pinifolic acid (LXVII) already mentioned, a bicyclic diterpene, abienol (LXXXV) had been isolated from *Pinus silvestris* (Pigulevskii *et al.*, 1961). This group of compounds also includes arixyl acetate (LXXXVI) obtained by Wienhaus (1947) from larch balsam (*Larix europeae* D.C.), for which Haeuser (1961) has also demonstrated the structure of a bicyclic diterpene. It can be seen from the examples given that the bicyclic diterpenes are also widespread in the family of the *Pinaceae*.

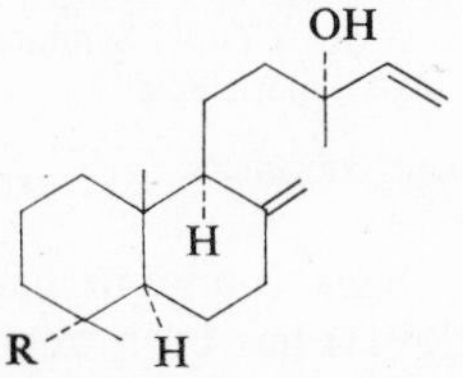

LXXXI. R=CH₃; Epimanool
LXXXII. R=CH₂OH; Hydroxyepimanool

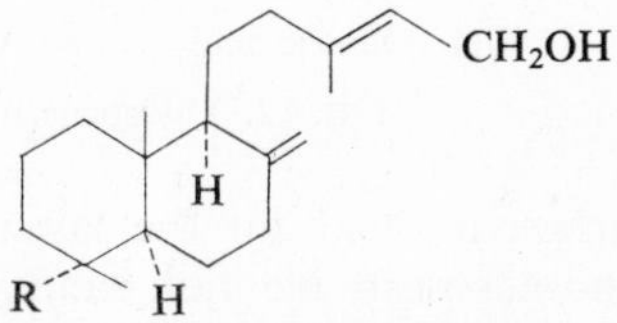

LXXXIII. R=CHO; Contortolal
LXXXIV. R=CH₂OH; Contortadiol

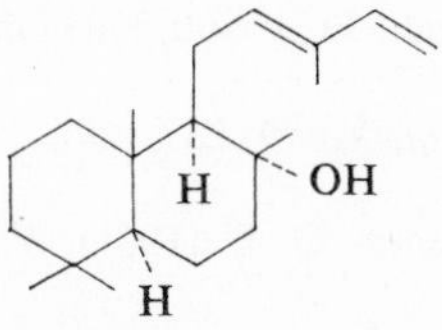

LXXXV. Abienol

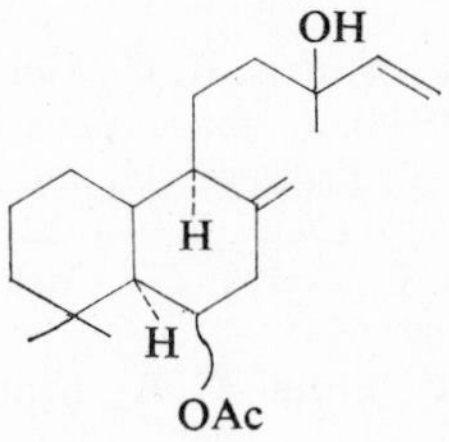

LXXXVI. Larixyl acetate

Alcohols and aldehydes from the series of tricyclic diterpenes, the biological precursors of the pine resin acids, have recently been isolated by Erdtman and Westfelt (1963) from the wood of *Pinus silvestris*. Besides the previously known aldehydes of the dextro- and isodextropimaric type, the corresponding aldehydes of the abietic and dehydroabietic acids were found, as well as various resin alcohols and hydrocarbons of the dextro- and isodextropimaric type (Fig. 12). Consequently, all the oxidation stages from hydrocarbon to

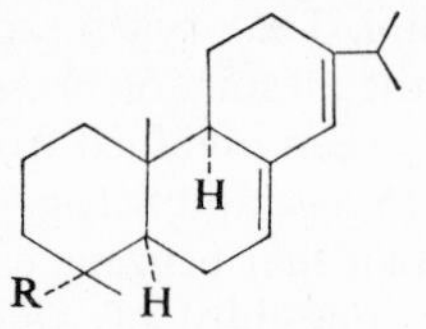

R = CH₂OH; Abietinol   R = CHO; Dehydroabietinal
R = CHO; Abietinal   R = COOH; Dehydroabietic acid
R = COOH; Abietic acid 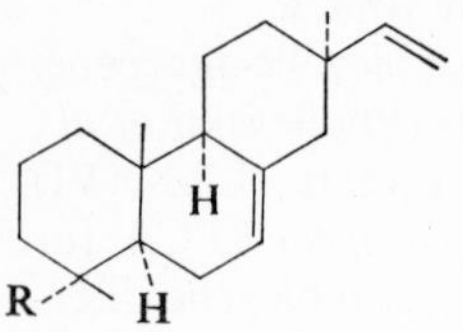

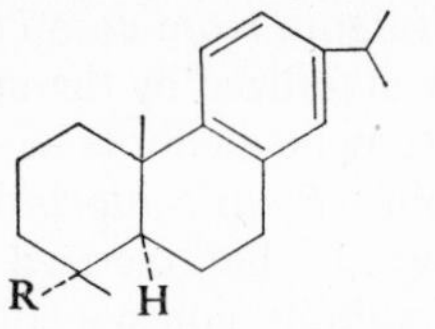

R = CH₃; Pimaradiene   R = CH₃; Isopimaradiene   LXXXVII. R = CH₂OH
R = CH₂OH; Pimarinol   R = CH₂OH; Isopimarinol   Sandaracopimarinol
R = CHO; Pimarinal   R = CHO; Isopimarinal   R = COOH Sandaracopi-
R = COOH; Pimaric acid   R = COOH; Isopimaric acid   maric acid

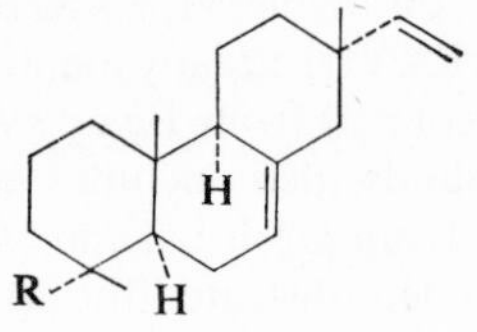

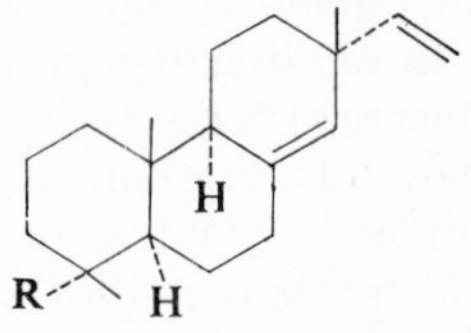

FIG. 12. Diterpene resinacids and their biogenetic congeners.

carboxylic acid for the last-mentioned acids have been found in nature. Another resin alcohol, sandaracopimarinol (LXXXVII) has been obtained by Nagahama from the wood of sugi (*Cryptomeria japonica* D. Don.) (Nagahama, 1964).

REFERENCES

Adams, R., Morris, R. C., Geissman, T. A., Butterbaugh, D. J. and Kirkpatrick, E. C. (1938). *J. Amer. chem. Soc.* **60**, 2193.
Akiyoshi, S., Erdtman, H. and Kubota, T. (1960). *Tetrahedron* **9**, 237.
Anon. (1964). *Chem. Techn.* **12**, 17, 343.
Archer, B. L., Ayrey, G., Cockbain, E. G. and McSweeney, G. P. (1961). *Nature, Lond.* **189**, 663.
Arya, V. P., Enzell, C. R., Erdtman, H. and Kubota, T. (1961). *Acta chem. scand.* **15**, 225.
Bates, R. B. and Hendrickson, E. K. (1962). *Chem. & Ind.* 1759.

Battaile, J. and Loomis, D. (1961). *Biochim. biophys. Acta* **51**, 545.
Bredenberg, J. B.-son and Erdtman, H. (1961). *Acta chem. scand.* **15**, 685.
Bruns, K. (1964). Dissertation Universität Hamburg.
Cambie, R. C. and Simpson, W. R. J. (1962). *Chem. & Ind.* 1757.
Carman, R. M. (1964). *Aust. J. Chem.* **17**, 393.
Carman, R. M. and Dennis, N. (1964). *Aust. J. Chem.* **17**, 390.
Chandra, G., Clark, J., McLean, J., Pauson, P. L. and Watson, J. (1964). *J. chem. Soc.* 3648.
Chow, Y.-L. and Erdtman, H. (1962). *Acta chem. scand.* **16**, 1291.
Crowley, K. J. (1962). *Proc. chem. Soc.* 245.
Enzell, C. R. (1960). *Svensk Kem. Tidskr.* **72**, 602.
Enzell, C. R. and Theander, O. (1962). *Acta chem. scand.* **16**, 607.
Enzell, C. R. and Thomas, B. R. (1964). *Tetrahedron Letters* No. 8, 391.
Erdtman, H. (1952). *Prog. Org. Chem.* **1**, 22.
Erdtman, H. and Westfelt, L. (1963). *Acta chem. scand.* **17**, 1826.
Haeuser, M. J. (1961). *Bull. Soc. chim. Fr.* 1490.
Hendrickson, J. B. (1959). *Tetrahedron* **7**, 82.
Heinstein, P. F., Smith, F. H. and Tove, S. B. (1962). *J. biol. Chem.* **237**, 2643.
Hirose, Y. and Nakatsuka, T. (1959). *Bull. Agr. Chem. Soc. Japan* **23**, 143.
Kremers, R. E. (1922). *J. biol. Chem.* **50**, 31.
Lynen, F. and Henning, U. (1961). *Angew. Chem.* (Int.) 9.
Lynen, F., Agranoff, B. W., Eggerer, H., Henning, U. and Möslein, E. M. (1959). *Angew. Chem.* **71**, 657.
Lynen, F., Eggerer, H., Henning, U. and Kessel, I. (1960). *Angew. Chem.* **70**, 738.
Lynen, F. (1960). *Chem. Weekbl.* **43**, 581.
Mangoni, L. and Belardini, M. (1964a). *Gazz. Chim. Ital.* **94**, 1108.
Mangoni, L. and Belardini, M. (1964b). *Tetrahedron Letters* No. 37. 2643.
Mayer, R. (1961). *Z. Chem.* **1**, 161.
Miller, J. A. and Wood, H. C. S. (1964). *Angew. Chem.* **76**, 301.
Moritz, O. (1958). *In* "Handbuch der Pflanzenanalyse", Vol. X, p. 49. Springer-Verlag, Berlin-Gottingen-Heidelberg.
Motl, O., Chow, W. Z. and Šorm, F. (1961). *Chem. & Ind.* 207.
Nagahama, S. (1964). *Bull. Chem. Soc. Japan* **37**, 886.
Nakatsuka, T. and Hirose, Y. (1956). *Bull. Agr. Chem. Soc. Japan* **20**, 215.
Nayak, U. R. and Dev, S. (1963). *Tetrahedron Letters*, 243.
Ourisson, G. (1955). *Bull. Soc. chim. Fr.* 895.
Pigulevskii, G. V., Kostenko, V. G. and Kostenko, L. D. (1961). *Zh. Obshch Khim.* **31**, 3143.
Popjak, G., Goodman, D. S., Cornforth, J. W., Cornforth, R. H. and Ryhage, R. (1961). *J. biol. Chem.* **236**, 1934.
Porsch, F. and Farnow, H. (1962a). *Dragoco-Report* **2**, 23.
Porsch, F. and Farnow, H. (1962b). *Dragoco-Report* **3**, 54.
Rao, G. S. K., Dev, S. and Guha, P. C. J. (1952). *Indian chem. Soc.* **29**, 721.
Reitsema, R. H. (1958). *J. Amer. Pharm. Ass. Sci. Ed.* **47**, 267.
Rowe, J. W. and Scroggins, J. H. (1964). *J. org. Chem.* **29**, 1554.
Rudloff, E. von and Erdtman, H. (1962). *Tetrahedron* **18**, 1315.
Rudloff, E. von and Nair, G. V. (1964). *Canad. J. Chem.* **42**, 421.
Ruzicka, L. (1938). *Angew. Chem.* **51**, 5.
Ruzicka, L. (1953). *Experientia* **9**, 357.
Ruzicka, L. (1963). *Pure appl. Chem.* **6**, 493.
Sandermann, W. (1938). *J. prakt. Chem.* **151**, 161.

120   G. Weissmann

Sandermann, W. (1959). 4th Internat. Congr. Biochem. 2, Pergamon Press.
Sandermann, W. (1962a). Holzforschung 16, 65.
Sandermann, W. (1962b). In "Comparative Biochemistry", (M. Florkin, and H. S. Mason eds.), Vol. III. Academic Press, New York and London.
Sandermann, W. and Schweers, W. (1962). Tetrahedron Letters No. 7, 257.
Sandermann, W. and Bruns, K. (1962a). Naturwissenschaften 49, 258.
Sandermann, W. and Bruns, K. (1962b). Chem. Ber. 95, 1863.
Shaykin, S., Law, J., Phillips, A. H., Tchen, T. T. and Bloch, K. (1958). Proc. nat. Acad. Sci. (Wash.) 44, 998.
Smith, G. M. (1955). "Cryptogramic Botany" 2nd ed., Vol. I, McGraw-Hill, London and New York.
Smith, R. H. (1964). US-Forest Serv. Res. Paper, PSW-15.
Tanenbaum, S. W., Basset, E. W. and Kaplan, M. (1959). Arch. biochem. Biophys. 81, 169.
Tanenbaum, S. W. and Basset, E. W. (1962). Biochim. biophys. Acta 59, 524.
Wallach, O. (1887). Annalen 239, 1.
Weissmann, G. and Bruns, K. (1965). Naturwissenschaften 52, 185.
Weissmann, G., Bruns, K. and Grützmacher, H. Fr. (1965). Tetrahedron Letters (In Press).
Wienhaus, H. (1947). Angew. Chem. 59, 248.
Yosioka, I., Hikino, H. and Sasaki, Y. (1959a). Chem. Pharm. Bull. 7, 817.
Yosioka, I., Takahashi, S., Hikino, H. and Sasaki, Y. (1959b). Chem. Pharm. Bull. 7, 319.
Zavarin, E. and Bicho, J. (1963). 144th Meeting. Amer. Chem. Soc. 10d.

CHAPTER 7

The Carotenoids

T. W. GOODWIN

Department of Biochemistry and Agricultural Biochemistry,
University College of Wales, Aberystwyth

I. DISTRIBUTION OF CAROTENOIDS IN PHOTOSYNTHETIC TISSUES

A. HIGHER PLANTS

All photosynthetic tissues so far examined always contain both chlorophylls and carotenoids which are confined to the photosynthetic organelles; they are present in the chloroplasts of higher plants and algae, where they accumulate in the grana, and in the chromatophores of photosynthetic bacteria (Pardee *et al.*, 1952). This distribution appears to be invariable and suggests a fundamental role for carotenoids in photosynthesis. On this view one would expect a regularity in the carotenoids of the chloroplasts of higher plants and this is found to be so. Although there are minor quantitative and qualitative variations, the major pigment components are always β-carotene (I), lutein (II), violaxanthin (III) and neoxanthin (probably, but not unequivocally, IV). Quantitative results (Table 1) show that while β-carotene is always the major

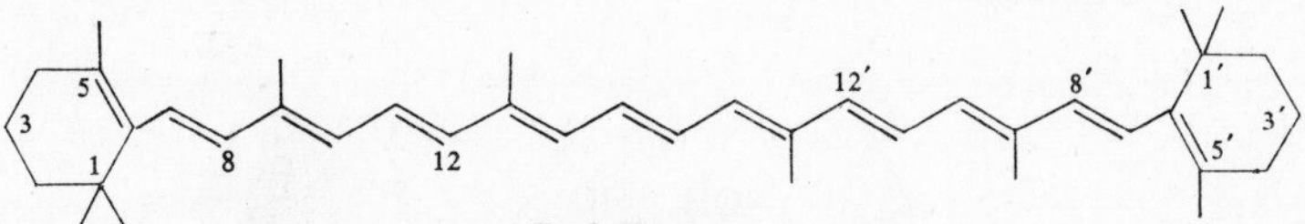

I. β-Carotene

In all other structural formulae the isoprenoid chain between C_8 and C_8' is omitted except where different from that in β-Carotene.

121

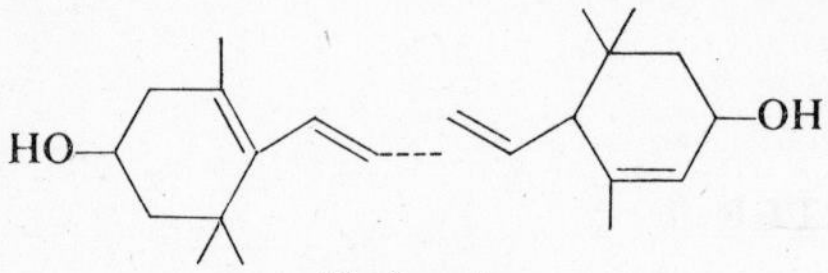

II. Lutein 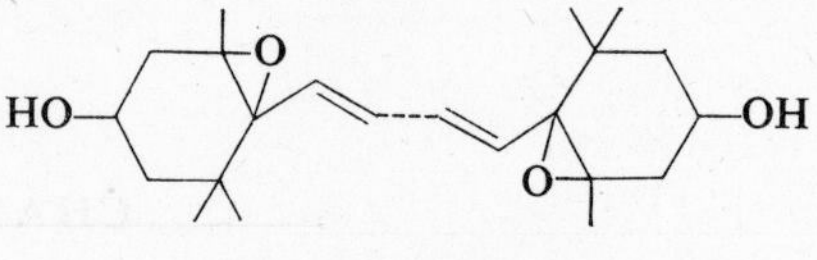   III. Violaxanthin

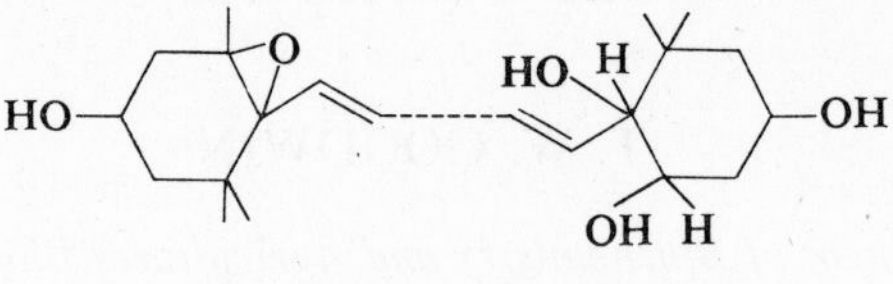

IV. Neoxanthin

carotene present (α-carotene (V) occurs sporadically in small amounts), lutein, II; (3,3′-dihydroxy-α-carotene) is always the major xanthophyll present. In this apparent immutability the carotenoids in the photosynthetic tissues of

TABLE 1. A typical distribution of carotenoids in leaf tissue

Pigment	% of Total pigments
β-Carotene I	25
Lutein II	40
Violaxanthin III	15
Neoxanthin IV	15

Varying amounts (0 to 2%) of α-carotene V, cryptoxanthin (VI 3-hydroxy-β-carotene), zeaxanthin (VII 3,3′-dihydroxy-β-carotene), antheraxanthin (VIII 5,6-epoxyzeaxanthin).

higher plants resemble the chlorophylls which exist together as chlorophylls *a* and *b* with no variants. Taxonomically this can mean only that all higher plants have evolved from the same common ancestor.

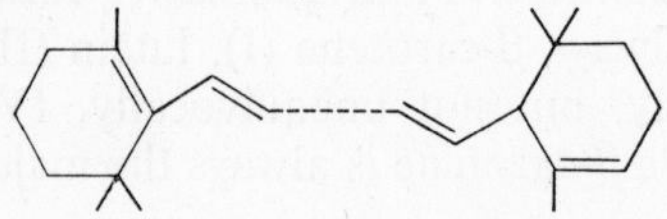

V. α-Carotene 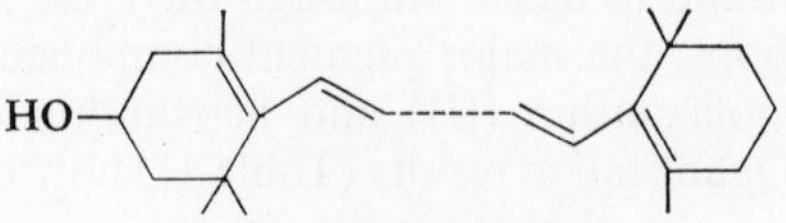   VI. Cryptoxanthin

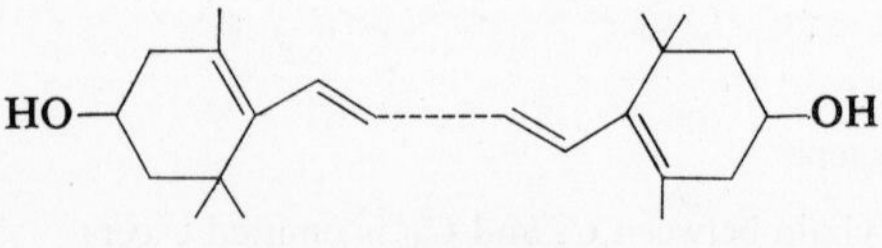

VII. Zeaxanthin 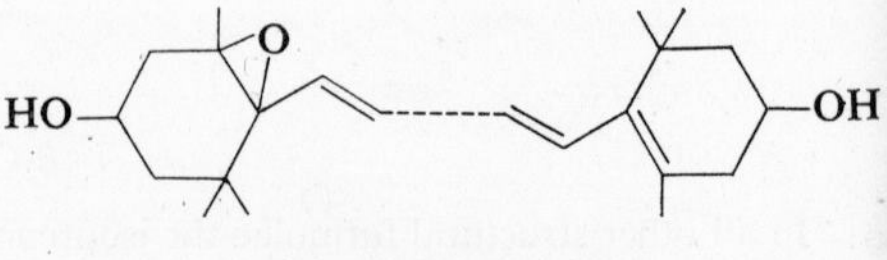   VIII. Antheraxanthin

B. ALGAE

The different classes of algae show a much wider spectrum of variation of plastid carotenoids and, to a lesser extent, chlorophylls than do higher plants. The increased availability during recent years of the more exotic algae in axenic (bacteria-free) culture and the development of rapid and reliable means of identifying the constituent carotenoids has led to a number of studies from which a general picture of carotenoid distribution is emerging (Table 2, Goodwin, 1965a). Based on such studies a pattern of algal evolution can be

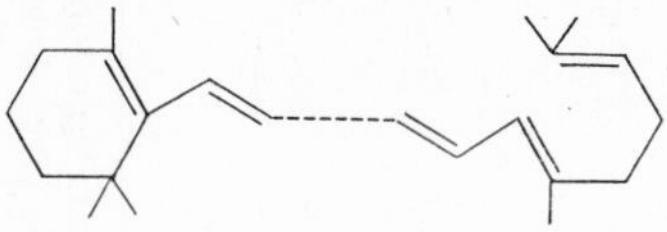

IX. γ-Carotene

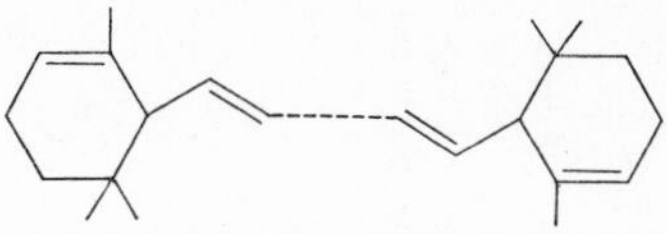

X. ε-Carotene

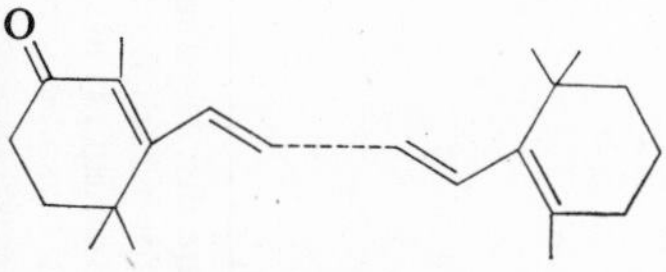

XI. Echinenone

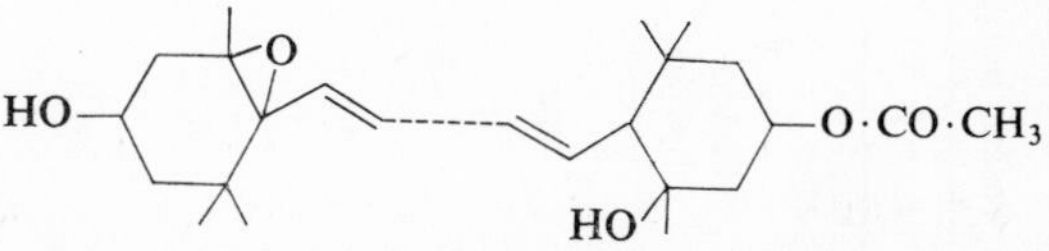

XII. Fucoxanthin

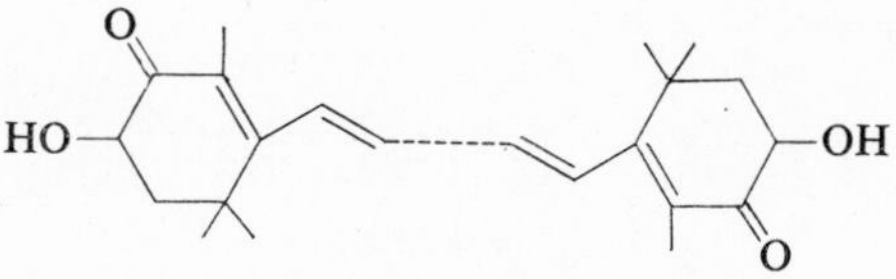

XIII. Astaxanthin

drawn (Fig. 1, Goodwin, 1962, 1963) which, with one or two exceptions, fits in well with the views based on more conventional assessments of evolution in algae.

According to Dougherty and Allen (1960) protists fall into three major levels of structural organization representing three major evolutionary steps. The lowest level is represented by the monerans (bacteria, blue-green algae), the second level by the mesoprotists (red algae) and the highest level by the meta-protists (all higher organisms except the metaphytes and metazoa). These levels are considered to be monophyletic, that is, the evolutionary pathway is blue-green algae (monerans) → red algae (mesoprotists) → metaprotists.

The blue-green algae are unique in synthesizing only β-carotene derivatives of which one is a xanthophyll, echinenone (XI; 4-oxo-β-carotene), in which the oxygen function is at C_4 (Goodwin, 1956). The red algae represent an increase in complexity in that they synthesize both α- and β-carotene derivatives (Strain, 1958). Their xanthophylls are also characteristic of the higher plants

TABLE 2. Major carotenoid distribution in various algal classes[1]

(+ = present; — = absent; ? possibly present in trace amounts)

Pigment[2]	Chlorophyta		Phaeophyta				Rhodophyta (Rhodophyceae)	Pyrrophyta (Dinophyceae)	Euglenophyta (Euglenineae)	Archephyta (Cyanophyceae)	Cryptophyta (Cryptophyceae)
	Charophyceae[3]	Chlorophyceae	Xanthophyceae (Heterokontae)	Bacillariophyceae (Diatomophyceae)	Chrysophyceae	Phaeophyceae					
Carotenes											
α-Carotene (V)	—	+	—	—	—	—	+	—	—	—	+
β-Carotene (I)	+	+	+	+	+	+	+	+	+	+	+
γ-Carotene (IX)	—	+[4]	—	—	—	—	—	—	—	—	—
ε-Carotene (X)	—	?[5]	—	+	+	+	+	—	—	—	—
Flavacene	—	—	—	—	—	—	—	—	—	—	—
Xanthophylls											
Echinenone (XI)	—	—	—	—	—	+?	+	—	+	+	—
Lutein (II)	+	+	—[8]	—	—	+?	+	—	—	?	—
Zeaxanthin (VII)	+	+	+?	—	—	—	+?	—	+	+	+[10]
Violaxanthin (III)	+	+	+?	—	—	+?	+?	—	—	—	—
Flavoxanthin	—	—	—	—	?	?	—	—	—	—	—
Neoxanthin (IV)	+	+	+?	—	—	—	—	—	+	—	—
Antheraxanthin (VIII)	—	—	+	—	—	?	?	—	+	+	—
Fucoxanthin (XII)	—	—	—	+	+	+	—	—	—	—	+
Diatoxanthin	—	—	+	+	+	?	—	?	—	—	—
Diadinoxanthin	—	—	—	+	+	+	—	+	+	—	—
Dinoxanthin	—	—	—	—	—	—	—	+	—	—	—
Peridinin	—	—	—	—	—	—	—	+	—	—	—
Myxoxanthophyll	—	—	—	—	—	—	—	—	—	+	—
Siphonaxanthin	—	+[6]	—	—	—	—	—	—	—	—	—
Astaxanthin (XIII)	—	+[7]	—	—	—	—	—	—	+[9]	—	—
Oscillaxanthin	—	—	—	—	—	—	—	—	—	+[9]	—

[1] From Goodwin (1965a). Occasional variations from this general picture are discussed in the text. No information exists on the carotenoids of the Chloromonadophyta (Chloromonadineae). [2] Structures of pigments for which no formula given unknown. Flavoxanthin is believed to be an isomer of XXII; diatoxanthin is related to zeaxanthin (VII); diadinoxanthin is a 5,6-epoxide related to lutein (II). [3] Only one species (*Chara fragilis*) studied; lycopene also reported present. [4] Present in traces in some species. [5] Present in one marine species. [6] The main pigments of the Siphonales. [7] The main extra-plastidic pigment (haematochrome) of some encysted flagellates. [8] Lutein epoxide and possibly

in that substitution occurs at C_3 and apparently never at C_4; thus they appear to occupy a central position in the evolutionary development of algae. It must be concluded that either the blue-green algae evolved the enzymes for the synthesis of their special xanthophylls after the red line split off, or in evolving systems for their xanthophylls the red line simultaneously lost the systems for synthesizing echinenone (XI) and myxoxanthophyll (structure unknown). Morphologically the most primitive metaprotists are the Cryptomonads and it is considered that these represent the link between the red algae and the most primitive class of the Phaeophyta, the Chrysophyceae. The cryptomonads so far examined synthesize α- and β-carotene derivatives, but are unique among

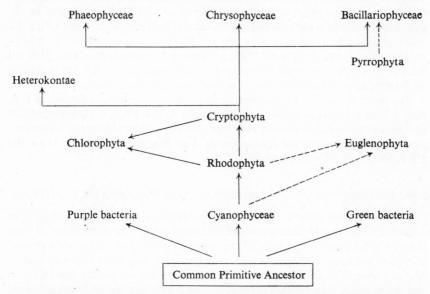

FIG. 1. The pattern of algal evolution as indicated by carotenoid distribution studies.

algae in that the major carotene is α-carotene and the major xanthophyll, a β-carotene derivative, zeaxanthin (VII; 3,3′-dihydroxy-β-carotene) or diatoxanthin (see Goodwin, 1965a). The chrysomonads synthesize mainly β-carotene (I) and fucoxanthin (see Goodwin, 1965a) (probable structure, XII, Bonnett *et al.*, 1964), a β-carotene derivative of higher oxidation state than zeaxanthin; fucoxanthin is also the characteristic xanthophyll of two other main classes of the Phaeophyta, the Phaeophyceae (brown algae) and Bacillariophyceae (diatoms). The diatoms and Chrysophyceae are also related in that they both synthesize diatoxanthin and dinoxanthin, pigments of unknown structure; the Phaeophyceae, on the other hand, do not appear to contain these pigments (see Goodwin, 1965a). However, the Heterokontae, which are always included in the Phaeophyta, do not produce fucoxanthin (XII), diatoxanthin or dinoxanthin, but a series of pigments which include antheraxanthin (VIII; zeaxanthin-

5,6-epoxide) and lutein epoxide (Thomas and Goodwin, 1965). Their position in the Phaeophyta could on this criterion be questioned. A reasonably close relationship between the Pyrrophyta and diatoms is suggested by the common synthesis of diadinoxanthin (unknown structure), but the Pyrrophyta never synthesize fucoxanthin and diatoms never synthesize peridinin (unknown structure) the unique pigment of the Pyrrophyta (see Goodwin, 1952).

The Chlorophyceae have, in general, the same carotenoid distribution as leaves of higher plants (see Table 1) which suggests that they are very close to the red algae; they do not synthesize any of the highly oxidized xanthophylls found in the Phaeophyta. A few green algae synthesize biliproteins (Allen, 1959), in particular *Cyanidium caldarum* which contains zeaxanthin (VII), and not lutein as the main xanthophyll (Allen *et al.*, 1960). Furthermore in the group Siphonales, which do not synthesize biliproteins, α-carotene (V) is the main hydrocarbon pigment (the nature of the main xanthophyll siphon-axanthin is not known) (Strain, 1951). These and other observations all recall the pigment situation in the Cryptophyta, and suggest that this latter class is an intermediate between the Rhodophyta and the Chlorophyta.

The pigments of the Euglenophyta are allied to the Cyanophyceae in that they are all β-carotene derivatives (the major xanthophyll is antheraxanthin) and that echinenone is present in traces; however, their chlorophyll distribution allies them with the Chlorophyta and Rhodophyta rather than the Cryptophyta and Cyanophyceae.

II. Distribution of Carotenoids in Non-Photosynthetic Tissues

It is in their flowers and fruit that higher plants exert their individuality in regard to carotenoid synthesis and accumulation. The remainder of this chapter will be concerned with an examination of the taxonomic importance, if any, of the apparently capricious distribution of carotenoids in the fruit of higher plants. The fruits can be divided into seven main groups according to the carotenoids they contain: (1) those which produce insignificant amounts of carotenoids, (2) those which produce the usual chloroplast carotenoids described above; (3) those in which there is a marked synthesis of the acyclic carotene lycopene (XIV), together with its partly saturated precursors,

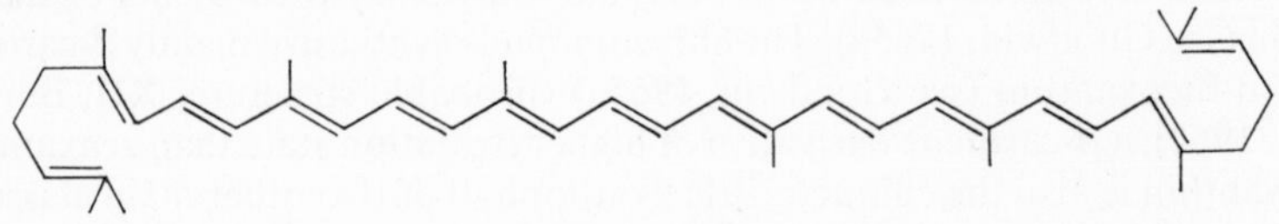

XIV. Lycopene

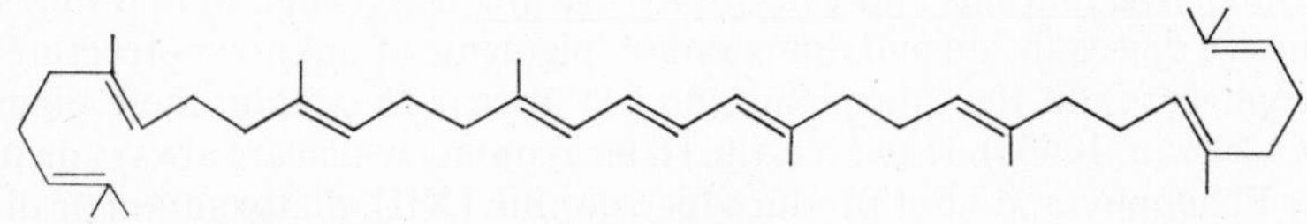

XV. Phytoene

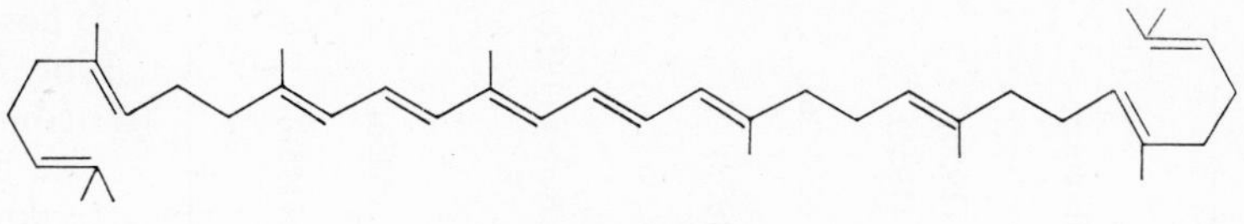

XVI. Phytofluene

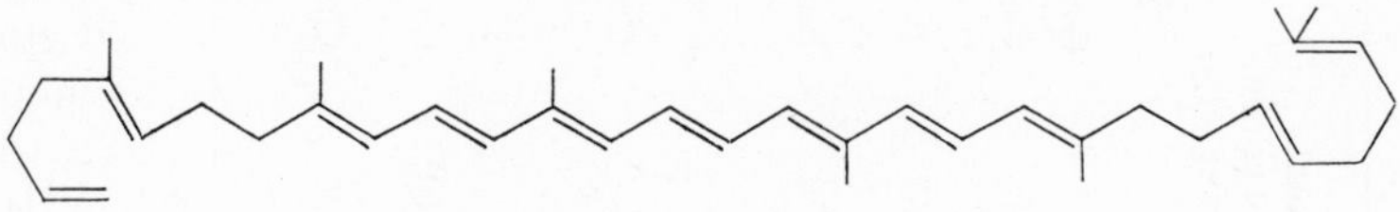

XVII. ζ-Carotene

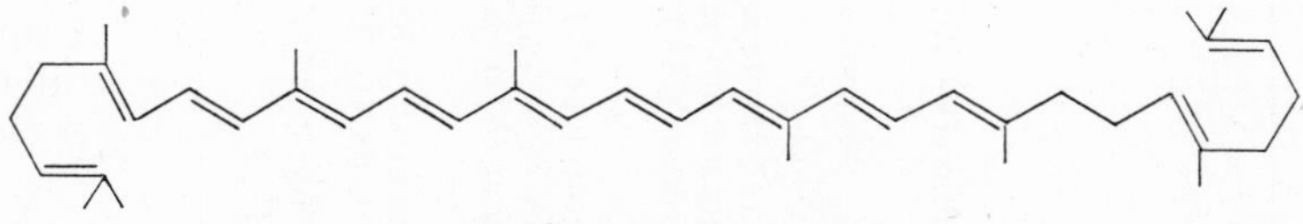

XVIII. Neurosporene

phytoene (XV), phytofluene (XVII), ζ-carotene (XVII), neurosporene (XVIII);
(4) those which produce large amounts of β-carotene and/or its derivatives
such as cryptoxanthin (VI; 3-hydroxy) and zeaxanthin (VII; 3,3′-dihydroxy);
(5) those that synthesize large amounts of epoxides; (6) those that synthesize
large amounts of allegedly unique pigments such as capsanthin (XIX),
rubixanthin (XX), rhodoxanthin (XXI); and (7) those which synthesize mainly
poly *cis*-carotenes, such as pro-γ-carotene and prolycopene (see also Goodwin,
1962, 1965a). Examples of each group are given in Table 3.

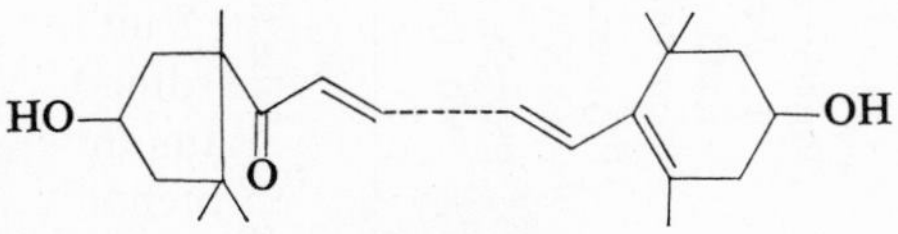

XIX. Capsanthin

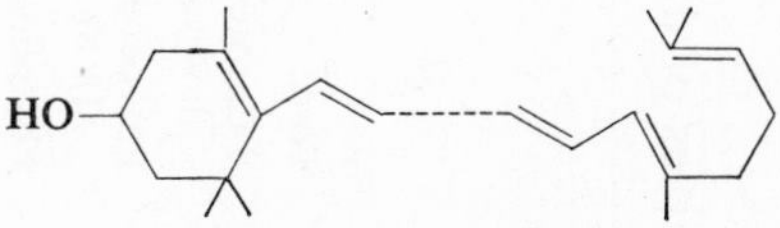

XX. Rubixanthin

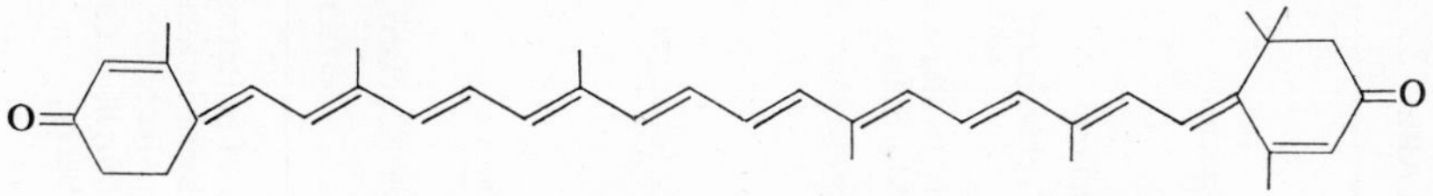

XXI. Rhodoxanthin (3,3′-Dioxo-*retro*-β-carotene)

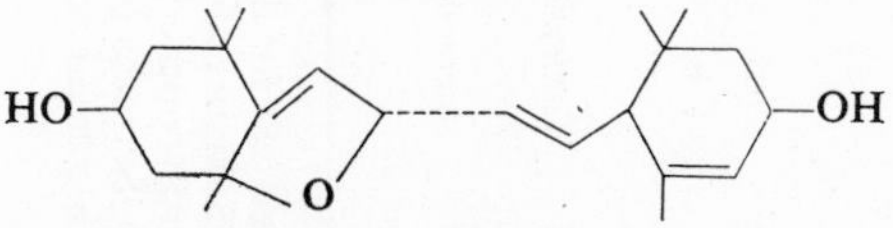

XXII. Chrysanthemaxanthin (Flavoxanthin)

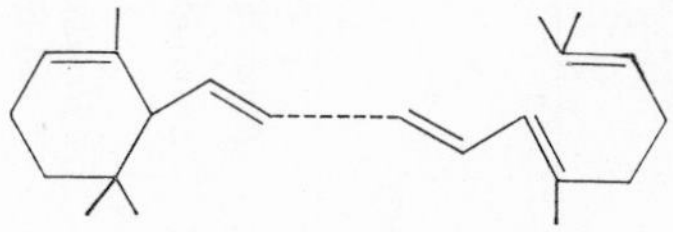

XXIII. δ-Carotene

TABLE 3. Examples of different carotenoid distribution patterns in fruit

Pattern[1]	Species	Main pigments[2]	Total pigment concn. (μg/g wet wt)	Reference
1	*Pyracantha rogeriana*	Trace	—	Goodwin (1956)
2	*Sambucus nigra* (elderberry)	β-Carotene (I), 19; lutein (II), 6·2; neoxanthin (IV), 10; flavoxanthin (XXII), etc., 10	16	Goodwin (1956)
3	*Citrullus vulgaris* (water-melon; red-fleshed—Mardella)	Phytoene (XV), 11; phytofluene (XVI), 4; β-carotene (I), 8; δ-carotene (XXIII), 2; lycopene (XIV), 75[4]	65	Tomes *et al.* (1963)
4	*Physalis alkekengi* (winter cherry)	Phytofluene (XVI), 1; β-carotene (I), 1; ζ-carotene (XVII), 1; cryptoxanthin (VI), 26; mutatochrome (XXIV), 2; zeaxanthin (VII), 50; luteochrome (XXV), 18	570	Baraud (1958)
5	*Citrus aurantium* (orange)	Phytoene (XV), 4; phytofluene (XVI), 1; α-carotene (V), 1; β-carotene (I), 1; δ-carotene (XXIII), 5; cryptoxanthin (VI), 5; lutein (II), 3; zeaxanthin (VII), 4; epoxides, 75	—	Curl and Bailey (1956)
6	*Taxus baccata* (yew)	Phytoene (XV), 1; β-carotene (I), 4; mutatochrome (XXIV), 3; lutein (II), 1·5; rhodoxanthin (XXI), 75	10[5]	Goodwin (1956)
7	*Arum maculatum* (Cuckoo pint)	Phytofluene (XVI), 2; β-carotene (I), 16; prolycopene,[3] 40; lycopene (XIV) and neolycopene, 40[4]	200	Goodwin (1956)

[1] See text for definition of patterns.
[2] The figure after a pigment indicates the % of the total pigments present.
[3] Prolycopene is poly *cis*-lycopene.
[4] Only traces of xanthophylls present.
[5] Calculated on whole berry.

XXIV. Mutatochrome
(5,8-Epoxy-β-carotene)

XXV. Luteochrome
(5,6,5′,8′-Diepoxy-β-carotene)

What taxonomic significance can be attached to these generalizations? It is very difficult at the moment to see. In some cases the various species of any one genus present a qualitatively homogeneous and unique pattern; for example, the three *Rosa* species examined by Goodwin (1956) have a qualitatively similar pattern and all contain the rather specific rubixanthin (XX) (Table 4).

TABLE 4. Carotenoid distribution in *Rosa* spp.[1]

Pigment	% of Total pigments[2]		
	R. canina	*R. moyesii*	*R. rubrifolia*
Phytoene (XV)	Trace	4·5	2·5
Phytofluene (XVI)	0·3	3·0	2·7
β-carotene (I)	16·5	14·5	28·5
ζ-carotene (XVII)	0·2	3·5	Trace
γ-carotene (IX)	1·4	—	Traces
Prolycopenes	0·5	12·5	—
Lycopene (XIV)	6·0	21·0	16·5
Mutatochrome (XXIV)	—	2·0	—
Cryptoxanthin (VI)	1·8	11·0	4·0
Zeaxanthin (VII)	6·0	4·5	2·5
Rubixanthin (XX)	42·0	14·0	41·0

[1] Goodwin (1956).
[2] Minor pigments are not recorded.

TABLE 5. Carotenoid present in *Cotoneaster* spp.[1]

Pigment	Species		
	C. bullata	*C. frigida*	*C. lebephylla*
β-Carotene (I)	+	+	+
Mutatochrome (XXIV)	+	+	+
Cryptoxanthin (VI)	+	+	+
Zeaxanthin (VII)	Trace	+	—
Flavoxanthin[2] or Chrysanthemaxanthin (XXII)	—	+	+
Aurochrome	+	—	—

[1] Goodwin, 1956.
[2] Believed to be an isomer of (XXII).

TABLE 6. Major carotenoid pigments in some Caprifoliaceae[7]

Pigment	*Lonicera japonica*	*Lonicera periclymenum*	*Sambucus nigra*	*Viburnum opulus*
Phytofluene (XVI)	+	?	−	+
β-Carotene (I)	+	+	+	+
ζ-Carotene (XVII)	+	+	−	+
γ-Carotene (IX)	+	−	−	+
Lycopene (XIV)	+	+	−	+
Mutatochrome (XXIV)	−	−	−	+
Cryptoxanthin (VI)	+	−	Trace	−
Zeaxanthin (VII)	+	+	−	−
Lutein (II)	+	−	+	−
Flavoxanthin[2] or	−	−	+	−
Chrysanthemaxanthin (XXII)	−	−	+	−
Auroxanthin (XXVI)	+	+	−	−

[1] Goodwin (1956).
[2] Believed to be an isomer of (XXII).

Similar observations have been made with *Cotoneaster* spp. (Table 5) although aurochrome is specific to *C. bullata* (Goodwin, 1956); two species of *Lonicera* (Table 6), as well as with a number of *Berberis* species (Bubicz and Wierzchowski, 1959).

However, if one were presented with a pattern of carotenoid distribution and asked for a taxonomic comment, nothing could be made of a mixture of pigments similar to that found in *Cotoneaster* spp., *Lonicera* spp. or *Berberis*

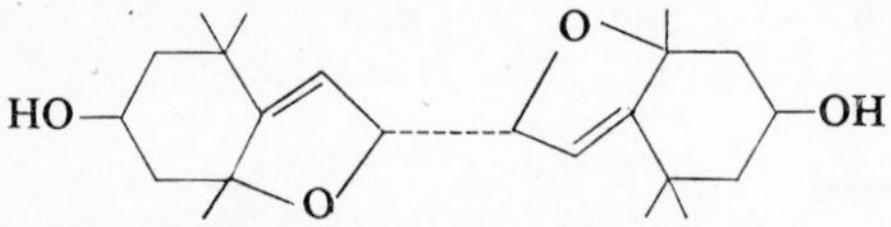

XXVI. Auroxanthin
(5,8,5′,8′-Diepoxyzeaxanthin)

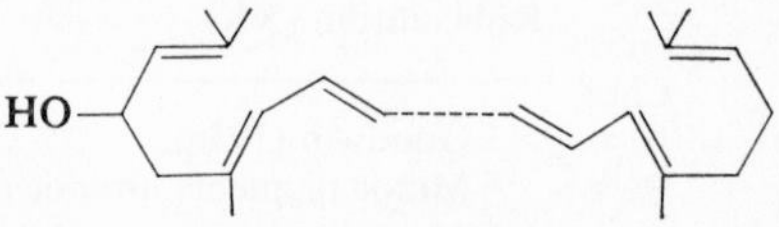

XXVII. Lycoxanthin

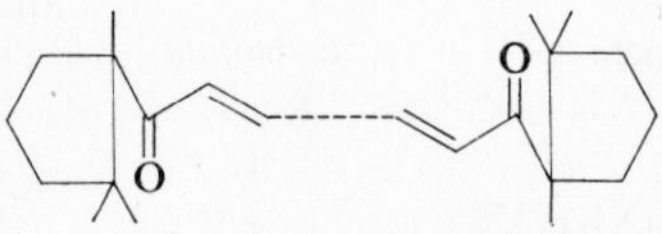

XXVIII. Capsorubin

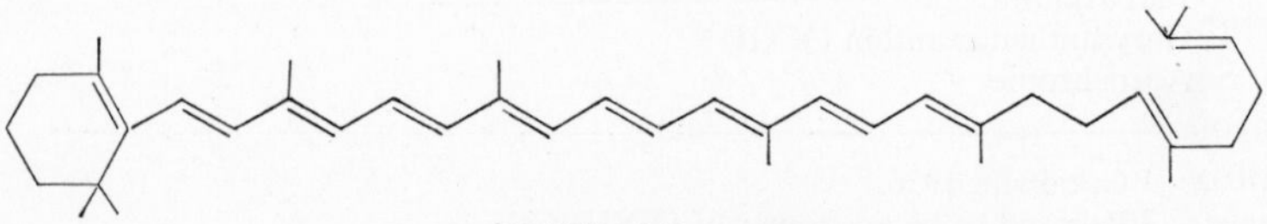

XXIX. β-Zeacarotene

spp., because similar patterns are found in many other genera. If the carotenoid mixture contained significant amounts of rubixanthin (XX), then one could say with some confidence that the source was the hips of a species of *Rosa*. However, the generalization could not be widened to cover the family Rosaceae, because genera other than *Rosa* do not synthesize rubixanthin, for example *Cotoneaster* spp. (Table 5), *Crataegus* spp., *Pyracantha* spp., *Sorbus aucuparia* (Goodwin, 1956).

The diversity is even more marked amongst the Caprifoliaceae (Table 6) and Solanaceae (Table 7) so far examined, but a rather consistent picture is emerging for the family Elaeagnaceae (*Hippophae rhamnoides* (Goodwin, 1956)); *Shepherdia canadensis* (Stabursvik, 1954); *Elaeagnus longipes* (Geiger-Vifian

TABLE 7. Major carotenoid pigments in some Solanaceae

Pigment	*Capsicum annuum* red[1] var.	*Capsicum annuum* yellow[2,3] var.	*Atropa belladonna*	*Solanum dulcamara*[1]	*Physalis alkekengi*[4]
β-Carotene (I)	+	+	+	Trace	+
Lycopene (XIV)	−	−	−	+	+
Cryptoxanthin (VI)	+	+	+	−	+
Mutatochrome (XXIV)	−	−	−	−	+
Zeaxanthin (VII)	+	+	−	−	+
Lutein (II)	−	+	−	+	−
Luteochrome (XXV)	−	−	−	−	+
Lycoxanthin (XXVII)	−	−	−	+	−
Capsanthin (XIX)	+	−	−	−	−
Capsorubin (XXVIII)	+	−	−	−	−

[1] Goodwin (1956).
[2] Cholnoky *et al.* (1958).
[3] Also significant amounts of violaxanthin (III) and antheraxanthin (VIII).
[4] Baraud (1958).

and Müller, 1945). Marked differences can also be observed amongst species of the same genus; for example the European *Pyracantha rogeriana* (Goodwin, 1956) and *P. coccinea* (Karrer and Rutschmann, 1945) produce no significant amounts of carotenoids, while *P. flava* falls into Group 4 (Table 3) (comparatively large amounts of β-carotene derivatives) (Goodwin, 1956). On the other hand *P. angustifolia*, grown in California produces large amounts of prolycopenes (Group 7) (Zechmeister and Sandoval, 1945). An interesting parallel is that a somewhat similar difference between new and old world species exist in the petals of *Mimulus* spp. The Californian *M. longiflorus* produces considerable amounts of γ-carotene (IX) and some lycopene (XIV) (Zechmeister and Schroeder, 1942) whilst the European *M. cupreus* synthesizes mainly β-carotene, and the main components of *M. tigrinus* are xanthophylls, with cryptoxanthin (VI) and taraxanthin (structure unknown) as the major components (Goodwin and Thomas, 1964).

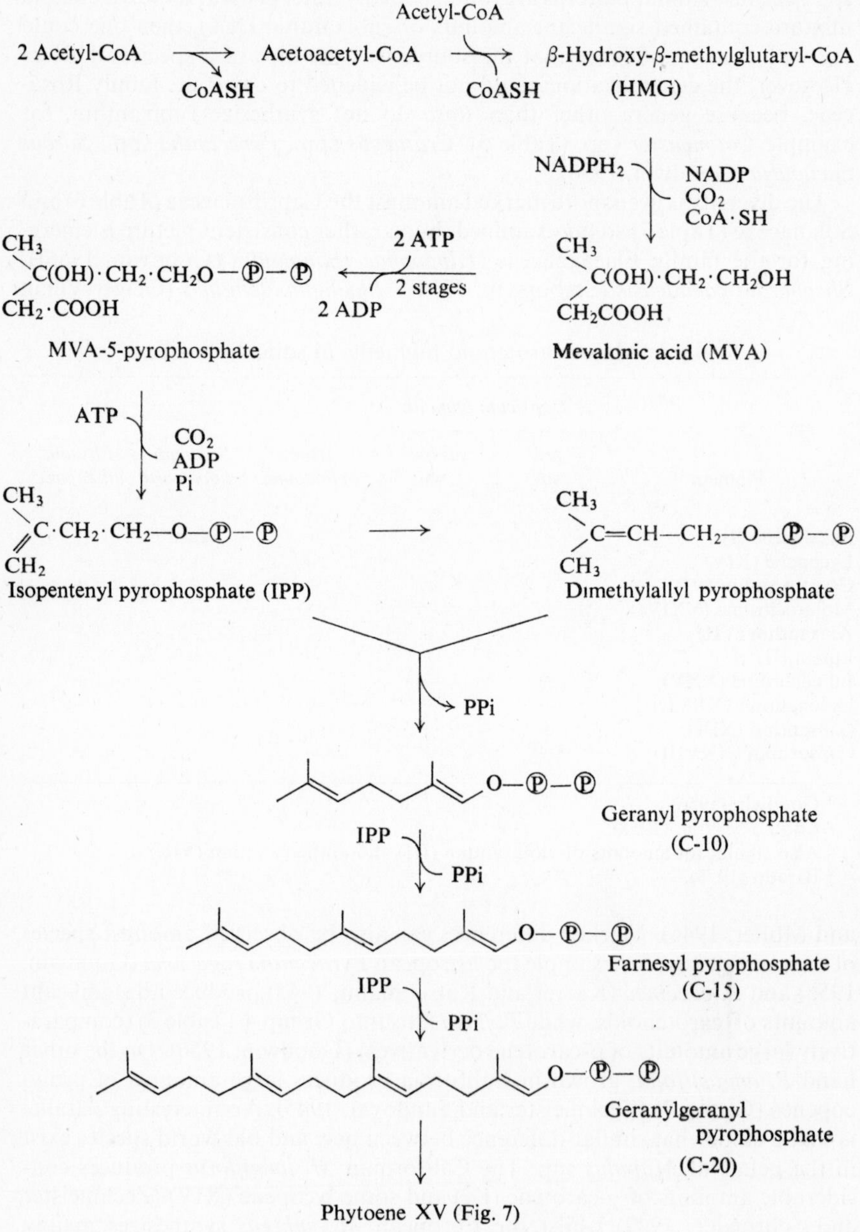

FIG. 2. The formation of isopentenyl pyrophosphate (IPP) and its conversion into geranylgeranyl pyrophosphate.

III. Biochemical Aspects of the Distribution of Fruit Carotenoids

A. GENERAL PATTERN OF SYNTHESIS

It is not necessary here to discuss in detail the formation of the biological isoprenoid precursor, isopentenyl pyrophosphate and its conversion into geranylgeranyl pyrophosphate (C-20) which dimerizes to form the first C-40 carotenoid precursor phytoene (XV); the pathway is summarized in Fig. 2. Phytoene is then stepwise dehydrogenated to neurosporene (XVIII); at this stage the pathway bifurcates, one path to lycopene (XIV) and the other to α-, (V) β-, (I) and γ- (IX) carotenes (see Goodwin, 1965a, b) (Fig. 3); both

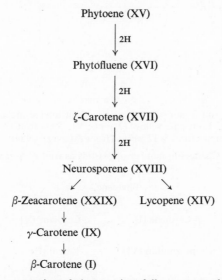

FIG. 3. The conversion of phytoene into fully unsaturated carotenoids.

pathways can exist in the same tissue, e.g. tomato fruit. The fully unsaturated carotenes may be hydroxylated at position 3 and 3′, and form epoxides, first the 5,6-epoxides (e.g. III) and then 5,8-epoxides (e.g. XXVI). Epoxide formation can only occur across the double bond of a β-ionone residue; however, β-carotene itself can be converted into its expoxide before hydroxylation. All these basic biosynthetic steps appear to occur to a greater or lesser extent in all carotenogenic fruit and this accounts for many of the similarities between different species. Only when unique steps occur, such as capsanthin (XIX) formation, can carotenoid studies be of any real taxonomic significance.

B. PATTERN IN CHLOROPLASTS

Most of the fruit mentioned in this chapter are green when immature and contain chloroplasts. As the fruit ripens the number of chloroplasts increases to a certain point and then remains constant for some time as with the case of *Sorbus* (Zurzycki, 1954) (Fig. 4). At the green stage (e.g. III-IV in Fig. 4) the carotenoids are very much the same in leaf and fruit as illustrated by work

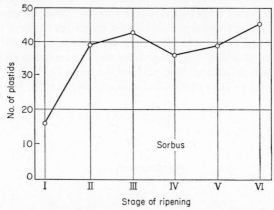

I. Full bloom; II. Fruit 3 mm pale green; III. Fruit intense green (5 mm); IV. Fruit pale green (7 mm); V. Fruit pale yellow (7 mm); VI. Ripe fruit (7 mm). Change from stage I to stage VI occupies from 8–12 weeks (from Zurzycki, 1954).

Fig. 4. Changes in number of plastids as fruit of *Sorbus* ripen.

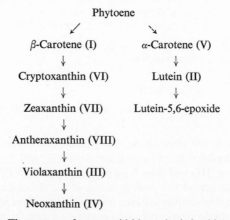

Fig. 5. The pattern of carotenoid biosynthesis in chloroplasts.

on peppers (Table 8). Electron microscope studies have shown that as fruit ripen further, the chloroplasts degenerate into chromoplasts (Frey-Wyssling and Kreutzer, 1958). The chlorophylls rapidly disappear and with the disappearance of photosynthetic ability the control of carotenoid synthesis is removed. This results is a rapid increase, generally in an overall oxidative fashion (Fig. 5); this is particularly true of the β-carotene pathway, as for example, in red peppers (Table 8), *Physalis alkekengi* (Table 3) *Cotoneaster* spp. (Table 5) and *Lonicera* spp. (Table 6). Rarely, if ever, are members of the α-carotene series, in particular lutein (II), synthesized in developing fruit.

TABLE 8. The major carotenoids of leaves, unripe and ripe fruit of *Capsicum annum* v. *lycopersiciforme rubrum*[1]

Pigment	Amount mg/100 g fresh wt		
	Leaves	Unripe fruit	Ripe fruit
β-Carotene (I)	7·92	0·095	2·35
Cryptoxanthin (VI)	0·45	0·027	1·10
Lutein (II)	13·99	0·276	—
Zeaxanthin (VII)	—	—	1·75
Antheraxanthin (VIII)	1·14	0·031	0·99
Violaxanthin (III)	8·27	0·042	0·70
Foliaxanthin[2]	5·66	0·058	—
Capsanthin (XIX)	—	—	9·60
Capsorubin (XXVIII)	—	—	1·46

[1] Adapted from Cholnoky *et al.*, 1956.
[2] Unknown 5,6-epoxide.

TABLE 9. The effect of temperature on carotenoid synthesis in tomatoes[1]

Pigment	Rutgers' strain (amount μg/g fresh wt) Temperature of maturation	
	23·5°	32°
Phytoene (XV)	29·0	9·6
Phytofluene (XVI)	8·2	0·4
β-Carotene (I)	4·9	5·5
γ-Carotene (IX)	1·4	0·6
Lycopene (XIV)	43·6	6·9
Total	87·1	23·0

[1] Tomes, 1963.

Occasionally there is a massive synthesis of β-carotene (I), and to a lesser extent α-carotene (V), without any appreciable xanthophyll synthesis; a typical example is the red palm (Argoud, 1958). More frequently, non-oxidative pathways involve the accumulation of the acyclic lycopene (XIV) and its precursors: phytoene (XV), phytofluene (XVI) etc. Typical examples are the commercial tomato, red water melon (Table 3) and *Viburnum opulus* (Table 6). Frequently β-carotene occurs alongside the lycopene series as in tomatoes and

Lonicera spp. (Table 6). There is, however, an important biosynthetic difference; in tomatoes ripened above 30° synthesis of the lycopene series is inhibited while that of β-carotene is unaffected; the lycopene effect is reversible (Table 9, Goodwin and Jamikorn, 1952; Tomes, 1963). In tomatoes containing the B gene the β-carotene level is increased to that of lycopene in normal tomatoes, whilst the lycopene level is concomitantly reduced. In this case β-carotene synthesis is temperature-sensitive.

Neoxanthin (IV) which is always present in chloroplasts and which is thought to be a β-carotene derivative, never appears during fruit maturation; this might point to the view that neoxanthin is not a β-carotene derivative.

The conversion of 5,6-epoxides into 5,8-epoxides is very easily achieved chemically by traces of mineral acids; since organic acids accumulate in the chromoplasts during ripening the formation of 5,8-epoxides in fruit could be due to the effect of the increasing acidic environment on the 5,6-epoxides rather than to an enzymatic reaction. Alternatively on disruption of the chromoplasts, the fragility of which varies from fruit to fruit, the pigments will be released into an acid environment. This is almost certainly the case with citrus fruit. In the variety Sarah, a pink sport of the Shamonti (Jaffa) orange, lycopene (XIV) is the main pigment and is concentrated in the pericarp; in contrast to normal oranges but similar to grapefruit, the filtered juice is colourless (Monselise and Halevy, 1961), which suggests that the carotenoids are not liberated from the chromoplasts. Furthermore when fresh pulp of pineapples is homogenized there is a rapid formation of furanoid (5,8-) epoxides (Singleton *et al.*, 1961).

Somewhat similar changes are observed in necrosing (autumn) leaves, when the degenerating chloroplasts lose their carotenoids by oxidative processes involving the formation of epoxides (Glover and Redfearn, 1953). In certain species, e.g. sycamore, the xanthophylls are esterified during this period before their complete destruction (Goodwin, 1958); many xanthophylls are also esterified in fruit, but never in chloroplasts. These observations suggest that the esterification occurs only in chloroplasts which have lost their chlorophylls, or outside the chloroplast following on its fragmentation.

C. CONCLUSIONS

The previous discussion attempts to provide an explanation at the biochemical level for the various pigment patterns observed in fruit. It follows that as there are only two main consequences of the breakdown of the carotenoid regulatory system in chloroplasts—formation of the acyclic lycopene series and of β-carotene and its oxidation products—a study of carotenoid distribution in fruit cannot have profound taxonomic significance. However, certain rather unique systems do arise such as rubixanthin (XX) in *Rosa* spp., rhodoxanthin (XXI) in *Taxus* spp. and capsanthin (XIX) and capsorubin (XXIX) in some *Capsicum* spp. As such these pigments have a limited use as taxonomic markers.

After all, carotenoid accumulation is merely one aspect of a fundamental process in fruit ripening—the conversion of chloroplasts into chromoplasts.

REFERENCES

Allen, M. B. (1959). *Arch. Mikrobiol.* **32**, 270.
Allen, M. B., Goodwin, T. W. and Phagpolngarm, S. (1960). *J. gen. Microbiol.* **23**, 93.
Argoud, S. (1958). *Oléagineux* **13**, 249.
Baraud, J. (1958). *Rev. gen. Bot.* **65**, 221.
Bonnett, R., Spark, A. A., Tee, J. L. and Weedon, B. C. L. (1964). *Proc. chem. Soc.* 419.
Bubicz, M. and Wierzchowski, Z. (1959). *Ann. Univ. Marie-Curie-Sklodoska Lublin*, **18**, 383.
Cholnoky, L., Gyorgyfy, C., Nagy, E. and Pancel, M. (1956). *Nature, Lond.* **178**, 410.
Cholnoky, L., Gyorgyfy, Nagy, E. and Pancel, M. (1958). *Acta Chim. Acad. Sci. Hung.* **16**, 227.
Curl, A. L. and Bailey, G. F. (1956). *Agr. Fd. Chem.* **4**, 156.
Dougherty, E. C. and Allen, M. B. (1960). *In* "Comparative Biochemistry of Photoreactive Pigments". (M. B. Allen, ed.). Academic Press New York and London.
Frey-Wyssling, A. and Kreutzer, E. (1958). *J. Ultrastructure Res.* **1**, 397.
Geiger-Vifian, A. and Müller, B. (1945). *Ber. schweiz. bot. Ges.* **55**, 320.
Glover, J. and Shah, P. P. (1957). *Biochem. J.* **54**, viii.
Goodwin, T. W. (1952). "The Comparative Biochemistry of the Carotenoids." Chapman and Hall, London.
Goodwin, T. W. (1956). *Biochem. J.* **62**, 346.
Goodwin, T. W. (1958). *Biochem. J.* **68**, 503.
Goodwin, T. W. (1962). *In* "Comparative Biochemisty", Vol. IV, B, (M. Florkin and H. S. Mason, eds.), p. 643. Academic Press, New York and London.
Goodwin, T. W. (1963). *Proc. 5th Int. Cong. Biochem.* **3**, 300.
Goodwin, T. W. (1965a). *In* "Chemistry and Biochemistry of Plant Pigments" (T. W. Goodwin, ed.), p. 127. Academic Press, London and New York.
Goodwin, T. W. (1965b). *In* "Biosynthetic Pathways in Higher Plants" (J. B. Pridham and T. Swain, eds.), Academic Press, London and New York.
Goodwin, T. W. and Jamikorn, M. (1952). *Nature, Lond.* **170**, 104.
Goodwin, T. W. and Thomas, D. M. (1964). *Phytochemistry* **3**, 47.
Karrer, P. and Rutschmann, J. (1945). *Helv. chim. acta.* **28**, 1528.
Monselise, S. P. and Halevy, A. H. (1961). *Science* **133**, 1478.
Pardee, A. B., Schachman, H. K. and Stanier, R. Y. (1952). *Nature, Lond.* **169**, 282.
Singleton, V. L., Gortner, W. A. and Young, H. Y. (1961). *J. Fd. Sci.* **26**, 49.
Stabursvik, A. (1954). *Acta chem. scand.* **8**, 1305.
Strain, H. H. (1961). *In* "Manual of Phycology" (G. M. Smith, ed.), Waltham, Massachusetts.
Strain, H. H. (1958). *Chloroplast Pigments and Chromatographic Analysis.* 32nd Annual Priestley Lecture. Pennsylvania State University, Pennsylvania.
Thomas, D. M. and Goodwin, T. W. (1965). *Phycology* (In press).
Tomes, M. L. (1963). *Bot. Gaz.* **124**, 180.
Tomes, M. L., Johnson, K. W. and Hess, M. (1963). *Proc. Amer. Soc. Hort. Sci.* **82**, 460.
Zechmeister, L. and Schroeder, W. A. (1942). *Arch. Biochem.* **1**, 231.
Zechmeister, L. and Sandoval, A. (1945). *Arch. Biochem.* **8**, 425.
Zurzycki, J. (1954). *Acta polsk. Tow. botan.* **23**, 161.

CHAPTER 8

The Natural Distribution of Plant Polysaccharides

ELIZABETH PERCIVAL

Royal Holloway College, University of London, Englefield Green, Surrey

I. INTRODUCTION

The majority of polysaccharides fulfil one of two main functions in plants. They either constitute the skeletal structure or form a major food reserve. There are, of course, polysaccharides which have other functions such as the gum exudates of the higher plants and the sulphated algal polysaccharides. All seaweeds synthesize at least one polysaccharide which carries sulphate ester groups and it seems to the author that these compounds must have some function related to the environment of the plant. As our knowledge of biology increases it is very probable that new and diverse functions for polysaccharides will also be discovered.

While certain polysaccharides are characteristic of a particular class of plant, others, such as cellulose and starch, are found in a wide variety of classes. Aspinall (1964a) has drawn attention to the fact that galacturonorhamnans may be found in exudate gums, pectic substances and plant mucilages. Only in a few instances have polysaccharides so far proved of value to the taxonomist. However, it should be remembered that the detailed structure of few, if any, polysaccharides has been completely determined. It is slowly becoming apparent that in many instances, such as the hemicelluloses of the higher plants, the galactans of the red algae, and the water-soluble sulphated polysaccharides of the green algae, there exist families of polysaccharides based on an essentially similar basic pattern, but which differ in the finer details. It is possible that as our knowledge increases, however, we shall find generic and even species differences in the individual polysaccharides.

II. CELLULOSE

The β-1,4-linked glucan, cellulose, is of almost universal occurrence throughout the plant world, and it seems that the only differences from class to class are the relative quantity, and possibly the degree of polymerization. The higher plants synthesize the largest amount and there are certain algae, e.g. the green seaweed genera *Codium, Caulerpa* and *Ulva*, and the diatom *Phaeodactylum tricornutum* which appear to be devoid of cellulose (see below). It should perhaps be emphasized that there are still many thousands of algae and other lower plants whose polysaccharides have not yet been investigated. It is said for instance that only about 10 per cent of all the algae have so far been identified and characterized.

III. HEMICELLULOSES

The rather loose term hemicellulose implies any of those polysaccharides found in close association with cellulose, especially in lignified tissue (Aspinall, 1959). In Angiosperms and Gymnosperms the hemicelluloses are divided into three main families: (*a*) those based on chains of D-xylose, the *xylans*, (*b*) polysaccharides comprising D-mannose usually in association with D-glucose and also occasionally with D-galactose, the *glucomannans*, and (*c*) polymers in which the major unit is D-galactose often associated with L-arabinose, the *arabinogalactans*.

In the lower plants, insofar as they have been studied, we find quite different types of polysaccharides. In the Phaeophyceae, cellulose is present, if at all, in very small amounts (Percival and Ross, 1949), its place apparently being taken by the β-1,4-linked mannuronoguluran, alginic acid (Hirst *et al.*, 1964; Hirst and Rees, 1965), closely associated with which is the highly sulphated fucan, fucoidin (Conchie and Percival, 1950; O'Neill, 1954; Côté, 1959) based mainly on sulphated fucose units, and sulphated polysaccharide material comprising D-xylose, L-fucose and D-glucuronic acid (Percival, unpublished results). Neither chemical studies nor X-ray investigations have provided definite evidence for the presence of cellulose in the Rhodophyceae (Myers *et al.*, 1956; Cronshaw *et al.*, 1958; Kreger, 1960; Frei and Preston, 1961). The major structural polysaccharides in this class are galactans of varying degrees of sulphation (Percival, 1963; Rees, 1962), sometimes accompanied, as will appear later, by mannans and xylans. From X-ray evidence (Cronshaw *et al.*, 1958) cellulose is considered to be present in some of the Chlorophyceae,

—4-D-Xyl p β1—4-D-Xyl p β1···4-D-Xyl p β1—
      3   2   3
      | | |
      1   1α 1
L-Ara f 4Me-D-GlcUA   R—L-Ara f

FIG. 1. General formula for xylans.

Abbreviations in this and all subsequent figures are those recommended in Appendix B to IUPAC Bulletin 12. Monosaccharides are represented by the first three letters of their names (Glc = glucose, GlcUA = glucuronic acid); α-, β-, D- and L- have the usual meaning; p and f show pyranose and furanose forms; numerals indicate position of links; Me indicates an *O*-methyl group; R, another chain.

type. On the whole the proportion of the 4-*O*-methylglucuronic acid is higher in softwood xylans (15–20%) than those from hardwoods (8–15%) (Jones and Painter, 1959). The view has been advanced (Thornber and Northcote, 1962) that the ratio of xylose to 4-methyl-*O*-glucuronic acid in the xylan decreases as the cell-wall material is laid down during the differentiation of the cell. While some wood xylans contain a small proportion of L-arabinofuranose linked at C-3, the xylans from cereals and grasses are generally characterized by a high proportion of single unit side chains of L-arabinose residues linked at C-3 to the xylose backbone, and in many cases a small proportion of D-glucuronic acid or its 4-*O*-methylether or both are present (Aspinall and Ferrier, 1957; Aspinall and Ross, 1963) (Fig. 1). There is, in fact, no marked structural division between these two groups of xylans, and in both, the side chains of ara$f$(1—3) and the D-glucuronic acid (1—2) show the same preferred mode of linkage. Recent work has shown that there is no simple regular arrangement of the side chains (Ewald and Perlin, 1959; Aspinall and Greenwood, 1962; Goldschmid and Perlin, 1963; Timell, 1962b; Aspinall and Ross, 1963). In the native 4-*O*-methylglucuronoxylans *O*-acetyl groups are present attached only to xylose residues (Bouveng *et al.*, 1960; Bouveng, 1961b).

142 *Elizabeth Percival*

In the lower classes of plants where xylans have been found they are structurally distinct from those of the higher plants. The only chemical investigation of a xylan reported from the red algae is that from *Rhodymenia palmata* (Chanda and Percival, 1950), which apparently is a homopolysaccharide essentially unbranched and comprising 80% of 1,4-linked and 20% of 1,3-linked units. Xylans constitute the structural material of the *Caulerpa* (Mackie and Percival, 1959, 1961), *Bryopsis, Halimeda* and *Chlorodesmis* genera (Iriki *et al.*, 1960) of the Chlorophyceae. Again these are essentially linear homopolysaccharides, but consisting entirely of β-1,3-linked units. X-ray studies have indicated a similar xylan in the cell walls of the red alga, *Porphyra umbilicalis* (Frei and Preston, 1964), but until chemical studies on this have been carried out it is not certain that it is devoid of 1,4-links. Although xylose has been reported as a polysaccharide constituent of a number of brown seaweeds no structural studies have so far been reported.

B. MANNANS AND GLUCOMANNANS

Pure essentially linear mannans comprising β-1,4-linked mannose units which appear to constitute the structural polysaccharide have been found in the red seaweed *Porphyra umbilicalis* (Jones, 1950), and in the green seaweeds, *Codium* (Love and Percival, 1964b), *Derbesia, Acetabularia* and *Halicoryne* (Iriki and Miwa, 1960). Although the mannan from *Codium fragile* appeared to contain *ca.* 5% of 1,4-linked glucose units.

The hemicellulose mannans also all comprise linear chains of 1,4-β-D-mannose, but with 1,4-β-D-glucose as part of the main structural features, in close association with cellulose in coniferous woods where it constitutes half the hemicellulose fraction (Timell, 1963), and may be once or twice branched (Timell, 1962c). The ratio of mannose:glucose varies, and is usually about 3:1 from softwood (Fig. 2) (Timell, 1961; Thornber and Northcote, 1962), and 2:1 from hardwood (Timell, 1965). Although it has been shown that there may be as many as four mannose and two glucose adjoining units (Aspinall *et al.*, 1962) it appears probable that there is a statistically random arrangement of glucose and mannose along the chains. In addition galactose

$$\text{—4-D-Man } p \text{ } \beta1\text{—4-D-Glc } p \text{ } \beta1\text{—4-D-Man } p \text{ } \beta1\text{—4-D-Man } p \text{ } \beta1\text{—}$$
$$6$$
$$|$$
$$1$$
$$\text{D-Gal } p$$

FIG. 2. Softwood mannan. (See Fig. 1 for symbols.)

occurs in nearly all mannans from softwood and ferns (Meier, 1960; Perila and Bishop, 1961; Mills and Timell, 1963; Schwarz and Timell, 1963) joined by 1,6-links to the other hexoses and some may be 1,3-linked.

Two unique mannans have recently been investigated. That from birch sap (Bishop, 1964) has the approximate normal composition of a galactoglucomannan but contains some mannofuranose units. Secondly the sulphated glucuronomannan from *Phaeodactylum tricornutum* which is apparently the structural material of this diatom, has a backbone of 1,3-linked mannose units to which are attached side chains of D-glucuronosyl(1—3)-D-mannosyl (1—2)-D-mannose (Ford and Percival, 1964). This is the first mannan of this type to be investigated, and it is possible that further studies on botanically related diatoms will reveal the presence of similarly constituted mannans.

C. GALACTANS AND ARABINOGALACTANS

Galactans and arabinogalactans are water-soluble polysaccharides found in many coniferous woods, but in largest proportion in larches. They are highly branched polysaccharides comprising 1,6- and 1,3-linked D-galactopyranose units (Aspinall *et al.*, 1958; Bouveng, 1959a, 1961a). A possible formula for

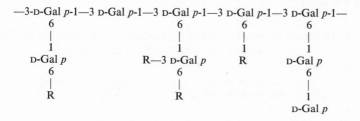

FIG. 3. Proposed structure for Japanese larch arabinogalactan.

(R = L-Ara *f*-1···; L-Ara *p*-1—3 L-Ara *f*-1···; or D-Gal *p*-1. Other symbols see Fig. 1.)

the Japanese larch, *Larix leptolepsis* (Aspinall, 1964b), arabinogalactan is shown in Fig. 3.

A 1,3-linked galactopyranose backbone to which are attached a variety of side chains comprising L-arabofuranose and D-galactopyranose is typical of the larch arabinogalactans. Recent work has shown that maritime pine (Roudier and Eberhard, 1963), mountain larch (Jones and Reid, 1963), and tamarack (Urbas *et al.*, 1963) arabinogalactans contain a small proportion of D-glucuronic acid residues linked 1,6 to D-galactose, and it is possible that most arabinogalactans from species of larch contain similar structural units.

Very similar arabinogalactans have been found in gum exudates, for example, from the Australian bunya pine (*Araucaria bidwilli*) (Aspinall, and Fairweather, 1965) (Fig. 4). It is becoming increasingly apparent that there is no clear division between the wood arabinogalactans and the exudate gums of the gum arabic type (Aspinall *et al.*, 1963). In the majority of cases the galactan skeleton has been shown to contain 1—3 linkages in the inner and 1—6 linkages in the outer chains. Both types of polysaccharide contain the same basal skeletons of galactose residues with the exudate gums carrying in general more highly ramified arrangements of L-arabinose, D-glucuronic acid and

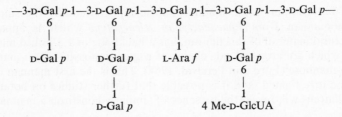

FIG. 4. Proposed structure for *Araucaria bidwilli* gum. (See Fig. 1 for symbols.)

L-rhamnose in the outer chains of the molecular structure. Table 1 illustrates the structural units common to the two types of polysaccharide.

Although the main structural features have been determined the points of attachment of the branches to the galactan framework have been established in very few cases.

TABLE 1. Different types of side-chain attached to the galactan skeleton in wood arabinogalactans and exudate gums

Source of wood (Arabinogalactans)	Neutral units	Acidic units
European, Western and Japanese larches[1]	A, B	—
Tamarack[2]	A, B	C
Mountain larch[3]	A, B	C
Scots pine[4]	A	
Maritime pine[5]	A, B	C
Maple sap[6]	A, E	
Source of gum		
Araucaria bidwilli[7] (Bunya pine)	A, E	C, F
Asafoetida[8]	A	C, F
Khaya senegalensis[9] (minor component)	A	C, F, G
Acacia senegal[10] (gum arabic)	A, B, E, H	C
Acacia pycnantha	A, B, E	C

Key: A=Ara f 1···; B=Ara p 1—3 Ara f 1—; C=GlcUA 1—6-D-Gal p 1—; E=L-Rha 1—; F=4MeGlcUA 1—6 D-Gal p 1—; G=4MeGlcUA 1—4 D-Gal 1; H=Gal 1—3 Ara f 1. (Symbols see Fig. 1.)

[1] Bouveng, 1959a, 1959b; Bouveng and Lindberg, 1958; Jones, 1953.
[2] Adams, 1960; Urbas *et al.*, 1963; Haq and Adams, 1961.
[3] Jones and Reid, 1963.
[4] Aspinall and Wood, 1963.
[5] Roudier and Eberhard, 1963.
[6] Adams and Bishop, 1960.
[7] Aspinall and Fairweather, 1965.
[8] Jones and Thomas, 1961.
[9] Aspinall, *et al.*, 1965b.
[10] Smith and Montgomery, 1959; Aspinall *et al.*, 1963.
[11] Aspinall *et al.*, 1959.

Arabinogalactans are also found in the green seaweeds belonging to the genera *Cladophora* (Fisher and Percival, 1957), *Chaetomorpha* (Mackie and Percival, 1965), *Codium* (Love and Percival, 1964a), and *Caulerpa* (Mackie and Percival, 1961) (Table 2). Although these polysaccharides are, as are the galactans of the Rhodophyceae, water soluble mucilages they are extremely difficult to remove completely from the algal cells; repeated aqueous, followed by acidic and alkaline extraction of the disintegrated cells fail to effect complete extraction in a number of instances, and it is possible that these polysaccharides have something of the same function as the hemicelluloses (Bean *et al.*, 1953; Percival, 1963). Apart from the presence of sulphate ester groups arabino-galactans from the Chlorophyceae are structurally similar to the equivalent polysaccharides shown in Table 1 in that they contain 1,3- and 1,6-linked

TABLE 2. Water-soluble polysaccharides from green seaweeds.

	Species			
Constituent sugars	*Cladophora rupestris*	*Chaetomorpha capillaris*	*Caulerpa filiformis*	*Codium fragile*
Arabinose	+ + + + +[1]	+ + + + +	+ + (?)	+ + + +
Galactose	+ + + + +	+ + + + +	+ + + + +	+ + + + +
Xylose	+ +	+ +	+ + +	+ +
Mannose	—	—	+ + +	—
Glucose	+	+	—	—
3,6-Anhydrohexose	+	+	+	+
SO_4^{2-} %	c. 16	15	17·5	15
$[\alpha]_D H_2O$	+53°	+70°	+13°	+46°

[1] Number of + indicates amount present.

galactose units. All contain galactose 6-sulphate units, and *Cladophora* and *Chaetomorpha* also have sulphate ester groups on C-3 of the arabinose units. Evidence for galactose units also sulphated at C-4 and for 1,3-linked arabinose units has been obtained for *Codium*. In spite of the presence of 6-sulphated and 1,3-linked galactose in the arabinogalactans of both *Cladophora* and *Codium* they are apparently quite different polysaccharides. For although each polymer contains alkali labile ester sulphate, this is mainly linked to arabinose units in *Cladophora* and to galactose units in the *Codium* polysaccharide. Galactans devoid of arabinose have been extracted from species of *Caulerpa* (Mackie and Percival, 1961). The polysaccharides of all four genera contain an appreciable quantity of xylose and little, if any, uronic acid. Those from *Cladophora* and *Chaetomorpha* contain a small amount of glucose, and that from *Caulerpa* appreciable quantities of mannose. Although these materials have defied fractionation into more than a single polysaccharide, no unequivocal proof that they are single heteropolysaccharides has so far been obtained.

The major polysaccharides of the Rhodophyceae contain 1,3-linked galactose units, but there the similarity to the galactans of the higher plants ends. The red algal galactans have varying proportions of 3,6-anhydrogalactose residues and ester sulphate groups within the macromolecule. In the different genera of the red weeds we find differences in the structure of the individual galactan. Agar is obtained from species of *Gelidium, Gracilaria, Ahnfeltia, Phyllophora* and *Pterocladia*, whereas *Chondrus, Eucheuma, Gigartina* and *Irideae* synthesize carrageenan (Smith and Montgomery, 1959). Both these materials consist of at least two polysaccharides agarose and agaropectin, and κ- and λ-carrageenan respectively. Agarose consists of alternate units of 1,3-linked β-D-galactose (some methylated at C-6) and 1,4-linked 3,6-anhydro-α-L-galactose units (Araki, 1959; 1965), whereas in κ-carrageenan, 3,6-anhydro-α-D-galactose replaces the L-sugar present in agarose. Furthermore in κ-carrageenan the galactose units are sulphated at C-4 and there is evidence for

$$—3\text{-D-Gal } p\ \beta1—4\ \text{L-Agal } \alpha1—$$
$$(a)$$

$$—3\text{-D-Gal } p\ \beta1—4\ \text{D-Agal } \alpha1—$$
$$4$$
$$SO_3H$$
$$(b)$$

FIG. 5. Units in (*a*) agarose and (*b*) κ-carrageenin (See Fig. 1 for symbols, Agal=3,6-anhydrogalactose.)

$$—3\text{-D-Gal } p\text{-}1—\qquad—4\text{-D-Gal } p\text{-}1—$$
$$4\qquad\qquad\quad 2\qquad 6$$
$$SO_3H\qquad\quad SO_3H\ \ SO_3H$$

FIG. 6. Major units present in λ-carrageenin.

branching at C-6 of every tenth D-galactose residue and for a 3,4- or 3,6-disulphated galactose as the terminal non-reducing unit (O'Neill, 1955) (Fig. 5).

The structures of agaropectin and λ-carrageenan are not so well established. The former contains the structural units found in agarose, a small proportion of sulphate, D-glucuronic acid and 1% of pyruvic acid in acetal linkage to D-galactose residues (Hirase, 1957). In contrast λ-carrageenan comprises a highly sulphated galactan containing α-1,3-linked D-galactose partly sulphated at C-4, *ca.* 40% of 1,4-linked 2,6-disulphated D-galactose and a small proportion of the 3,6-anhydro-D-derivative (Morgan and O'Neill, 1959; Rees, 1961b) Fig. 6. There is some evidence that the ester sulphate groups are distributed between different sites on the polysaccharide depending upon the species, and if further work substantiates this, it should be of considerable value to the taxonomist. Alcohol fractionation separates a small proportion of polysaccharide containing L-galactose units (Smith *et al.*, 1955). Analyses of samples of carrageenan from different environment show a very variable composition, that from *G. stellata* having a lower λ-content (no matter what the season of collection) than that from American *Chondrus*. Samples of American λ-carrageenan were practically devoid of 3,6-anhydrogalactose, whereas three samples of British λ-carrageenan contained appreciable quantities, and corresponding lower proportions of 2,6-disulphate (Rees, 1963), the apparent precursor of the anhydro sugar.

Recent studies (Black *et al.*, 1965) have revealed genus and species differences in carrageenan, particularly in the proportion of the λ-polysaccharide. Whereas that from *Gigartina acicularis* and *Gigartina pistillata* can be readily fractionated into κ- and λ-carrageenan, the polysaccharide from *G. stellata* and *G. radula* is poorly fractionated and the carrageenan from *Euchema spinosum* resists fractionation and comprises a more highly sulphated κ-carrageenan. Furthermore that from *Polyides rotundis* differs from the other carrageenans in that it contains all three types of sulphated residues, but negligible amounts of 3,6-anhydrogalactose. The situation is further complicated in the case of *C. crispus* by seasonal and environmental variations.

A carrageenan-like material can be extracted from *Agardhiella tenera* (Smith and Montgomery, 1959), *Furcellaria fastigiata* (Clancy *et al.*, 1960; Painter, 1960), and species of *Hypnea* (Yaphe, 1959).

The galactan (porphyran) from *Porphyra umbilicalis*, again thought to be a mixture of related polysaccharides, resembles agarose in containing 3,6-anhydro-L-galactose units, and is similar to carrageenan in that it contains both D- and L-galactose and C-6 ester sulphate, the latter predominantly linked to L-galactose. Porphyran differs from carrageenan and resembles agar in containing 6-*O*-methylgalactose residues (Peat *et al.*, 1961; Turvey and Rees, 1961). Similar galactans are synthesized by a variety of *Porphyra* species (Nunn and von Holdt, 1957; Rees and Conway, 1962a), *Bangia fuscopurpurea* (Wu and Ho, 1959) *Laurencia* and three species of *Polysiphonia* (J. R. Turvey, private communication). It appears that these algae synthesize a DL-galactan in which some of the 1,3-linked D-galactose units carry a methyl group on C-6, and the ester sulphate occurs predominantly as 1,4-linked D-galactose 6-*O*-sulphate (Turvey, 1960).

The composition of the polysaccharide extracts of different species of *Porphyra* collected at different seasons and from different environments has been shown to vary (Rees and Conway, 1962b): e.g. 3,6-anhydro-L-galactose 5–19%; ester sulphate 6–11%; 6-*O*-methyl-D-galactose 3–28%; galactose 24–45%. In all the samples examined, however, 3,6-anhydro-L-galactose and L-galactose-6-sulphate equal 50% of the sugar units, the rest comprising D-galactose and 6-*O*-methyl-D-galactose. In all these galactans the galactose-6-sulphate unit is thought to be the biological precursor of the 3,6-anhydro-derivative and in equilibrium with it, and evidence of this enzymic transformation has been reported (Rees, 1961a). In spite of the differences in the fine structure of these galactans they are all based on alternate units of α-1,3- and β-1,4-linked galactose or modified galactose. In fact the Edinburgh School (Anderson *et al.*, 1965) have recently converted, by desulphation and methylation, porphyran into methylated agarose. These authors suggest that the

6

difference in the fine structure is appropriate for a particular species growing in a particular environment.

Apart from galactans based on 1,3-linked units there are galactans characterized by β-1,4- with in some cases 1,6-linkages. These occur as neutral galactans and as integral constituents of acidic polysaccharides (Dutton and Unrau, 1963). In birchwood a galactan containing 1,4-linked units occurs (Gillham *et al.*, 1958), and in Norway spruce compression wood, an acidic polysaccharide is present which on partial hydrolysis yields an homologous series of 1,4-linked galactose oligosaccharides (Bouveng and Meier, 1959). Beech tension wood contains an even more complex polysaccharide (Meier, 1962) comprising a neutral fraction of 1,4- and 1,6-linked D-galactose units, and an acidic fraction containing D-galactose, D-glucuronic acid, D-galacturonic acid and L-rhamnose units. It is interesting to compare the structural

TABLE 3. Structural units found in beech tension wood galactan and in *Combretum leonense* gum

Source	Structural unit present[1]
Beech tension wood galactan	C, D, E, F, J (trace), K.
Combretum leonense gum	A, B, C, D, F, G, H, J.

A=Ara 1—3 Ara; B=Gal 1—3 Ara; C=Gal 1—6 Gal 1—6 Gal; D=Gal 1—4 Gal; E=Gal 1—6Gal 1—4Gal; F=GalUA1—2 Rha; G=Gal 1—? GalUA1—2Rha; H=GalUA 1—2 Rha 1—4 Gal; J=GlcUA 1—6 Gal; K=4 MeGlcUA 1—6 Gal.

units found in this latter polysaccharide with those of the gum exudate from the West African Tree, *Combretum leonense* (Aspinall and Bhavanandan, 1965) Table 3.

IV. PECTIC ACID AND GUM EXUDATES

Brief mention only will be made of the complex polysaccharides known as pectic acid and the highly branched acidic gum exudates, since the structural chemistry of these materials is still very far from complete. It has been known for over thirty years (Ehrlich, 1932) that pectic substances are composed of residues of L-arabinose, D-galactose and D-galacturonic acid, the last, at least in part, as its methyl ester. Other sugars, including D-xylose, L-rhamnose and L-fucose may also be present, and in recent years small amounts of 2-*O*-methyl-D-xylose and 2-*O*-methyl-L-fucose have been isolated from the hydrolysis products of various pectic materials. The different polysaccharides all contain some contiguous residues of D-galacturonic acid, but considerable variations are encountered in the nature of the units in the side chains. These side chains in turn may be attached to inner chains containing residues either of both D-galacturonic acid and L-rhamnose or of D-galacturonic acid alone. It may well be that as the detailed structure of these complex materials is determined a species difference will emerge. Table 4 shows some of the structural units present in pectic materials from different sources.

There is an essential similarity between the gum exudates of *Khaya* (Aspinall *et al.*, 1956; 1960), *Sterculia* (Hirst *et al.*, 1958; Aspinall and Fraser, 1965), *Cochlospermum gossypium* (Aspinall *et al.*, 1965c) and *Combretum leonense* (Aspinall and Bhavanandan, 1965). With the exception of the last (which also contains arabinose) all are composed of the same four sugar residues: D-galactose, L-rhamnose, D-galacturonic acid and D-glucuronic acid or its 4-methyl ether. They differ in the nature of the branching points and in the detailed sequence of sugars. The most striking difference so far recognized between *Khaya* and *Sterculia urens* (Aspinall and Nasir-ud-din, 1965) lies in the site of attachment of D-glucuronic acid end groups: in the former they are linked to

TABLE 4. Structural units found in a number of pectins

Source of Pectin	Units Found[1]
Lucerne[2]	A, B, C.
Citrus[3]	A, B, C, H, J.
Soybean hulls[4]	A, B, C, (G).
Soybean meal[5]	A, B, C, (D), E, F.
Tragacanthic acid[6]	A, B, D, K, L.

[1] Key to Units: A=GalUA 1—4 GalUA; B=GalUA 1—4 GalUA (1—4 GalUA; C=GalUA 1—2 Rha; D= Xyl 1—3 GalUA; E=GalUA 1—6 Gal; F=GlcUA 1—4 Gal; G=GlcUA 1—? Fuc; H=Gal 1—4 Gal; J=Gal 1—4 Gal 1—4 Gal; K=Fuc 1—2 Xyl; L=Gal 1—2 Xyl.
[2] Aspinall and Fanshawe, 1961.
[3] Aspinall, G. O. and Whyte, J. L., unpublished results.
[4] Aspinall and Hunt, unpublished results.
[5] Aspinall, G. O. and Whyte, J. N. C., unpublished results.
[6] Aspinall and Baillie, 1963.

D-galactose and in *Sterculia* to D-galacturonic acid. Each of these gums contains O-α-D-GalUA 1—2 L-Rha as a structural unit.

The gum exudates from *Virgilia araboides* (Stephen, 1962, 1963) and *Anogeissus latifolia* (gum Ghatti) (Aspinall *et al.*, 1965a) and *A. schimperi* (Aspinall and Christensen, 1961) are very similar. They differ from the previous gums in that they are devoid of galacturonic acid and rhamnose and they contain mannose units. They consist of long chains of 1,6-linked β-D-galactose with a small proportion of 1,3-links. The macromolecule is highly branched and contains 40% of arabinose. The glucuronic acid is linked to mannose and to galactose (i.e. GlcUA(1—2)Man, and GlcUA(1—6)Gal.) and there is a surrounding sheath of L-arabinofuranose units.

V. URONIC ACID-CONTAINING POLYSACCHARIDES OF THE CHLOROPHYCEAE

Acidic polysaccharides which differ in many respects from the acidic polysaccharides of the higher plants are synthesized by the genera, *Ulva*

(Percival and Wold, 1963; Haq and Percival, 1966), *Enteromorpha* (McKinnell and Percival, 1962b) and *Acrosiphonia centralis* (*Spongamorpha arcta*) (O'Donnell and Percival, 1959) of the Chlorophyceae. Before discussing these I should like to digress to remind you that *Acrosiphonia* is botanically very closely related to *Cladophora* and *Chaetomorpha*, yet its polysaccharide synthesis is quite different and definitely of the same family of polysaccharides as those found in *Ulva* and *Enteromorpha*. In the polysaccharides of these three genera the major sugar is L-rhamnose and this is accompanied by xylose, glucose, *ca.* 20% glucuronic acid and ester sulphate groups varying from 7–20% (Table 5).

There is evidence of a high proportion of 1,3-linkages and/or of branch points and a structural unit common to the three genera is the aldobiouronic acid (McKinnell and Percival, 1962a) shown below. A high proportion of the sulphate groups are linked to C-2 of rhamnose and some of the xylose units

TABLE 5. Uronic acid containing polysaccharides of the Chlorophyceae

	Species		
Constituent sugars	*Acrosiphonia centralis*	*Enteromorpha compressa*	*Ulva lactuca*
L-Rhamnose	+ + + +	+ + + + +	+ + + + +
D-Xylose	+ + +	+ + +	+ + +
D-Glucose	+ +	+	+ +
D-Glucuronic acid %	19	18	20
SO_4^{2-}%	7·8	16	17·5
4-*O*-Glucuronosyl-L-rhamnose	+	+	+
$[\alpha]_D H_2O$	−31°	−49°	−47°

are also sulphated at C-2. All attempts to fractionate these materials have proved unsuccessful, and recent partial hydrolyses on a desulphated reduced sample of *Ulva* polysaccharide have led to the characterization of *O*-L-Rha1—4-*O*-D-Xy1—3-D-Glc, as a structural unit in the polysaccharide (Haq and Percival, 1966). It is very probable that the glucose unit is a glucuronic acid residue in the native polysaccharide. Evidence of direct linkage between glucuronic acid and xylose and between xylose and rhamnose in the polysaccharide has also been obtained; providing evidence of the heterogeneity of this polysaccharide.

VI. FOOD STORAGE POLYSACCHARIDES

A. STARCHES

The best known food storage polysaccharide is of course starch which is widespread among plants. The green algae also synthesize starch, but whereas in the higher plants the starch is laid down as a complex granule containing molecules of amylose and of amylopectin, the algal starches are much less organized and appear to have smaller molecular weights. It is of interest from the taxonomic point of view to consider the ratio of amylose to amylopectin. While the majority of higher plants contain *ca.* 20 % of amylose, the wrinkled pea has as much as 60 % whereas waxy maize starch is practically devoid of amylose. The green algal starches so far examined, with the exception of that from *Ulva lactuca* fall into the general pattern of *ca.* 20 % amylose (Love *et al.*, 1963). In contrast the red algae appear to synthesize starch devoid of amylose (Fleming *et al.*, 1956; Peat *et al.*, 1959; Meeuse and Kreger, 1954, 1959) which in some respects resembles glycogen, the α-1,4-linked glucan synthesized by animals. Recently the Bangor School (J. R. Turvey, private communication) have examined *Ceramium rubrum, Girgartina stellata, Gracilaria confervoides, Porphyra umbilicalis* and *Corallina officinalis* and found that they have a shorter average chain-length (11–13; one 15, glucose units) than the higher plant amylopectins (~20).

B. β-1,3-LINKED GLUCANS

β-1,3-Linked glucans also occur widely distributed in plants, ranging from cytoplasmic deposits or vacuolar inclusions which may consist wholly of β-1,3-linked glucose units to cell-wall components where they may be associated with considerable proportion of β-1,4- or β-1,6-linkages, other polysaccharides, proteins or lipids (Clarke and Stone, 1963). The largest quantity of a β-1,3-glucan is found in the Phaeophyceae in the form of laminarin. It is an essentially linear polymer with occasional branch points at C-6 and with a variable proportion of the glucose chains terminated at the potential reducing end with a molecule of mannitol (Peat *et al.*, 1957; Handa and Nisizawa, 1961). The soluble and insoluble forms of laminarin are considered to differ only in the degree of branching (Manners and Fleming, 1965).

β-1,3-Linked glucans devoid of mannitol occur in very diverse plant tissues: in the higher plants as callose (Aspinall and Kessler, 1957) and in, for example, sieve tubes, young tracheids, and various parts of the pollen mechanism; as chrysolaminarin or leucosin (Beattie *et al.*, 1961; Archibald *et al.*, 1958; Ford and Percival, 1964), the characteristic polysaccharide of the Chrysophyta; as paramylon in the form of cytoplasmic granules in the Euglenophyta (Clarke and Stone, 1960) and other flagellates (Cunningham *et al.*, 1962).

An essentially linear glucan, lichenin, present in the thallus of several lichens contains β-1,3-glucosidic linkages in the molecules together with a relatively high proportion of β-1,4-glycosidic linkages. Structural studies indicate it to be built mainly of cellotriose units joined through 1,3-linkages (Peat *et al.*, 1957; Cunningham and Manners, 1961; Perlin and Suzuki, 1962). Similar glucans have also been isolated from oat and barley (Parrish *et al.*, 1960; Goldstein *et al.*, 1959). Studies on different samples of barley indicate that it is possible that the variable proportion of 1,4-linked units may reflect a real difference in the composition of the glucan from different barley varieties.

C. FRUCTANS (Percival and Percival, 1962)

A characteristic polysaccharide of the Compositeae which completely replaces starch as a reserve food, is inulin. This consists of straight chain of *ca* 35 molecules of 2,1-linked β-fructofuranose residues terminated by a sucrose unit. A different type of fructan, levan, comprising 6,2-linked β-fructofuranose residues again thought to be terminated by sucrose, is the characteristic storage material of the leaves and stems of many monocotyledons. These are comparatively small mainly linear molecules of 20–50 units in keeping with their function as readily available storage products.

Highly branched fructans containing both the 2,1-linkages of inulin and the 6,2-linkages of levan have been isolated from the underground stems of *Triticum repens* (Arni and Percival, 1951), from the stem of *Agave vera cruz*, Mill (Aspinall and Das Gupta, 1959), and *Cordyline terminalis* (Boggs and Smith, 1956) and from a number of cereals (Schlubach, 1953, 1958). The basic structure of all of these fructans is similar but there are differences in their finer details, and as techniques in polysaccharide studies advance it is possible that here, too, we may find a species difference. It is worthy of note that no polysaccharide containing fructose has been reported from the algae, although many bacteria synthesize 6,2-linked fructans, highly branched at C-1 of the fructose units.

VII. MANNANS, GALACTOMANNANS AND GLUCOMANNANS (Kooiman and Kreger, 1960)

A. MANNANS

Ivory nut (*Phytelephas macrocarpa*) (Aspinall *et al.*, 1953), seeds of the ornamental palm (*Phoenix canariensis*) (Courtois and Dizet, 1964), date seed (*P. dactylifera*) (Meier, 1958; Mukherjee, 1962), Huacra Pona palm seed (*Iriartea ventricosa*) (Sowa and Jones, 1964) and *Cocos nucifera* (Mukherjee and Rao, 1962, 1964) consist of essentially linear mannans comprising β-1,4-linked mannose units only.

B. GALACTOMANNANS

Seeds of a number of leguminous plants (Smith and Montgomery, 1959; Courtois *et al.*, 1958) have been found to contain galactomannans. In those from locust bean (carob gum, *Ceratonia siliqua*) (Smith, 1948; Hirst and Jones, 1948), guar gum (*Cyamopsis* sp.) (Whistler and Durso, 1952), and Kentucky coffee bean (Larson and Smith, 1955), the galactomannan occurs as a vitreous layer on the inner side of the seed coat. The ratio of galactose to mannose varies from 45:55 to 14:86. The polysaccharide consists of β-1,4-linked mannopyranose backbone to which single α-D-galactopyranose units are attached at C-6 (Fig. 7). The degree of branching is a measure of the molecular proportions of galactose:mannose; almost equimolecular proportions of the two sugars have been reported recently for the galactomannan of the seeds of *Trifolium repens* L. (Courtois and Dizet, 1963; Horvei and Wickstrom, 1964).

—4-D-Man *p* β1—4-D-Man *p* β1—4-D-Man *p* β1—4-D-Man *p* β1—
      6   6
      | |
     1α   1α
    Gal   Gal

Galactomannan from leguminous seed.

—4-D-Man *p* β1—4-D-Man *p* β1—4-D-Man *p* β1—4-D-Glc *p* β1—4-D-Man—

Structural unit in *Orchis* glucomannan.

FIG. 7. Structural units in galacto- and gluco-mannans.

C. GLUCOMANNANS

Mucilagenous polysaccharides composed of variable proportions of glucose and mannose are found in the tubers of species of *Amorphophallus* (Iles mannan) (Smith and Srivastava, 1956), the seeds of species of *Iris* and members of the Liliaceae and in some orchid bulbs (*Cremastra*). These polysaccharides consist of essentially linear chains of β-D-1,4-linked glucose and mannose residues although side chains of single glucose and mannose residues (ratio Man:Glc = 1·5:1) attached to C-3 of glucose have been reported for *Amorphophallus konjak* (Srivastava and Smith, unpublished results). The glucomannan extracted from bulbs of french *Orchis* species (Daloul *et al.*, 1963; Petek *et al.*, 1963) has a linear molecule of β-1,4-linked mannose and glucose units in the ratio of 3:1 (see Fig. 7), terminated at the non-reducing end by D-mannopyranose or D-glucopyranose residues (cf. glucomannans of softwoods). From two species of *Orchis*, branching at C-6 of glucose has been established (Courtois *et al.*, 1963). On the other hand the glucomannans from the seeds of *Lilium candidum*, *L. henryii* (Andrews *et al.*, 1956) and *Scilla nonscripta* (Thompson and Jones, 1964) have a ratio of Man:Glc of *ca* 2:1 for the first two and of 1·3:1 for the last species, and the linear chains of β-1,4-linked units are terminated at the non-reducing end by a D-glucopyranose unit.

VIII. *Plantago* MUCILAGES (Smith and Montgomery, 1959)

Mucilages extracted from seeds of *Plantago* sp. are composed of D-xylose, L-arabinose, D-galacturonic acid and in some cases L-rhamnose or D-galactose (Fig. 8).

D-GalUA α1—2-L-Rha

In *P. ovata* and *P. arenaria*.

O-D-GalUA—L-Ara.

In *P. psyllum* and *P. fastigiata*.

FIG. 8. Structural units in *Plantago* mucilages.

At least two polysaccharides differing in their uronic acid content are present, and in this they may be compared with linseed mucilage (Hunt and Jones, 1962), with gum tragacanth, and with pectin all of which contain a neutral polysaccharide and a polyuronide.

IX. CONCLUSIONS

There are a number of other plant polysaccharides that it is impossible to include in an account of this nature (see Hegnauer, 1962). All fungal and bacterial polysaccharides have been omitted and the other main groups of polysaccharides only outlined in such a manner as to illustrate their essential similarities and differences.

It is very apparent from this brief survey that our knowledge of the detailed structure of these complex molecules is very incomplete and that this can only be regarded as an interim report. From what has been said, however, it is clear that although different types of polysaccharide may resemble one another and have the same basic structure, they also differ from one another in the types and the arrangement of the side chains and in the mode of linkage of these chains to one another and to the basic structure. It is in these differences that generic and even species differences may be found. Only in a few instances, as, for example, in the Rhodophyceae galactans, is our knowledge sufficiently complete for this to be apparent, and even in these cases much remains to be investigated.

This chapter would not be complete without a few words on polysaccharide biosynthesis. Recent investigations have shown that the phosphorylases formerly thought to catalyse the synthesis, as well as the breakdown, of starch, function in the cell mainly as degradative enzymes. Although they can be used to synthesize saccharides in the laboratory, their natural role is predominantly a catabolic one. Considerable evidence has accumulated in the last few years that glycosyl esters of nucleotides act as glycosyl donors in polysaccharide synthesis (Neufeld and Hassid, 1963; Ginsberg, 1964), and it is generally the case that the monosaccharide residues present in a polysaccharide are also found in the same plant as the glycosyls of sugar nucleotides. The reported enzymic synthesis of cellulose from guanosine diphosphate D-glucose (I)

I. *Guanosine glucopyranosyl pyrophosphate*

(Elbein *et al.*, 1964) is one of the major advances in our knowledge of polysaccharide synthesis. The biosynthetic pathway in the synthesis of the nucleotide is also of considerable importance, and it has recently been shown (Loewus, 1964) that *myo*inositol is a precursor of the pentans and pectins.

From the evidence the suggested pathway for pectin synthesis is: D-glucose—
*myo*inositol—L-bornesitol (1-methyl *myo*inositol)—methyl D-glucuronate—
methyl D-glucuronate nucleotide—methyl D-galacturonate nucleotide—
pectin. The role of glucose in this pathway has been confirmed by the use of
^{14}C-glucose in biosynthetic studies in polysaccharide synthesis (Andrews *et
al.*, 1965).

Proof of the final steps in the synthesis of the macromolecule is lacking.
In the complex highly branched polymers the different branched units may be
synthesized by a template mechanism and subsequently polymerized, or, as is
more generally believed, the main chain is synthesized by sequential trans-
glycosylation and the various side chains added by gradual apposition.

The author would like to record her gratitude to Dr. G. O. Aspinall for his
generous loan of unpublished manuscripts on the arabinogalactans and gum
exudates, and Drs. J. R. Turvey and W. A. P. Black for permission to use
unpublished results.

REFERENCES

Adams, G. A. (1960). *Canad. J. Chem.* **38**, 280.
Adams, G. A. and Bishop, C. T. (1960). *Canad. J. Chem.* **38**, 2380.
Andrews, P., Hough, L. and Jones, J. K. N. (1956). *J. chem. Soc.* 181.
Andrews, P., Hough, L. and Picken, J. M. (1965). *Biochem. J.* **94**, 75.
Anderson, N. S., Dolan, T. C. S. and Rees, D. A. (1965). *Nature, Lond.* **205**, 1060.
Araki, C. (1959). "Carbohydrate Chemistry of Substances of Biological Interest"
  (M. L. Wolfrom, ed.), Pergamon Press, 1959, p. 24.
Araki, C. (1965). 5th Internat. Seaweed Symposium (1965), Halifax, Nova Scotia,
  Abstracts, p. 3.
Archibald, A. R., Manners, D. J. and Ryley, J. F. (1958). *Chem. & Ind.* 1516.
Arni, P. C. and Percival, E. G. V. (1951). *J. chem. Soc.* 1822.
Aspinall, G. O. (1959). *Advances in Carbohydrate Chemistry* **14**, 429. Academic
  Press, New York and London.
Aspinall, G. O. (1964a). "Symposium International sur la Chemie et la Biochemie
  de la lignine, de la cellulose et des hemicelluloses.", p. 421. Grenoble, July.
Aspinall, G. O. (1964b). Recent Developments in Chemistry of Arabinogalactans.
  "International Symposium sur la Chemie et la Biochemie de la lignine, de la
  cellulose et des hemicelluloses.", p. 89. Grenoble, July.
Aspinall, G. O. and Ferrier, R. J. (1957). *J. chem. Soc.* 4188.
Aspinall, G. O. and Kessler, G. (1957). *Chem. & Ind.* 1296.
Aspinall, G. O. and Das Gupta, P. C. (1959). *J. chem. Soc.* 718.
Aspinall, G. O. and Christensen, T. B. (1961). *J. chem. Soc.* 3461.
Aspinall, G. O. and Fanshawe, R. S. (1961). *J. chem. Soc.* 4215.
Aspinall, G. O. and Greenwood, C. T. (1962). *Rev. J. Inst. Brewing* LXVIII, 167.
Aspinall, G. O. and Ballie, J. (1963). *J. chem. Soc.* 1702.
Aspinall, G. O. and Ross, K. M. (1963). *J. chem. Soc.* 1681.
Aspinall, G. O. and Wood, T. M. (1963). *J. chem. Soc.* 1686.
Aspinall, G. O. and Bhavanandan, V. P. (1965). *J. chem. Soc.* 2685; 2693.
Aspinall, G. O. and Fairweather, R. M. (1965). *Carbohydrate Res.* I, 83.
Aspinall, G. O. and Fraser, R. N. (1965). *J. chem. Soc.* 4318.
Aspinall, G. O. and Nasir-ud-din (1965). *J. chem. Soc.* 2710.
Aspinall, G. O., Hirst, E. L., Percival, E. G. V. and Williamson, I. R. (1953).
  J. chem. Soc. 3184.

Aspinall, G. O., Hirst, E. L. and Mahomed, R. S. (1954). *J. chem. Soc.* 1734.
Aspinall, G. O., Hirst, E. L. and Matheson, N. K. (1956). *J. chem. Soc.* 989.
Aspinall, G. O., Hirst, E. L. and Ramstad, E. (1958). *J. chem. Soc.* 593.
Aspinall, G. O., Hirst, E. L. and Nicolson, A. J. (1959). *J. chem. Soc.* 1697.
Aspinall, G. O., Johnston, M. J. and Stephen, A. M. (1960). *J. chem. Soc.* 4918.
Aspinall, G. O., Begbie, R. and McKay, J. E. (1962). *J. chem. Soc.* 214.
Aspinall, G. O., Charlson, A. J., Hirst, E. L. and Young, R. (1963). *J. chem. Soc.* 1696.
Aspinall, G. O., Bhavanandan, V. P. and Christensen, T. B. (1965a). *J. c hem. Soc.* 2677.
Aspinall, G. O., Johnston, M. J. and Young, R. (1965b). *J. chem. Soc.* 2701.
Aspinall, G. O., Fraser, R. N. and Sanderson, G. R. (1965c). *J. chem. Soc.* 4325.
Bean, R. C., Putman, E. W., Trucco, R. E. and Hassid, W. Z. (1953). *J. biol. Chem.* **204**, 169.
Beattie, A., Hirst, E. L. and Percival, Elizabeth (1961). *Biochem. J.* **79**, 531.
Black, W. A. P., Blakemore, W. R., Colquhoun, J. A. and Dewar, E. T. (1965). 5th Internat. Seaweed Symposium (1965), Halifax, Nova Scotia, Abstracts, p. 4.
Bishop, C. T. (1964). "International Symposium sur la Chemie et la Biochemie de la lignine, de la cellulose et des hemicelluloses." Grenoble, July.
Boggs, L. A. and Smith, F. (1956). *J. Amer. chem. Soc.* **78**, 1880.
Bouveng, H. O. (1959a). *Acta. chem. scand.* **13**, 1869.
Bouveng, H. O. (1959b). *Acta. chem. scand.* **13**, 1877.
Bouveng, H. O. (1961a). *Acta. chem. scand.* **15**, 78.
Bouveng, H. O. (1961b). *Acta. chem. scand.* **15**, 87; 96.
Bouveng, H. O. and Lindberg, B. (1958). *Acta chem. scand.* **12**, 1977.
Bouveng, H. O. and Meier, H. (1959). *Acta chem. scand.* **13**, 1884.
Bouveng, H. O., Garegg, P. J. and Lindberg, B. (1960). *Acta chem. scand.* **14**, 742.
Chanda, S. K. and Percival, E. G. V. (1950). *Nature, Lond.* **166**, 787.
Clancy, M. J., Walsh, K., Dillon, T. and O'Colla, P. S. (1960). *Proc. roy. Dublin Soc.* **AI**, 197.
Clarke, A. E. and Stone, B. A. (1960). *Biochim. biophys. Acta* **44**, 161.
Clarke, A. E. and Stone, B. A. (1963). "Chemistry and Biochemistry of β-1,3-Glucans." *Rev. Pure appl. Chem.* **13**, 134.
Conchie, J. and Percival, E. G. V. (1950). *J. chem. Soc.* 827.
Coté, R. H. (1959). *J. chem. Soc.* 2248.
Courtois, J. E. and Dizet, L. (1963). *Bull. Soc. Chim. biol.* **45**, 731.
Courtois, J. E. and Dizet, L. (1964). *Bull. Soc. Chim. biol.* **46**, 535.
Courtois, J. E., Anagnistopoulos, C. and Petek, F. (1958). *Bull. Soc. Chim. biol.* **40**, 1277.
Courtois, J. E., Daloul, M. and Petek, F. (1963). *Bull. Soc. Chim. biol.* **45**, 1255.
Cronshaw, J., Myers, A. and Preston, R. D. (1958). *Biochim. biophys. Acta* **27**, 89.
Cunningham, W. L. and Manners, D. J. (1961). *Biochem. J.* **80**, 42P.
Cunningham, W. L., Manners, D. J. and Ryley, J. F. (1962). *Biochem. J.* **82**, 12P.
Daloul, M., Petek, F. and Courtois, J. E. (1963). *Bull. Soc. Chim. biol.* **45**, 1247.
Dutton, G. G. S. and Unrau, A. M. (1963). *Canad. J. Chem.* **41**, 1417.
Eddy, B. P., Fleming, I. D. and Manners, D. J. (1958). *J. chem. Soc.* 2827.
Ehrlich, F. (1932). *Biochem. Z.* **250**, 525.
Elbein, A. D., Barber, G. A. and Hassid, W. Z. (1964). *J. Amer. chem. Soc.* **86**, 309.
Ewald, C. M. and Perlin, A. S. (1959). *Canad. J. Chem.* **37**, 1254.
Fisher, I. S. and Percival, Elizabeth (1957). *J. chem. Soc.* 2666.
Fleming, I. D., Hirst, E. L. and Manners, D. J. (1956). *J. chem. Soc.* 2831.
Ford, C. W. and Percival, Elizabeth (1964). "Third International Carbohydrate Symposium." Münster. July.

Frei, E. and Preston, R. D. (1961). *Nature, Lond.* **192**, 939.
Frei, E. and Preston, R. D. (1964). *Proc. roy. Soc. B.* (**160**), 293.
Gillham, J. K., Perlin, A. S. and Timell, T. E. (1958). *Canad. J. Chem.* **36**, 1741.
Ginsberg, J. (1964). *Advanc. Enzymol.* **26**, 35.
Goldschmid, H. R. and Perlin, A. S. (1963). *Canad. J. Chem.* **41**, 2272.
Goldstein, I. J., Hay, G. W., Lewis, B. A. and Smith, F. (1959). *Amer. Chem. Soc. Meeting Bos. Mass. Abs.* 135.
Handa, N. and Nisizawa, K. (1961). *Nature, Lond.* **192**, 1078.
Haq, S. and Adams, G. A. (1961). *Canad. J. Chem.* **39**, 1563.
Haq, N. Q. and Percival, Elizabeth (1966). *In* "Some Contemporary Studies in Marine Science" (H. Barnes, ed.), p. 365, Allen & Unwin, London.
Hegnauer, R. (1962). "Chemotaxonomie der Pflanzen," Band I. Birkhäuser, Basel und Stuttgart.
Hirase, S. (1957). *Bull. Chem. Soc. Japan* **30**, 70.
Hirst, E. L. and Jones, J. K. N. (1948). *J. chem. Soc.* 1278.
Hirst, E. L. and Rees, D. A. (1965). *J. chem Soc.* 1182.
Hirst, E. L., Percival, Elizabeth and Williams, R. S. (1958). *J. chem. Soc.* 1942.
Hirst, E. L., Percival, Elizabeth and Wold, J. K. (1964). *J. chem. Soc.* 1493.
Hirst, E. L., Mackie, W. and Percival, Elizabeth (1965). *J. chem. Soc.* 2958.
Horvei, K. F. and Wickstrom, A. (1964). *Acta chem. scand.* **18**, 833.
Hunt, K. and Jones, J. K. N. (1962). *Canad. J. Chem.* **40**, 1266.
Iriki, Y. and Miwa, T. (1960). *Nature, Lond.* **185**, 178.
Iriki, Y., Suzuki, T., Nisizawa, K. and Miwa, T. (1960). *Nature, Lond.* **187**, 82.
Jones, J. K. N. (1950). *J. chem. Soc.* 3292.
Jones, J. K. N. (1953). *J. chem. Soc.* 1672.
Jones, J. K. N. and Painter, T. J. (1959). *J. chem. Soc.* 573.
Jones, J. K. N. and Thomas, G. H. S. (1961). *Canad. J. Chem.* **39**, 192.
Jones, J. K. N. and Reid, P. E. (1963). *J. Polymer Sci.* Pt. C., No. 2, 63.
Kooiman, P. and Kreger, D. R. (1960). *Proc. Akad. Sci. Amst.* Series C. 63. No. 5, 634.
Kreger, D. R. (1960). *Proc. Akad. Sci. Amst.* Series C. 63, 613.
Larson, E. B. and Smith, F. (1955). *J. Amer. chem. Soc.* **77**, 429.
Loewus, F. A. (1964). *Nature, Lond.* **203**, 1175.
Love, J. and Percival, Elizabeth (1964a). *J. chem. Soc.* 3338.
Love, J. and Percival, Elizabeth (1964b). *J. chem. Soc.* 3345.
Love, J., Mackie, W., McKinnell, J. P. and Percival, Elizabeth (1963). *J. chem. Soc.* 4177.
Mackie, I. M. and Percival, Elizabeth (1959). *J. chem. Soc.* 1151.
Mackie, I. M. and Percival, Elizabeth (1961). *J. chem. Soc.* 3010.
McKinnell, J. P. and Percival, Elizabeth (1962a). *J. chem. Soc.* 2082.
McKinnell, J. P. and Percival, Elizabeth (1962b). *J. chem. Soc.* 3141.
Manners, D. J. and Fleming, M. (1965). *Biochem. J.* **94**, 17P.
Meeuse, B. J. D. and Kreger, D. R. (1954). *Biochim. biophys. Acta* **13**, 593.
Meeuse, B. J. D. and Kreger, D. R. (1959). *Biochim. biophys. Acta* **35**, 26.
Meier, H. (1958). *Biochim. biophys. Acta* **28**, 229.
Meier, H. (1960). *Acta chem. scand.* **14**, 749.
Meier, H. (1962). *Acta chem. scand.* **16**, 2275.
Mills, A. R. and Timell, T. E. (1963). *Canad. J. Chem.* **41**, 1389.
Morgan, K. and O'Neill, A. N. (1959). *Canad. J. Chem.* **37**, 1201.
Mukherjee, A. K. (1962). *J. Indian Chem. Soc.* **39**, 71.
Mukherjee, A. K. and Rao, C. V. N. (1962). *J. Indian Chem. Soc.* **39**, 687.
Mukherjee, A. K. and Rao, C. V. N. (1964). *Bull. Soc. Chim. biol.* **46**, 505.

Myers, A., Preston, R. D. and Ripley, G. W. (1956). *Proc. roy. Soc.* B. **144**, 450.
Neufeld, E. F. and Hassid, W. Z. (1963). *Advanc. Carbohydrate Chem.* **18**, 340.
Nunn, J. R. and von Holdt, Mrs. M. M. (1957). *J. chem Soc.* 1095.
O'Donnell, J. J. and Percival, Elizabeth (1959). *J. chem. Soc.* 2168.
O'Neill, A. N. (1954). *J. Amer. chem. Soc.* **76**, 5074.
O'Neill, A. N. (1955). *J. Amer. chem. Soc.* **77**, 6324.
Painter, T. J. (1960). *Canad. J. Chem.* **38**, 112.
Parrish, F. W., Perlin, A. S. and Reese, E. T. (1960). *Canad. J. Chem.* **38**, 2094.
Peat, S., Whelan, W. J. and Roberts, J. G. (1957). *J. chem. Soc.* 3916.
Peat, S., Whelan, W. J. and Lawley, H. G. (1958). *J. chem. Soc.* 729.
Peat, S., Turvey, J. R. and Evans, J. M. (1959). *J. chem. Soc.* 3223.
Peat, S., Turvey, J. R. and Rees, D. A. (1961). *J. chem. Soc.* 1590.
Percival, E. G. V. and Ross, A. G. (1949). *J. chem. Soc.* 3041.
Percival, E. G. V. and Percival, Elizabeth (1962). "Structural Carbohydrate Chemistry", p. 272. J. Garnet Miller Ltd.
Percival, Elizabeth (1963). "Algal Polysaccharides and their Biological Relationships." 4th Internat. Seaweed Symposium, p. 18. Pergamon Press, Oxford.
Percival, Elizabeth and Wold, J. K. (1963). *J. chem. Soc.* 5459.
Perila, O. and Bishop, C. T. (1961). *Canad. J. Chem.* **39**, 815.
Perlin, A. S. and Suzuki, S. (1962). *Canad. J. Chem.* **40**, 50.
Petek, F., Courtois, J. E. and Daloul, M. (1963). *Bull. Soc. Chim. biol.* 1261.
Rees, D. A. (1961a). *Biochem. J.* **81**, 347.
Rees, D. A. (1961b). *J. chem. Soc.* 5168.
Rees, D. A. (1962). *British Phyco. Bull. Abs.* **2**, 180.
Rees, D. A. (1963). *J. chem. Soc.* 1821.
Rees, D. A. and Conway, E. (1962a). *Nature, Lond.* **195**, 398.
Rees, D. A. and Conway, E. (1962b). *Biochem. J.* **84**, 411.
Roudier, A. J. and Eberhard, L. (1963). *Bull. Soc. Chim. biol.* 844.
Schlubach, H. H. (1953). *Experientia* **9**, 230.
Schlubach, H. H. (1958). *Fortschr. Chem. Org. Naturf.* **15**, 1.
Schwarz, E. C. A. and Timell, T. E. (1963). *Canad. J. Chem.* **41**, 1381.
Smith, D. B., O'Neill, A. N. and Perlin, A. S. (1955). *Canad. J. Chem.* **33**, 1352.
Smith, F. (1948). *J. Amer. chem. Soc.* **70**, 3249.
Smith, F. and Montgomery, R. (1959). "The Chemistry of Plant Gums and Mucilages", Reinhold Publishing Corp., New York.
Smith, F. and Srivastava, H. C. (1956). *J. Amer. chem. Soc.* **78**, 1404.
Sowa, W. and Jones, J. K. N. (1964). *Canad. J. Chem.* **42**, 1751.
Stephen, A. M. (1962). *J. chem. Soc.* 2030.
Stephen, A. M. (1963). *J. chem. Soc.* 1974.
Thompson, J. L. and Jones, J. K. N. (1964). *Canad. J. Chem.* **42**, 1088.
Thornber, J. P. and Northcote, D. H. (1962). *Biochem. J.* **82**, 340.
Timell, T. E. (1961). *Tappi* **44**, 88.
Timell, T. E. (1962a). *Svensk. Papperstidn.* **65**, 266.
Timell, T. E. (1962b). *Svensk. Papperstidn.* **65**, 435.
Timell, T. E. (1962c). *Tappi* **45**, 734; 799.
Timell, T. E. (1965). "Cellular Ultrastructure of Woody Plants," p. 127, Syracuse University Press, Syracuse, N.Y.
Turvey, J. R. (1960). *Colloq. Intern. Centre. Nat. Rech. Sci. Paris* **103**, 29.
Turvey, J. R. and Rees, D. A. (1961). *Nature, Lond.* **189**, 831.
Urbas, B., Bishop, C. T. and Adams, G. A. (1963). *Canad. J. Chem.* **41**, 1522.
Whistler, R. L. and Durso, D. F. (1952). *J. Amer. chem. Soc.* **74**, 3795, 5140.
Wu, Y. C. and Ho, H. K. (1959). *J. Chinese Chem. Soc.* **6**, 84.
Yaphe, W. (1959). *Canad. J. Bot.* **37**, 451.

CHAPTER 9

The Asperulosides and The Aucubins

E. C. BATE-SMITH AND T. SWAIN

Low Temperature Research Station, Downing Street, Cambridge

I. INTRODUCTION

The extensive chemical investigations of plants of medicinal importance which were started at the end of the eighteenth century soon led to the isolation of a large number of crystalline compounds the structures and possible inter-relationships of which naturally remained unknown for several decades. In many instances, however, it was only in the last few years or so that the detailed structures of certain compounds or groups of compounds have been completely elucidated. There are of course many reasons why this was so: the three main ones are, (*a*) difficulties in obtaining the reputed source material (often due to faulty initial identification, but cf. the occurrence of glycyphyllin in *Smilax glycyphylla*, Chapter 17, p. 30); (*b*) extreme complexity of structure (for example, strychnine was first obtained in a crystalline state in 1818 (Pelletier and Caventou), but its structure (I), was not finally settled until

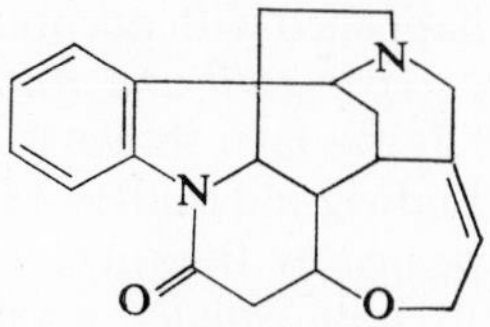

I. Strychnine

1945 (Robinson, 1952)); and (*c*) instability of the compounds, their simple derivatives or degradation products. Substances in this latter group were often difficult to purify and, after their properties had been recorded, were usually placed to one side and partially forgotten until rediscovered, often years later, by a different investigator from a different source and hence given an entirely new name. Compounds related to asperuloside (II) and aucubin (III) are prime

CH_3COOCH_2　OGlc

$HOCH_2$　OGlc

II. Asperuloside

III. Aucubin

examples of such substances: the majority of these compounds are glucosides, the aglycones of which are unstable and rapidly give intractable tars even under mild conditions of hydrolysis. It was not until the less labile members of the class were obtained and their structures determined that any real progress was made at all.

Asperuloside itself, for example, was first isolated in an impure state from madder root (*Rubia tinctorum*) in 1848 by Schunk and called chlorogenin. In 1851, Rochleder re-isolated the compound from the same source and called it rubichloric acid. In the same year Schwarz (1851) showed it to be present in *Asperula odorata*. Later the compound, as rubichloric acid, was reported in *Galium aparine, G. verum* (Schwarz, 1852), *G. mollugo* (Vielguth, 1865), *Gardenia grandiflora*, (Orth, 1855), *Morinda umbellata* (Rochleder, 1852), *Oldenlandia umbellata* (Schützenberger, 1867) and (as alstonin) in *Alstonia constricta* (Palm, 1863). Excluding the brief mention of its being present in *Oldenlandia umbellata* and *Morinda umbellata* by Perkin and Hummel (1893, 1894), there are apparently no further references (Briggs and Nichols, 1954) until Herissey isolated it in crystalline form in 1925 from *Asperula odorata* and called it asperuloside. Next year, Herissey (1926) showed that asperuloside and rubichloric acid were identical and went on to report its occurrence in several of the above plants! It is rather a pity that the original name of chlorogenin has been forgotten as it is certainly more descriptive of the compound's behaviour when heated in dilute acid solution.

A similar course of events happened with aucubin. It was first discovered in *Rhinanthus alectorolophus, R. crista-galli* and *Antirrhinum majus* and called rhinanthin by Ludwig in 1868. It was later shown to be present in *Melampyrum sylvaticum* and *M. cristatum* (Ludwig and Müller, 1872). A crystalline glucoside was isolated from *Aucuba japonica* by Bourquelot and Herissey in 1902 and called aucubin (it had actually been isolated a year earlier by Champenois, 1901). It was not until twenty years later that aucubin and rhinanthin were shown to be identical (Bridel and Braecke, 1922).

Both aucubin and asperuloside may be readily recognized in plants by the fact that on hydrolysis with dilute acid they produce first blue or green colours

(Trim and Hill, 1952; Briggs and Nichols, 1954) and subsequently, if present in sufficiently high concentration, black precipitates. Many compounds of this class, however, do not give this colour test or indeed any of the other characteristic tests devised by Trim and Hill (1952) except those for glucosides (e.g. benzidine-trichloroacetic acid (Duff *et al.*, 1965)). Thus they are less easily detected in plants or plant extracts than for example the leucoanthocyanins (Bate-Smith, 1954) and the alkaloids (Arthur and Cheung, 1960). The isolation of many compounds related to asperuloside and aucubin which do not give the typical colour reactions of these two substances has been, therefore, mainly due to the fortuitous examination of particularly rich sources. For

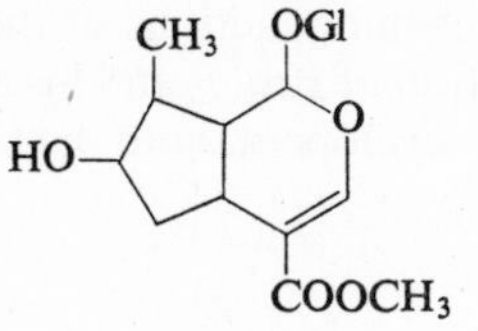

IV. Loganin

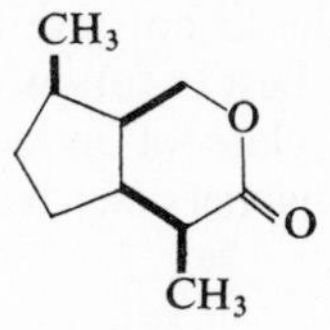

VI. Iridomyrmecin

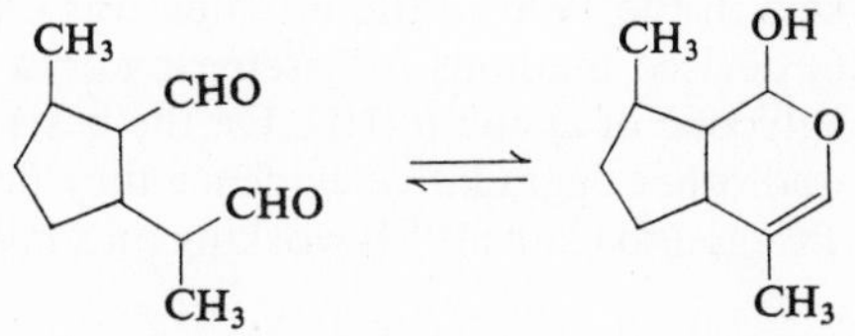

a

b

V. Iridodial

example, loganin (IV), was isolated from the pulp of the fruit of *Strychnos nux vomica* where it occurs in amounts of up to 25 per cent on a dry weight basis (Dunstan and Short, 1884). Attention has been mainly drawn to such plants either because of their supposed medicinal importance, or because of other observations, such as that the plant has a bitter taste (Korte *et al.*, 1959) or rapidly darkens on injury or death (Paris, 1954; Paris and Chaslot, 1955).

The introduction of paper and thin layer chromatography has not made the task of identifying these compounds in plant extracts any easier, although such methods have allowed ready distinction to be made between aucubin, asperuloside and other compounds which give the characteristic colour tests on paper (Paris and Chaslot, 1955; Winde, 1959; Winde and Hänsel, 1960, and Duff *et al.*, 1965), and for the separation and examination of their hydrolysis products on silica thin-layer chromatograms (Bobbitt *et al.*, 1962). It should be remembered, therefore, that when the distribution of this group of compounds is discussed below (Part IV), it is mainly the distribution of asperuloside and aucubin that is being considered.

Because aucubin and asperuloside yield blue colorations on heating in dilute acid, this class of compounds has been termed *pseudoindicans* by many workers (Hegnauer, 1964; Hänsel *et al.*, 1964; Winde, 1959). Other terms which have been proposed for the group are *melampyrosides* (J. Grimshaw, personal communication) and *iridoids* (Briggs *et al.*, 1963). The latter term was suggested on the basis of the name of the simplest compound of this class of substances, iridodial (V*a, b*), which was isolated from the common meat ant *Iridomyrmex detectus* (Cavill *et al.*, 1956). Other ants of this class (Dolichoderine ants) yield a variety of related cyclopentanoid lactones (Cavill, 1960); iridomyrmecin (VI) from *I. humulis* which was the first one to be isolated (Pavan, 1952, 1955) and have its structure elucidated (Fusco *et al.*, 1955). Since, as is pointed out above, many of the compounds isolated from plants which are related to aucubin and asperuloside do not give the chromogenic reactions of these substances, it would appear best to follow the suggestion of Briggs and his co-workers and call the whole class of monoterpenoid cyclopentanoid lactones and their biosynthetic congeners (see below) *iridoids*.

II. The Chemistry of the Iridoids

Although the chemistry of both asperuloside and aucubin was investigated by numerous workers in the 1930's little of value was obtained until Karrer and Schmidt (1946) devised methods for preparing crystalline derivatives of aucubinogen, the aglycone of aucubin (III). On the basis of the properties of these compounds and other degradative evidence they proposed the formula (VII) for aucubin. Briggs and Cain (1954) working on similar lines put forward

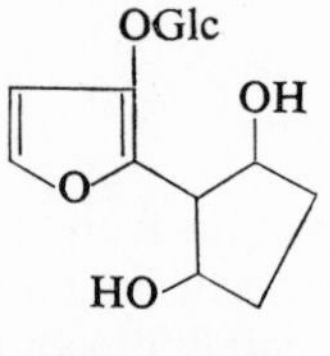

VII. Aucubin
(Karrer and Schmid, 1946)

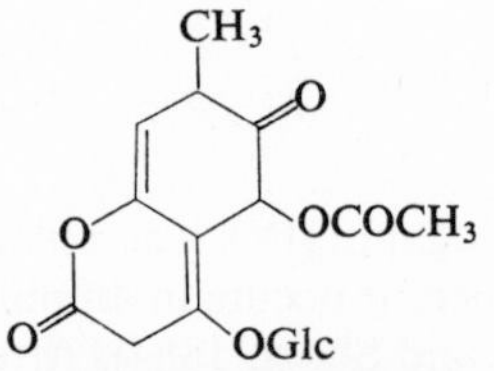

VIII. Asperuloside
(Briggs and Cain, 1954)

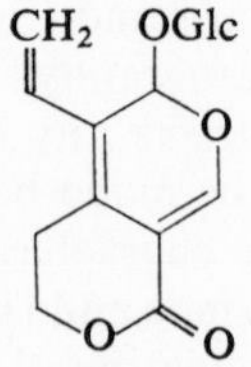

IX. Gentiopicrin

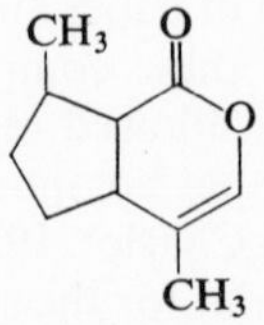

X. Nepetalactone

the structure (VIII) for asperuloside which is reminiscent of the biosynthetically related compound gentiopicrin (IX) from *Gentiana lutea* and related species whose structure was deduced in 1961 by Canonica *et al.* The now accepted formulae for asperuloside (II) and aucubin (III) were not in fact settled until 1961 (Grimshaw, 1961) and 1960 (Fujise *et al.*, 1960; Grimshaw and Juneja, 1960; Wendt *et al.*, 1960), respectively.

The difficulties in determining the structure of these two compounds (II and III) appear in retrospect somewhat laboured when we examine some of the earlier work carried out on their less labile congeners, but this only serves to reinforce the remarks made in the introduction.

We now know that the first real break-through in the chemistry of the iridoid compounds came in fact from the investigations of McElvain and his co-workers (1941) on the active principles in the oil of catmint, *Nepeta cataria*. They found that the major constituent was an acidic aldehyde $C_{10}H_{16}O_3$ which, in solution, enolized and then lactonized to give the compound nepetalactone (X). Although they deduced the essential chemistry of the lactone ring in (X), and suggested that the remainder of the molecule was a methylcyclopentane, the actual structure of this latter ring and hence of (X) itself was not settled for another thirteen years (Meinwald, 1954; McElvain and Eisenbraun, 1955; Cavill, 1960).

At about the same time, Schmidt and his co-workers were continuing their investigations mentioned earlier, on glucosides related to the aucubins, and turning their attention to *Plumeria* species which were known as a source of

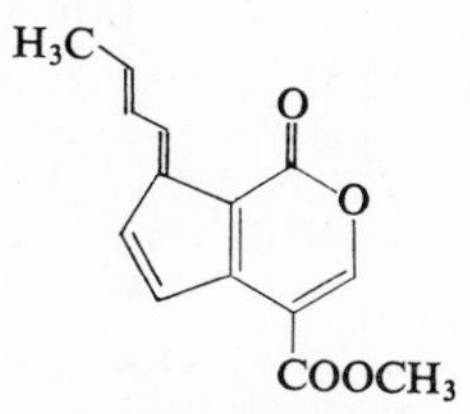

XI. Fulvoplumierin

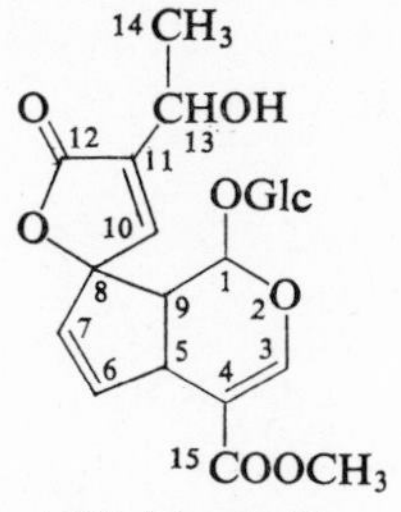

XII. Plumieride

(Numbering of side chains after Yeowell and Schmidt, 1964.)

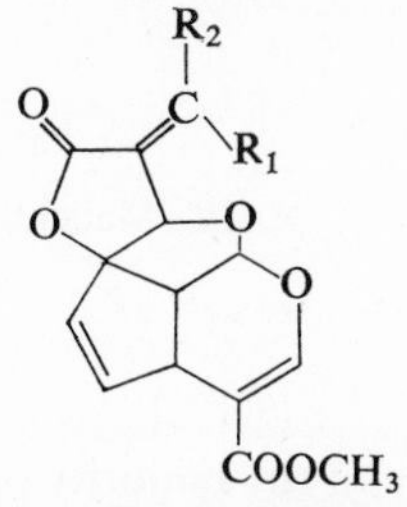

XIII. $R_1 = CH_3$, $R_2 = H$ Plumericin
XIV. $R_1 = H$, $R_2 = CH_3$ Isoplumericin

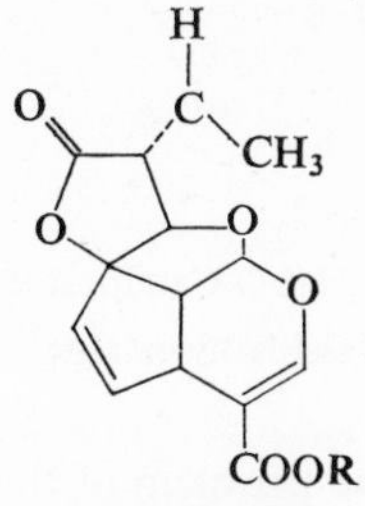

XV. $R = CH_3$
β-Dihydroplumericin
XVI. $R = H$
β-Dihydroplumericinic acid

the bitter principle agoniapicrin and of plumieride. They isolated several compounds from the bark of *P. rubra* var. *alba* and the first to have its structure determined was fulvoplumierin (XI) (Schmid and Bencze, 1953a, b; Albers-Schönberg *et al.*, 1962). The other compounds isolated were plumieride (XII), plumericin (XIII), isoplumericin (XIV), β-dihydroplumericin (XV) and β-dihydroplumericinic acid (XVI), (Halpern and Schmid, 1958; Albers-Schönberg and Schmid, 1961) and the stereochemistry of all this series has now been deduced.

The work on the Dolichoderine ant extractives has already been mentioned. Besides iridodial (V) and iridomyrmecin (VI), Cavill and his co-workers (Cavill *et al.*, 1956; Cavill, 1960) isolated isoiridomyrmecin (XVII) from *Iridomyrmex nitidus* and dolichodial (XVIII) from *Dolichoderus diceratoclinea scabridus* and related species.

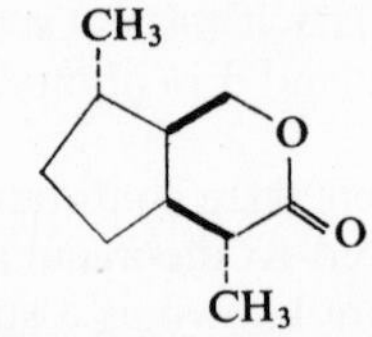

XVII. Isoiridomyrmecin

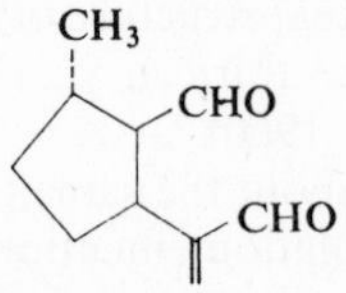

XVIII. Dolichodial

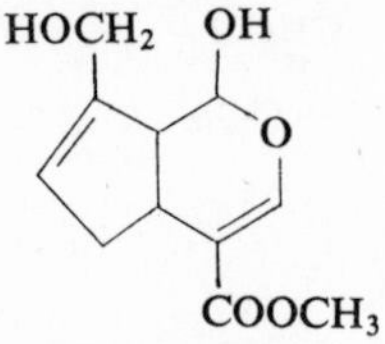

XIX. Genipin

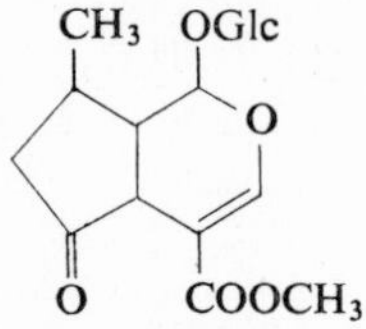

XX. Verbenalin

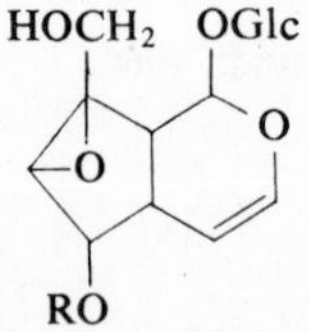

XXI. R=H, Catalpol
XXII. R=*p*-HOC$_6$H$_5$CO-Catalposide
XXIII. R=CH$_3$ 6-*O*-Methylcatalpol

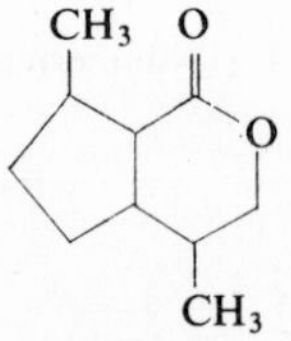

XXIV. Matatabilactone

By the year 1962 the problem of the structure of the majority of the iridoid compounds was finally settled. Besides the publications on asperuloside (II) and aucubin (III) already mentioned, structures were put forward for genipin (XIX) from the fruit of *Genipa americana* (Djerassi *et al.*, 1960, 1961); verbenalin (XX) the first of the class to be isolated (as cornin) from *Cornus florida* (Geiger,

1835) and whose structure was worked out by Büchi and Manning (1960); catalpol (XXI) and its *p*-hydroxybenzoyl ester catalposide (XXII) from *Catalpa ovata* (Bobbitt *et al.*, 1962) (incidentally *p*-hydroxybenzoic acid was first isolated from a related species and known as catalpic acid, Sardo, 1884); and loganin (meniantin) (IV) from *Strychnos nuxvomica* (Sheth *et al.*, 1961). The structure of matatabilactone (XXIV) from *Actinidia polygama*, is closely related to that of nepetalactone (X) and like the latter attracts cats and other Felidae (Sakan *et al.*, 1959). The structure of two related substances, oleuropein (probable structure XXV), the bitter blackening substance from olives (*Olea europaea*) was deduced by Panizzi and his co-workers (1960); Shasha and Leibowitz (1961), and that of swertiamarin (XXVI) from *Swertia japonica* by Kubota and Tomita (1961). The latter compound is closely related to gentiopicrin (IX) and on hydrolysis gives the dihydroisocoumarin erythrocentaurin (XXVII) (Kubota and Tomita, 1958) long believed to be a natural product (see Kariyone and Matsushima, 1927). It is interesting that gentiopicrin likewise gives rise to an artifact, the so-called alkaloid gentianin (XXVIII). It has recently been shown using [15]N (Floss *et al.*, 1964) that this compound results from the hydrolysis of gentiopicrin in the presence of ammonia. The mechanism for both these reactions has been discussed by Dean (1963). Bakankosin, which has the stucture (XXIX) according to Büchi (quoted in Büchi and Manning,

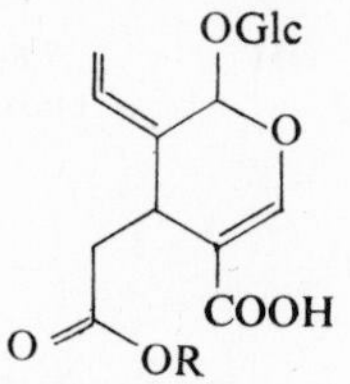

XXV. R = 3,4-diOHC$_6$H$_4$(CH$_2$)$_2$-CO-
Oleuropein

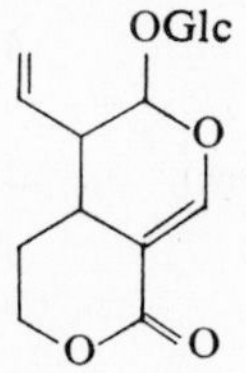

XXVI. Swertiamarin

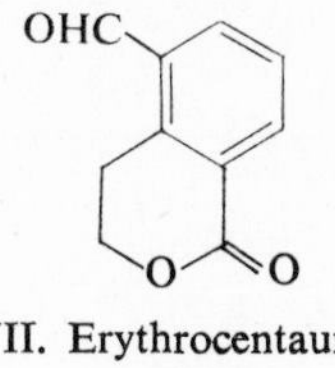

XXVII. Erythrocentaurin

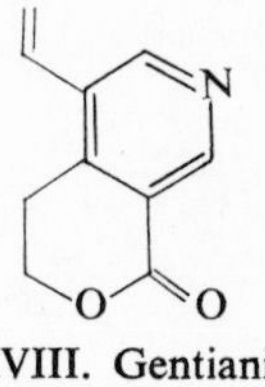

XXVIII. Gentianin

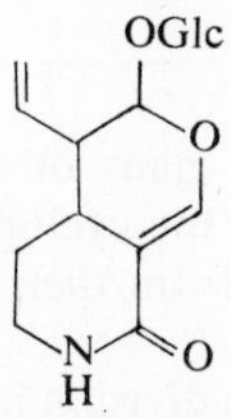

XXIX. Bakankosin

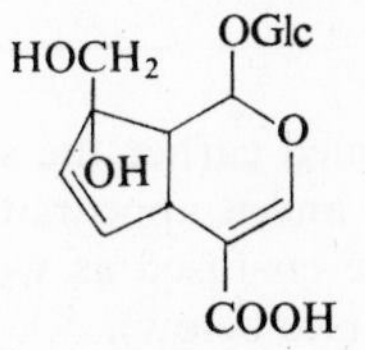

XXX. Monotropein

1960) may also arise in a similar way. In the last year or so a few other iridoids have been isolated or had their structures determined: monotropein from *Monotropa hypopitys* is (XXX) (Bobbitt *et al.*, 1964); agnuside, the *p*-hydroxybenzoate of aucubin (Winde and Hänsel, 1960) is believed to have the acid esterifying the primary alcoholic group (C_{10}, III, Hänsel *et al.*, 1964) (cf. catalposide, XII); a 6-*O*-methyl ether of catalpol (XXIII) (Duff *et al.*, 1965); and finally the part structure of harpagoside (XXXI) (Tunmann and Stierstorfer, 1964) from *Harpagophytum procumbens*.

TABLE 1. Substitution pattern of selected cyclopentane iridoids

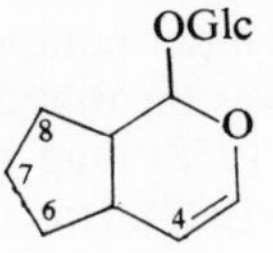

Compound	Double bond	Substitution at			
		4	6	7	8
Aucubin	Δ^7	—	OH	—	CH_2OH
Catalpol	Δ^7	—	OH	7,8-epoxy	CH_2OH
Genipin (aglycone)	Δ^7	CO_2Me	—	—	CH_2OH
Asperuloside	Δ^7	CO—O		—	CH_2OAc
Loganin	—	CO_2Me	—	OH	CH_3
Verbenalin	—	CO_2Me	=O	—	CH_3
Monotropein	Δ^6	CO_2Me	—	—	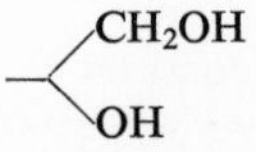 CH_2OH / OH
Plumieride	Δ^6	CO_2Me	—	—	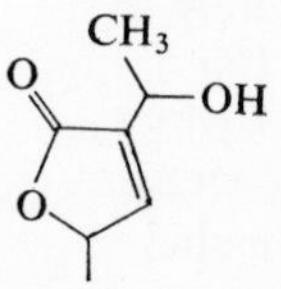 (ring structure)

As mentioned earlier the stereochemistry of many of these compounds is now known, and it appears that in the case of the cyclopentane iridoids the two rings are *cis*-fused as would be expected from their proposed mode of biosynthesis (see below).

Finally we should point out that although the division is purely artificial we can divide the cyclopentane iridoids into two groups depending on whether the substituent carbon atom at C_4 in the ring closed form of iridodial (V*b*) has been retained or not: these correspond approximately to the asperuloside (II) and aucubin (III) groups of compounds respectively. The substitution patterns in selected cyclopentane iridoids is shown in Table 1.

III. BIOSYNTHESIS OF IRIDOIDS

Although Wenkert (1962), and Wenkert and Bringi, (1959) have suggested that iridoids and related substances might be formed from prephenic acid (XXXII) by *retro* aldolization and addition of a formaldehyde unit to give finally XXXIII, it can now be confidently assumed that the compounds are in

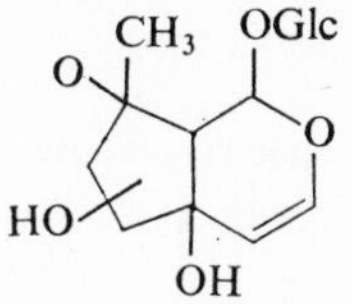

XXXI. Harpagoside

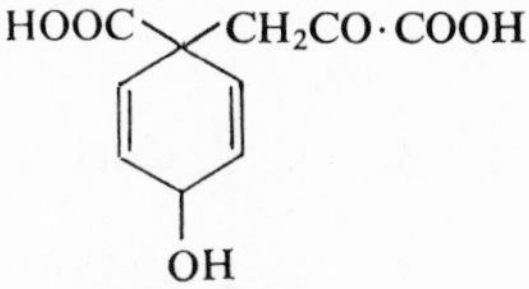

XXXII. Prephenic acid

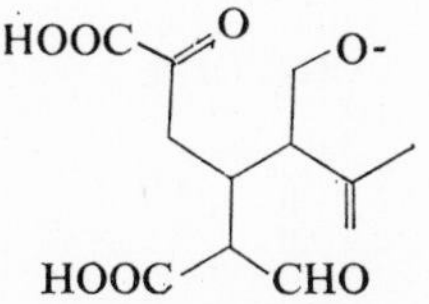

XXXIII. *Seco*-prephenate-
formaldehyde unit

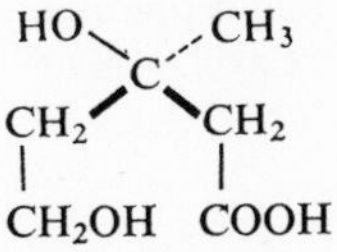

XXXIV. Mevalonic
acid

fact formed from the mevalonate pathway as for other mono-terpenoids (Thomas, 1961). Indeed, Meinwald (1954) decided between alternative structures of nepetalactone (X) on the basis of this hypothesis, and the pathway was finally proven by the use of ^{14}C-labelled mevalonate (XXXIV) in a study of the biosynthesis of plumieride (XII) by Yeowell and Schmid (1964) (Fig. 1). The side chain (C-10 to C-14) of this latter compound (XII) is synthesized from acetate as shown by the fact that the labelled atoms from acetate-1-^{14}C give activity at C-12 and C-13. Mevalonate-2-^{14}C gives half the activity in the expected C-3 and its equivalent C-15 positions and the other in the expected C-7 (although the degradation used did not distinguish between this carbon and those at C-6 and C-5).

The route from mevalonate probably goes via L-citronellal (XXXV) which can be transformed into iridodial (V) in the laboratory (Clark *et al.*, 1959). The steps from the latter compound to the asperuloside and aucubin series of compounds are unknown. Cleavage of the cyclopentane ring in iridodial (V) between carbon atoms 1 and 6 would lead to the related gentiopicrin (IX) and swertiamarin (XXVI). The outline of this pathway is shown in Fig. 1.

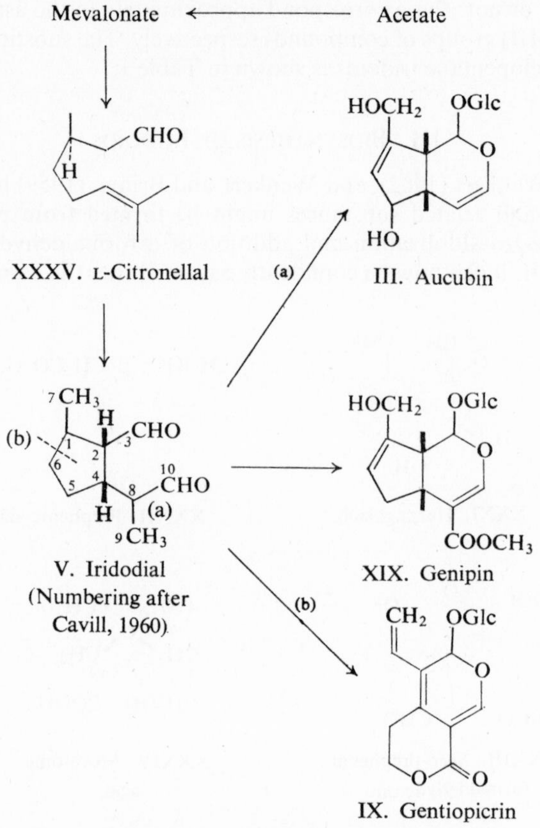

FIG. 1. Outline of the biosynthesis of the iridoids.

IV. CHEMICAL TAXONOMY OF THE IRIDOIDS

Without implying any prior assumption as to a division of the naturally occurring iridoids into two or more classes, it will be convenient to consider the systematic distribution of the two most commonly recorded of these compounds, asperuloside (II) and aucubin (III). One reason for the frequency of their recorded occurrence must be, as we have already mentioned, the ease with which their presence can be demonstrated in plant tissues by virtue of the pseudo-indican reaction, a property not possessed to the same degree by other members of this class of compounds.

There are three especially useful sources of information about the occurrences of aucubin and asperuloside: Winde (1959), Paris and Chaslot (1955) and Briggs and Nichols (1954). These and other more recent studies indicate that both of these substances are of especially frequent occurrence in a few particular families: asperuloside in the Rubiaceae, aucubin in the Scrophulariaceae, Plantaginaceae, and some other closely related families in the Sympetalae. Apart from these there are some rather widely scattered and apparently sporadic occurrences; these are represented in Table 2.

TABLE 2. The distribution of asperuloside and aucubin

Asperuloside	Aucubin
Rubiaceae	Cornaceae
Escallonia (Saxifragaceae) *Daphniphyllum* (Euphorbiaceae)	Scrophulariaceae Orobanchaceae Lentibulariaceae Globulariaceae
	Plantaginaceae
Fouquieria (Fouquieriaceae) (Asperocotillin)	*Buddleia* (Loganiaceae) *Garrya* (Garryaceae) *Eucommia* (Eucommiaceae) *Hippuris* (Hippuridaceae)

A. SUBSTANCES CLOSELY RELATED TO ASPERULOSIDE

Genipin (XIX), which occurs in the fruits of *Genipa americana* (Djerassi *et al.*, 1960, 1961), a member of the Rubiaceae, is closely similar to asperuloside (II) in structure, and from its occurrence in the same family can be presumed to have a similar biogenetic origin. Asperocotillin, which has not yet been isolated, but from colour reactions cannot be very different from asperuloside, occurs in *Fouquieria splendens* in the family Fouquieriaceae (Bate-Smith, 1954). The only way it differs from asperuloside is in its R_f value in a number of solvents. (In butanol-acetic acid-water (4:1:5), for instance, its R_f value is 0·45 compared with asperuloside 0·51.) The most likely explanation of its very close similarity to asperuloside, taking into account the difference in R_f value, is that asperocotillin has the structure (XXXVI), i.e. not acetylated at C-10 as in asperuloside. (The fact that asperocotillin is indistinguishable from asperuloside in its colour reactions raises the question as to whether many of the recorded occurrences of asperuloside might not in fact be attributable to asperocotillin or related derivatives.)

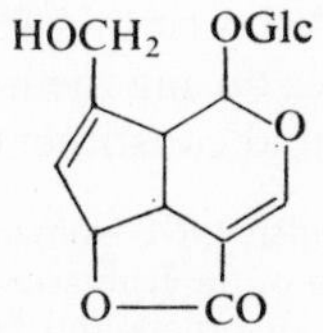

XXXVI. Asperocotillin

170   *E. C. Bate-Smith and T. Swain*

B. SUBSTANCES CLOSELY RELATED TO AUCUBIN

Catalpol (XXI) which co-occurs with aucubin in *Buddleia* and *Plantago* (Duff *et al.*, 1965), is clearly as closely related to aucubin (III) as genipin (XIX) is to asperuloside (II). Whether verbenalin (XX) is so closely related is possibly more questionable since it is rather different in structure from aucubin and catalpol, but it does co-occur with aucubin in *Cornus* species. The Verbenaceae, in which it is frequently found, are regarded as quite nearly related to the Scrophulariaceae. One species of *Verbena* in fact, contains an acylated aucubin, agnuside (Winde and Hänsel, 1960; Hänsel *et al.*, 1964).

Not so closely related either chemically or botanically are the iridoid-containing families of the Gentianales*: Gentianaceae, Loganiaceae, Menyanthaceae and Apocyanaceae; but loganin (XXIV) and plumieride (XII) may perhaps be regarded as not so very distant from aucubin (III) and verbenalin (XX). Even gentiopicrin (IX) can be seen as easily derivable from the iridoid structure by oxidative cleavage of the furan ring and lactonization with carboxyl at position C_4 as mentioned above (Fig. 1).

C. SUBSTANCES NOT CLOSELY RELATED TO EITHER ASPERULOSIDE OR AUCUBIN

As mentioned earlier nepetalactone (X) present in the Labiatae, seems to have the simplest structure of any of the plant iridoids. Monotropein (XXX), while it has chemical features found in both groups (e.g. the hydroxyl substitution at C_8 corresponding with the epoxide bond in catalpol; the absence of hydroxyl at C_5 corresponding with genipin and plumieride, and the carboxyl group at C_4 corresponding with asperuloside, see Table 1) cannot be placed in either group. By assigning it provisionally to a group of its own, the botanically isolated position of the Pyrolaceae (in the Ericales) from all other iridoid-containing families will be acknowledged.

D. THE PRESENT TAXONOMIC SITUATION

In considering the present taxonomic situation in the Angiosperms we are fortunate that a new (twelfth) edition of Engler's system has just been published (Engler, 1964). By comparing the new edition[1] with the seventh, which having been used by J. C. Willis in all editions of his Dictionary of the Flowering Plants and Ferns from 1919 up to the latest reprinting of the 6th edition in 1951 (Willis, 1960) has provided a certain stability in systematic usage, it is possible to evaluate the changes during the last fifty years in the taxonomy of the families containing the iridoid constituents.

* The Gentianales are formed (Engler, 1964) from the earlier Contortae by exclusion of the Oleaceae and *Buddleia* and inclusion of the Rubiaceae.
[1] The composition and sequence of orders and families in this edition are outlined in Hegnauer (1964) (cf. Chapter 13).

It is immediately noticeable that almost all the genera not enclosed within frames in Table 2 have had or are in prospect of having their positions revised. Two of them (*Daphniphyllum* and *Buddleia*) are now promoted into families. Several including the Rubiaceae, have been moved into different positions, as follows:

(i) Rubiaceae are included in a new order, Gentianales, comprising the previous Contortae less the Oleaceae. Thus the Rubiaceae are brought into closer relationship with other iridoid-containing families, but at the expense of separating them from the Caprifoliaceae. (ii) Fouquieriaceae from an association with the Tamaricaceae in the Parietales (now divided up and mainly replaced by Guttiferales and Violales) to a relationship with Polemoniaceae in the Tubiflorae; bringing this family also into the close assemblage of iridoid-containing families. (iii) *Buddleia* (together with 18 other genera) becomes a family Buddleiaceae, near the Scrophulariaceae in the Tubiflorae, this position being justified in part on biochemical grounds, including the presence of aucubin. (iv) *Garrya* from among the Apetalous orders with which the Syllabus begins to a position alongside the Cornaceae. (v) *Eucommia* from the Rosales to the Urticales (this would be a very good subject for further taxonomic study from a phytochemical standpoint). (vi) While *Hippuris* remains in the same position in the Myrtiflorae, there are authors who suggest a position for it in or near the Tubiflorae, which the presence of aucubin would support.

Besides the removal of *Buddleia* from the Loganiaceae, and the exclusion of the Oleaceae, another change has been made in the composition of the erstwhile Contortae: the creation of a new family, Menyanthaceae, for the genus *Menyanthes* previously in the Gentianaceae. This genus contains meniatin, which has recently been shown to be identical with loganin (XXIV).

These changes help in bringing several of the outlying genera into closer relationship with groups of similar phytochemical constitution. The only change which might be seriously questioned is moving the Rubiaceae into the Gentianales. It would seem reasonable on phytochemical grounds to look for possibilities of relationship with the other asperuloside-containing taxa—with *Escallonia* for instance—and in this way to try to identify links between the Archichlamydeae and the Sympetalae which undoubtedly must exist.

If the Rubiaceae were, in fact, removed from the Gentianales, the iridoid-containing families could be divided into three or four groups differentiated both chemically and botanically, no member of which would contain iridoid constituents found in any of the other groups. These would be: (1) The asperuloside group; (2) the aucubin-catalpol-verbenalin group; (3) the loganin-gentiopicrin-plumieride group; and possibly (4) the monotropein family, the Pyrolaceae.

These groups have certain distinguishing features over and above their different iridoid constitution. Group 1 mainly consists of woody plants, most of which contain leuco-anthocyanins. Groups 2 and 3 mainly consist of herbaceous plants, almost entirely without leuco-anthocyanins, the woody members, the Cornaceae excepted, probably (cf. Takhtajan, 1959) being secondarily woody. Alkaloids are present in many members of Group 3, these being almost absent from members of Groups 1 and 2. The Pyrolaceae and the Ericales generally are chemically distinct from the other three groups.

While these may be reasonable conclusions from the facts at present known, it seems likely that there is much more to be learnt about these constituents that will completely change the systematic picture. The chemical differences between the members of the different groups are not at all marked and they may represent a much more homogeneous assemblage when, as is to be expected, other members are discovered in many other species of plants. Hegnauer (1964) has pointed out, in fact, that with a little adjustment of Takhtajan's system (Takhtajan, 1959) the main iridoid-containing orders can be represented as a compact group stemming from an assemblage which he (Hegnauer) calls Rosiflorae. It would indeed be encouraging to the chemical taxonomist if the families in which the synthesis of iridoids occurs came to be recognized by the systematic botanist as a natural assemblage.

<h2 style="text-align:center">REFERENCES</h2>

Albers-Schönberg, G. and Schmid, H. (1961). *Helv. chim. acta* **44**, 1447.
Albers-Schönberg, G., Philipsborn, W.v., Jackman, L. M. and Schmid, H. (1962). *Helv. chim. acta* **45**, 1406.
Arthur, H. R. and Cheung, H. T. (1960). *J. Pharm., Lond.* **12**, 567.
Bate-Smith, E. C. (1954). *Biochem. J.* **58**, 122.
Bate-Smith, E. C. (1964). *Phytochemistry* 3, 623.
Bobbitt, J. M., Spiggle, D. W., Mahboob, S., Philipsborn, W. v., and Schmid, H. (1962). *Tetrahedron Letters* 321.
Bobbitt, J. M., Rao, K. V. and Kiely, D. E. *Chem. & Ind.* **1964**, 931.
Bourquelot, E. and Herissey, H. (1902). *C.R. Acad. Sci., Paris* **134**, 1441.
Bridel, M. and Braecke, M. (1922). *C.R. Acad. Sci., Paris* **175**, 532.
Briggs, L. H. and Cain, B. F. (1954). *J. chem. Soc.* 4182.
Briggs, L. H. and Nichols, G. A. (1954). *J. chem. Soc.* 3940.
Briggs, L. H., Cain, B. F. and LeQuesne, P. W. (1963). *Tetrahedron Letters* 69.
Büchi, G. and Manning, R. E. (1960). *Tetrahedron Letters* No. **26**, 5.
Canonica, L., Pelizzoni, F., Manitto, P. and Jommi, G. (1961). *Tetrahedron* **16**, 192.
Cavill, G. W. K. (1960). *Rev. Pure appl. Chem.* **10**, 169.
Cavill, G. W. K., Ford, D. L. and Locksley, H. D. (1956). *Aust. J. Chem.* **9**, 288.
Clark, K. J., Fray, G. I., Jaeger, R. H. and Robinson, R. (1959). *Tetrahedron*, 6, 217.
Champenois, G. (1901), *C.R. Acad. Sci., Paris* **133**, 885.
Dean, F. M. (1963). "Naturally Occurring Oxygen Ring Compounds". Butterworths, London.
Djerassi, C., Gray, J. D. and Kincl, F. (1960). *J. org. Chem.* **25**, 2174.
Djerassi, C., Nakano, T., James, A. N., Zalkow, L. H., Eisenbrau, E. J. and Schoolery, J. N. (1961). *J. org. Chem.* **26**, 1192.
Duff, R. B., Bacon, J. S. D., Mundie, C. M., Farmer, V. C., Russell, J. D. and Forrester, A. R. (1965). *Biochem. J.* (In press).
Dunstan, W. R. and Short, F. W. (1884). *J. chem. Soc.* **53**, 1409.
Engler, A. (1964). "Syllabus der Pflanzenfamilien", 12th Ed. Vol. 11. Borntraeger, Berlin.
Floss, H. G., Mothes, U. and Rettig, A. (1964). *Z. Naturf.* **19b**, 1106.
Fujise, S., Obara, H. and Uda, H. (1960). *Chem. & Ind.* **1960**, 289.

Fusco, R., Trave, R. and Vercellone, A. 1955 *Chim. Ind.* **37**, 958.
Geiger, P. L. (1835). *Liebigs Ann.* **14**, 206.
Grimshaw, J. (1961). *Chem. & Ind.* **1961**, 403.
Grimshaw, J. and Juneja, H. R. (1960). *Chem. & Ind.* **1960**, 657.
Halpern, O. and Schmid, H. (1958). *Helv. chim. acta* **41**, 1109.
Hänsel, R., Rimpler, H., Schoof, D. and Sirait, M. (1964). *Arch. Pharm.* **297**, 493.
Hegnauer, R. (1964). "Chemotaxonomie der Pflanzen", Vol. III, p. 29. Birkhäuser, Basel.
Herissey, H. (1925). *C.R. Acad. Sci., Paris* **180**, 1695.
Herissey, H. (1926). *J. Pharm. Chim.* **4**, 481.
Kariyone, T. and Matsushima, Y. (1927). *J. Pharm. Soc. Japan* **47**, 727.
Karrer, P. and Schmid, H. (1946). *Helv. chim. acta* **29**, 529.
Korte, F., Barkemeyer, H. and Korte, I. (1959). *Forschr. Chem. org. Naturst.* **17**, 124.
Kubota, T. and Tomita, Y. (1958). *Chem. & Ind.* 230.
Kubota, T. and Tomita, Y. (1961). *Tetrahedron Letters* 453.
Ludwig, H. (1868). *Arch. Pharm.* **186**, 64.
Ludwig, H. and Müller, H. (1872). *Arch. Pharm.* **199**, 6.
McElvain, S. M., Bright, R. D. and Johnson, P. R. (1941). *J. Amer. chem. Soc.* **63**, 1558.
McElvain, S. M. and Eisenbrau, E. J. (1955). *J. Amer. chem. Soc.* **77**, 1599.
Meinwald, J. (1954). *J. Amer. chem. Soc.* **76**, 4571.
Orth, H. V. (1855). *J. prakt. Chem.* **64**, 10.
Palm, J. (1863). *Vjschr. prakt. Pharm.* **12**, 161.
Panizzi, L., Scarpati, M. L. and Oriente, G. (1960). *Gaz. Chim. Ital.* **90**, 1449.
Paris, R. (1954). *Bull. Soc. Bot. Fr.* **93**, 159.
Paris, R. and Chaslot, M. (1955). *Ann. pharm. franç.* **13**, 648.
Pavan, M. (1952). *Trans. Ninth Int. Cong. Entomol.* **1**, p. 321.
Pavan, M. (1955). *Chim. Ind.* **37**, 625.
Pelletier, P. J. and Caventou, J. B. (1818). *Ann. Chim. Phys.* **8**, 323.
Perkin, A. G. and Hummel, J. J. (1893). *J. chem. Soc.* **63**, 1160.
Perkin, A. G. and Hummel, J. J. (1894). *J. chem. Soc.* **64**, 851.
Robinson, R. (1952). *Prog. Org. Chem.* **1**, 1.
Rochleder, F. (1851). *Ann. Chem.* **78**, 246.
Rochleder, F. (1852). *J. prakt. Chem.* **55**, 396.
Sakan, T., Fujimo, A., Murai, F., Burtsugan, Y. and Suzai, A. (1959). *Bull. Chem. Soc. Japan* **32**, 315.
Sardo, S. (1884). *Ber.* **17**, 583.
Schmid, H. and Bencze, W. (1953a). *Helv. chim. acta* **36**, 205.
Schmid, H. and Bencze, W. (1953b). *Helv. chim. acta* **36**, 1468.
Schützenberger, K. (1867). "Traité de Matières Colorants", Paris.
Schunk, W. (1848). *Ann. Chem.* **66**, 174.
Schwarz, R. (1851). *S. B. Wien Acad. Math-naturwiss.* **6**, 446.
Schwarz, R. (1852). *Ann. Chem.* **83**, 57.
Shasha, B. and Leibowitz, J. (1961). *J. org. Chem.* **26**, 1948.
Sheth, W., Ramstad, E. and Wolinsky, J. (1961). *Tetrahedron Letters* 394.
Takhtajan, A. (1959). "Die Evolution der Angiospermen". Fischer, Jena.
Thomas, R. (1961). *Tetradedron Letters*, 544.
Trim, A. R. and Hill, R. (1952). *Biochem. J.* **50**, 310.
Tunmann, P. and Stierstorfer, N. (1964). *Tetrahedron Letters* 1697.
Vielguth, P. (1865). *Vjschr. prakt. Pharm.* **5**, 187.
Wendt, M. W., Haegele, W., Simonitsch, E. and Schmid, H. (1960). *Helv. chim. Acta* **43**, 1440.

Wenkert, E. (1962). *J. Amer. chem. Soc.* **84**, 98.
Wenkert, E. and Bringi, N. V. (1959). *J. Amer. chem. Soc.* **81**, 1474.
Willis, J. C. (1960). "A Dictionary of the Flowering Plants and Ferns", 6th Edition, University Press, Cambridge.
Winde, E. (1959). Inaugural Dissertation, Free University of Berlin.
Winde, E. and Hänsel, R. (1960). *Arch. Pharm.* **293**, 556.
Yeowell, D. A. and Schmid, H. (1964). *Experientia* **20**, 250.

CHAPTER 10

The Distribution of Ranunculin and Cyanogenetic Compounds in the Ranunculaceae

H. W. L. RUIJGROK

Laboratorium voor Experimentele Plantensystematiek, Leiden, Netherlands

I. SYSTEMATICS OF RANUNCULACEAE

Most authorities agree with regard to the delimitation of the Ranunculaceae. There are only a few genera (*Paeonia, Hydrastis, Glaucidium, Kingdonia* and *Circaeaster*) whose affinities are in doubt and are much discussed in taxonomic literature; these will not be considered here. The remaining genera, which we shall call "true Ranunculaceae", are traditionally divided into two main groups, the Helleboreae and Anemoneae of Prantl (1891), the Helleboroideae and Ranunculoideae of Janchen (1949) or the Helleboraceae (including *Glaucidium* and *Hydrastis*) and the Ranunculaceae of Hutchinson (1959). The arrangement of the family according to Janchen is depicted in Table 1. Helleboroideae are characterized in the first instance by multiovulate carpels giving rise to follicular fruitlets; Ranunculoideae by uniovulate carpels producing dry achenes.

Langlet (1927, 1932) carried out extensive cytological work with the Ranunculaceae. He observed that the chromosomes fell into two sharply and markedly different groups based on both shape and size. The predominantly large, slender and bent chromosomes of many genera he called *Ranunculus*-type (R-type). The remainder of the genera characterized by short, kidney-shaped chromosomes he reunited in his *Thalictrum*-type (T-type). Gregory (1939–1941) extended the work of Langlet and proposed a classification of the family based primarily on caryotypes. He separated *Coptis* and *Xanthorrhiza* from the T-type group of genera and proposed the classification of the family shown in Table 2.

According to Gregory uniovulate carpels originated more than once from multiovulate carpels in the family. It is interesting to see that extensive morphological investigations of Ranunculaceae led Tamura, in recent years (1963) to

TABLE 1. The taxa of Ranunculaceae according to Janchen (1949)

Subfamilies	Tribes	Subtribes	Genera
Hydrastidoideae	Hydrastideae	Glaucidiinae	*Glaucidium, Hydrastis*
Paeonioideae	Paeonieae	Paeoniinae	*Paeonia*
Helleboroideae	Isopyreae	Cimicifuginae	*Xanthorrhiza, Coptis, Cimicifuga, Actaea, Anemonopsis*
		Helleborinae	*Helleborus, Eranthis*
		Isopyrinae	*Enemion, Isopyrum, Leptopyrum, Semiaquilegia, Aquilegia*
	Trollieae	Calthinae	*Caltha, Trollius*
		Nigellinae	*Nigella*
		Delphiniinae	*Aconitum, Delphinium*
Ranunculoideae	Clematideae	Thalictrinae	*Anemonella, Thalictrum*
		Anemoninae	*Anemone, Knowltonia*
		Clematidinae	*Clematis, Clematopsis*
		Kingdoniinae	*Kingdonia, Circaeaster*
	Ranunculeae	Ranunculinae	*Trautvetteria, Ranunculus, Ceratocephalus, Myosurus*
		Laccopetalinae	*Laccopetalum*
		Adonidinae	*Callianthemum, Adonis*

TABLE 2. The classification of Ranunculaceae according to caryotypes

(Gregory 1939–1941.)

Tribes	Characters	Genera
Helleboreae	R-type; $\times = 8$ (7, 6) Fruitlets follicular, capsular or rarely a berry	*Caltha, Trollius, Helleborus, Eranthis, Nigella, Actaea, Cimicifuga, Anemonopsis, Delphinium, Aconitum*
Anemoneae	R-type; $\times = 8$ (7) Fruitlets dry achenes	*Anemone, Clematis, Ranunculus, Callianthemum, Adonis*
Thalictreae	T-type; $\times = 7$ Fruitlets follicular or dry achenes	*Aquilegia, Isopyrum, Anemonella, Thalictrum*
Coptideae	C-type; $\times = 9$ Fruitlets follicular	*Coptis, Xanthorrhiza*

essentially the same conclusions. He summarized the results of his studies in the scheme shown in Fig. 1.

Being convinced that a thorough knowledge of chemical characters of plants may contribute in many instances valuable arguments for an ultimately satisfactory delimitation and classification of taxa, we started investigations on the distribution of distinct chemical compounds in Ranunculaceae. In the first instance we payed attention to the acrid principles and to cyanogenetic compounds.

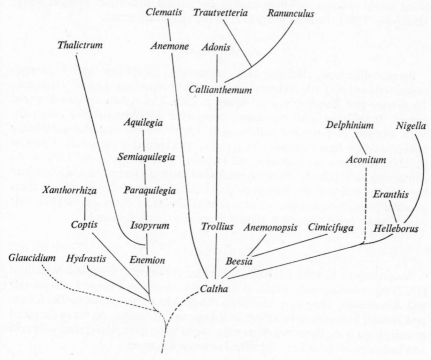

Fig. 1. The relationship of genera of Ranunculaceae according to Tamura (1962).

II. Chemical Characters of Ranunculaceae

At present no constituents characteristic for the family as a whole are known. However, besides many constituents of only sporadic occurrence, there are a number of compounds which seem to be highly characteristic for infrafamiliar taxa (genera, subtribes, tribes).

A. Phenolic Constituents

Kaempferol and quercetin are very widespread in the Ranunculaceae. They are usually accompanied by derivatives of caffeic acid (chlorogenic acid, 1-caffeoyl glucose) (Bate-Smith 1962; Harborne and Corner, 1961; Egger, 1959). Two flavonoid compounds, called ranunculetin and flavescetin occur consistently in hydrolysed extracts of the leaves of representatives of some genera (*Paeonia, Helleborus, Caltha, Anemone, Ranunculus*) but are rare or totally lacking in others (*Cimicifuga, Aquilegia, Trollius, Nigella, Aconitum, Delphinium, Adonis, Callianthemum*) (Egger, 1959); ranunculetin and flavescetin appear to be 7-glycosides of quercetin and kaempferol which originate from more complex 3,7-diglycosides by partial hydrolysis (Egger, 1961; Harborne, 1965). Tannins are absent in true Ranunculaceae.

B. ALKALOIDS

Isoquinoline-type alkaloids (e.g. berberine-, aporphine- and bisbenzyl-isoquinoline-types) are common in the genera *Aquilegia, Coptis, Hydrastis, Thalictrum* and *Xanthorrhiza* (compare p. 227). The zygomorphous-flowered genera *Delphinium* and *Aconitum* accumulate diterpene-derived alkaloids. Very recently Nijland and Uffelie (1965) have shown that the quaternary aporphine-type base magnoflorine is rather widespread in the family (compare p. 227); it occurs in *Aconitum* and in many genera considered hitherto as containing no alkaloids. Thus magnoflorine seems to form a biochemical link between the different alkaloid-structural types of Ranunculaceae (i.e. accumulators of phenylisoquinoline; of diterpene alkaloids; of damascenine [*Nigella*]; and of many genera in which alkaloids are apparently lacking).

C. SAPONINS

According to Greshoff (1909), Schneider (1930) and Gilg and Schürhoff (1932) saponins are widespread in the genera *Anemone, Clematis, Knowltonia* and *Ranunculus*. They have also been detected in some species of *Thalictrum* and *Trollius* but seem to be absent in *Adonis* and *Myosurus*. As far as chemical investigation goes, the sapogenins were found to be triterpenes; oleanolic acid and hederagenin have been identified in some instances.

D. OTHER CONSTITUENTS

Cardenolides occur frequently in *Adonis* and bufodienolides are present in *Helleborus*. Adonitol (= ribitol) seems to be a characteristic pentitol of *Adonis*, while *Delphinium* species accumulate the hexitol, mannitol. The distribution of ranunculin and cyanogenetic compounds is discussed in Section III.

III. THE DISTRIBUTION OF RANUNCULIN

Ranunculin (I) is the most characteristic compound of Ranunculaceae. So far it has never been found outside this family. Müller, (1850) and Beckurts (1892), demonstrated that acrid representatives of the Ranunculaceae on steam distillation of the fresh herb produce an acrid distillate which slowly loses acridity and at the same time crystals form which are called anemonin (III). Beckurts described such properties for some species of the genera *Anemone, Clematis*, and *Ranunculus*. Asahina and Fujita (1922) isolated the acrid precursor of anemonin from a steam distillate of *Ranunculus japonicus*

Langsd., called it protoanemonin and elucidated its structure (II). Proto-anemonin, the vesicant and acrid principle, was shown to be the lactone of γ-hydroxyvinylacrylic acid. Protoanemonin is also liberated on bruising fresh plants and this phenomenon was used by Bergmann (1944) to demonstrate the presence of vesicant principles in different species of *Ranunculus*. She supposed that protoanemonin itself is generated from a non-vesicant con-stituent present in the plant. Shearer (1938) elaborated a quantitative assay procedure for protoanemonin of buttercups. He steam-distilled fresh plants, extracted the distillate with ether and performed a lactone titration on the ether residue.

The presumed precursor of protoanemonin was isolated by Hill and van Heyningen (1951) from different species of *Ranunculus*. They showed it to be a glycoside and called it ranunculin (I). At the same time they demon-strated that protoanemonin is generated from ranunculin enzymatically or by steam distillation. According to Hill and van Heyningen ranunculin, protoanemonin and anemonin are interrelated as shown in Fig. 2. Hellström (1959) and Bredenberg (1961) confirmed the structure proposed for ranunculin by Hill and van Heyningen.

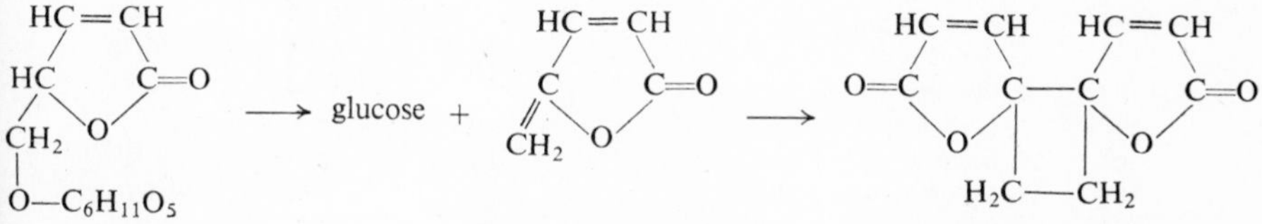

I. Ranunculin (bitter, crystalline substance; non-vesicant).

II. Protoanemonin (vesicant oil, volatile in steam, with an acrid taste).

III. Anemonin (non-vesicant crystalline substance).

FIG. 2. Relationship between ranunculin, protoanemonin and anemonin.

As ranunculin (I) is the most characteristic constituent of some members of Ranunculaceae it seemed worthwhile to us to study its distribution in the family systematically. To demonstrate its presence we used the procedures already described (Ruijgrok, 1963) which are summarized in footnote 1 to Table 3. All species investigated are shown in Table 3. Herbarium specimens were prepared at the same time from all plant samples collected; the latter were numbered and deposited in the herbarium of our laboratory.

Table 3 needs some comments. Method *b* is the most sensitive, but in a few instances traces of protoanemonin indicated to be present by this method could not be confirmed by either of the others. It also happened sometimes with small amounts of ranunculin that there was a discrepancy between method *a* and *b*. This was mainly due to the fact that ranunculin deteriorates slowly in fresh plants or in extracts even if they are kept in a refrigerator; method *b* was always carried out first, and in a few instances ranunculin detected in appreciable quantity (up to 0·13 per cent) could not be confirmed by method *a*.

7

TABLE 3. The distribution of ranunculin in Ranunculaceae

Herbarium no.	Plants investigated	Method of identification[1,2]		
		a	*b*	*c*
2090	*Aconitum altaicum*	—	0	—
2088	*A. anthora*	—	0	—
2089	*A. bicolor*	—	0	—
2200	*A. ferox*	0	0	—
2134	*A. fischeri* cv. "Wilson"	0	0	—
546	*A. lycoctonum*	0	0	—
2133	*A. lycoctonum*	—	0	—
2203	*A. lycoctonum*	—	0	—
2135	*A. napellus*	—	0	—
2136	*A. napellus*	—	0	—
2124	*A. napellus*	—	0	—
2137	*A. ranunculifolius*	—	0	—
547	*Actaea spicata*	—	0	—
550	*Adonis aestivalis*	0	0	0
552	*A. amurensis*	—	—	0
557	*A. vernalis*	—	—	0
2107	*Anemone alpina*	—	0·09	0·07
1384	*A. apennina*	+	0·73	0·78
1383	*A. apennina*	+	0·62	0·51
1590	*A. barbulata*	+	—	—
563	*A. fulgens*	+	—	—
564	*A. hepatica* var. *angulosa*	0	—	—
565	*A. hepatica*	0	—	—
2105	*A. hepatica*	—	0·07	0·07
567	*A. hupehensis* cv.	+	0·42	—
1412	*A. multifida*	+	1·52	2·10
578	*A. nemorosa*	+	1·50	—
589	*A. pulsatilla*	+	2·78	2·90
1381	*A. ranunculoides*	+	0·24	0·18
1388	*A. sylvestris*	+	0·18	0·46
2202	*A. vitifolia* var. *albadura*	—	2·07	1·82
2086	*A. vitifolia*	—	0·07	—
595	*Anemonella thalictroides*	—	0	—
601	*Aquilegia alpina* cv.	—	0	—
216	*A. vulgaris*	—	0	—
623	*Callianthemum anemonoides*	0	0	—
627	*Caltha leptosepala*	—	0	—
637	*C. palustris*	—	0	—
638	*C. palustris* var. *plena*	—	0	—
640	*C. palustris* var. *polypletala*	—	0	—
642	*Ceratocephalus falcatus*	—	0·22	—
643	*C. falcatus*	+	0·32	—
645	*Cimicifuga cordifolia*	0	0	—
646	*C. racemosa*	—	0	—
648	*C. simplex* cv. "White Pearl"	—	0	—
650	*Clematis* × *bonstedtii* cv.	+	0·31	0·40
2130	*C. fremonti*	+	4·96	4·37
2623	*C. hirsutissima*	+	2·74	2·98
2129	*C. integrifolia*	+	1·77	1·87

TABLE 3—*continued*

Herbarium no.	Plants investigated	Method of identification[1,2]		
		a	b	c
651	C. integrifolia	+	—	—
653	C. montana	+	0·26	0·27
2624	C. orientalis	+	0·82	0·93
2622	C. paniculata	+	4·80	—
2102	C. tangutica	+	0·35	0·33
662	C. vitalba	+	0·27	—
2935	C. vitalba	+	0·91	—
657	C. viticella cv.	+	0·15	0·23
2294	C. wilfordi	+	0·11	0·21
2295	C. wilfordi	+	0·28	0·34
1297	Clematopsis stanleyi	+	—	—
664	Coptis trifolia	—	0	—
2093	Delphinium ajacis	—	0	—
667	D. ambiguum	—	0	—
668	D. consolida	—	0	—
2094	D. cultorum	—	0	—
2096	D. dyctiocarpum	—	0	—
2097	D. elatum	—	0	—
2095	D. formosum	—	0	—
673	D. grandiflorum	—	0	—
3864	Eranthis hiemalis	0	0	—
686	Helleborus abchasicus	+	0·34	0·38
687	H. corsicus	+	8·75	—
688	H. corsicus	+	4·66	4·40
690	H. foetidus	+	1·40	1·69
693	H. niger	+	2·94	3·41
	Isopyrum biternatum[3]	—	0	—
	I. thalictroides[4]	—	0	—
698	Leptopyrum fumaroides	0	0	—
700	Myosurus minimus	+	0·23	—
1413	M. minimus	+	0·40	0·46
1414	M. minimus	+	0·24	0·34
702	Nigella damascena	—	0	—
2635	N. hispanica	—	0	—
922	N. sativa	—	0	—
704	Paeonia anomala	—	0	—
710	P. officinalis	—	0	—
	Ranunculus abortivus var. encyclus (Montreal)[5]	+	—	—
721	R. aconitifolius	0	—	—
1389	R. aconitifolius	+	—	—
2109	R. aconitifolius	0	0·12	0·10
	R. acris (701—Greenland)[5]	+	—	—
	R. acris (5005—USSR)[5]	+	—	—
	R. acris (702—Greenland)[5]	+	1·30	1·46
	R. acris (59—Sweden)[5]	+	—	—
	R. acris (60—Sweden)[5]	+	—	—
	R. acris (61—Sweden)[5]	+	—	—
723	R. acris	+	—	—

H. W. L. Ruijgrok

TABLE 3—*continued*

Herbarium no.	Plants investigated	Method of identification[1,2]		
		a	b	c
732	R. acris	—	1·37	1·66
736	R. acris	+	—	—
2069	R. aquatilis ssp. *aquatilis*	—	0·06	0·05
2070	R. aquatilis ssp. *aquatilis*	+	0·21	0·36
758	R. arvensis	+	1·46	1·44
765	R. auricomus	+	—	—
1380	R. auricomus	+	0·44	0·41
1382	R. auricomus	+	0·53	0·52
2080	R. baudotii	—	1·27	1·26
2081	R. baudotii	—	0·14	0·13
2082	R. baudotii	+	0·54	0·53
2083	R. baudotii	—	0·42	0·41
766	R. borealis	+	—	—
2120	R. breyninus	—	1·22	1·72
1417	R. bulbosus	+	2·60	3·17
	R. carpaticus[5]	+	—	—
787	R. chius	+	0·88	0·80
789	R. circinatus	0	0·13	—
2071	R. circinatus	0	0·04	—
2100	R. circinatus	+	0·06	0·06
791	R. ficaria	0	0·02	—
795	R. ficaria	+	0·08	—
2072	R. flammula	+	3·12	3·68
2121	R. flammula	—	2·48	2·90
818	R. fluitans	+	0·20	0·18
1387	R. fluitans	+	0·31	0·36
820	R. glacialis	+	1·05	1·20
	R. gouani[5]	+	—	—
1390	R. gramineus	+	3·47	3·27
823	R. hederaceus	+	0·50	—
1416	R. hederaceus	+	0·28	0·21
824	R. lanuginosus	+	—	—
1411	R. lanuginosus	+	0·77	0·84
2111	R. lanuginosus	+	0·11	0·09
826	R. lingua	+	1·19	—
2077	R. lingua	—	2·27	2·07
2112	R. montanus	—	0·36	0·36
2125	R. montanus	—	0·29	0·30
832	R. muricatus	+	0·30	—
834	R. neapolitanum	+	—	—
2078	R. ololeucos	+	0·60	0·64
	R. pedatifidus (Greenland)[5]	+	—	—
736	R. pedatifidus	+	0·33	—
	R. pseudohirculus (Leningrad)[5]	+	—	—
2126	R. pyrenaeus	—	0·66	0·54
	R. recurvatus (Montreal)[5]	+	0·41	0·38
856	R. repens	0	—	—
1418	R. repens	0	0·06	0·08
	R. seguierii (Austria)[5]	+	—	—

TABLE 3—*continued*

Herbarium no.	Plants investigated	Method of identification[1,2]		
		a	b	c
868	R. sardous	+	0·45	—
2079	R. sardous	+	2·38	2·00
870	R. seleratus	+	—	—
2065	R. seleratus	+	1·41	1·44
2084	R. stevenii	+	2·66	3·00
	R. sulfureus (Spitsbergen)[5]	+	—	—
1588	Semiaquilegia ecalcarata	—	0	—
889	Thalictrum aquilegifolium	—	0	—
886	T. angustifolium	—	0	—
891	T. flavum	—	0	—
893	T. galioides	—	0	—
2098	T. glaucum	—	0	—
894	T. minus	—	0	—
2201	Trollius asiaticus	—	0	—
903	T. europaeus	—	0	—
2108	T. europaeus	—	0	—
1585	T. pumilus	—	0	—
2085	T. yunnannensis	—	0	—
908	T. sinensis	—	0	—

[1] a: paper chromatographic identification of ranunculin; b: spectrophotometric estimation (extinction at 260 nm) of protoanemonin in steam distillates; c: colorimetric estimation (violet colour of ferric complex of hydroxamic acid) of protoanemonin in steam distillates.

[2] Numbers represent the calculated ranunculin content of fresh plants (%). 0: ranunculin or protoanemonin not detected; —: assay not performed; +: ranunculin present.

[3] Only sterile plants from U.S.A. available.

[4] These plants were obtained fresh from Prof. Widder (Graz) and used completely for chemical examination.

[5] These plants were kindly sent by Prof. Böcher (Copenhagen).

The ranunculin content may vary considerably within a species. This may be the result of genetic differentiation (e.g. *Ranunculus ficaria*; *R. repens*) or of variations during development. Species of the Batrachium section, for instance, always contain readily detectable amounts of ranunculin at the time they have immature fruits.

Paper chromatographic analysis (method a) demonstrated that extracts of *Anemone hepatica*, *A. sylvestris*, *Clematis wilfordi*, *Ranunculus auricomus*. *R. circinatus* and *R. repens* contained a ranunculin-like main constituent whose R_f value deviated from that of pure ranunculin. By co-chromatography of pure ranunculin with extracts of *Clematis wilfordi* and *R. circinatus*, it was proved that another ranunculin-like compound was present in these extracts. The structure of this glycoside is not yet known.

As well as ranunculaceous species, we investigated a few species not belonging to this family for the presence of ranunculin. The latter were chosen from other families of Ranunculales and from some families of monocotyledons. In no instance could the presence of ranunculin be demonstrated. The following species were concerned: Alismataceae: *Alisma lanceolatum* (2826); Araceae: *Arum maculatum* (3866); Berberidaceae: *Berberis vulgaris* (3865); Ceratophyllaceae: *Ceratophyllum submersum* (2828); Liliaceae: *Narthecium ossifragum* (3155).

IV. THE DISTRIBUTION OF CYANOGENETIC COMPOUNDS

All species of *Aquilegia* (*sensu lato*) and *Isopyrum* (*sensu lato*), so far investigated, contain cyanogenetic compounds. Furthermore some species of *Ranunculus* (especially *R. arvensis* and *R. repens*), *Thalictrum* (especially *T. aquilegifolium*, *T. dipterocarpum* and *T. polygamum*) and *Clematis* release appreciable amounts of HCN on bruising (for references see Hegnauer 1959, 1961).

We tested all species enumerated in Table 3 for the presence of cyanogenetic compounds. Our results are completely in line with the observations reported in literature. Of the plants investigated only the following ones gave a strongly positive reaction: 601 *Aquilegia alpina*, 605 *A. chrysantha*, 616 *A. vulgaris*, — *Isopyrum biternatum*, —*I. thalictroides*, 698 *Leptopyrum fumaroides*, 758 *Ranunculus arvensis*, 856/1418 *R. repens*, 1588 *Semi-aquilegia ecalcerata*, 888/889 *Thalictrum aquilegifolium*.

The only species we observed for the first time to be strongly cyanogenetic was *Isopyrum biternatum*. The chemical nature of the constituents of Ranunculaceae, which generate HCN, is still unknown.

V. DISCUSSION

We believe that the observations made so far by us prove that in Ranunculaceae some biochemical trends run parallel to morphological evolution.

Ranunculin has only been detected in the R-type group of genera (see Table 2). It occurs in high concentration in species of *Helleborus* and in many species of *Anemone* (sensu lato), *Clematis* and *Ranunculus*. It occurs also in *Ceratocephalus* and *Myosurus* and presumably in *Knowltonia*. This, at least, indicates a more intimate relationship between some genera with multiovulate pistils (*Helleborus*) and genera with mono-ovulate pistils (e.g. *Ranunculus*, *Anemone* and *Clematis*) than is assumed in the conventional arrangement of genera (Table 1). On the other hand this biochemical resemblance resulting in the production of a highly specific compound is in line with the ideas of Langlet and Gregory (Table 2).

Accumulation of cyanogenetic compounds seems to be a character of Tamura's evolutionary line *Enemion→Isopyrum→Paraquilegia→Semiaquilegia→Aquilegia* (see Fig. 1); most species of *Thalictrum* seem to have lost this character. In *Ranunculus* and *Clematis* cyanogenetic compounds are present only very sporadically. Whether HCN originates in the same way as in the *Aquilegia*-branch of Ranunculaceae is still unknown.

If we combine with our observations the results of chemical investigations reported in literature especially on alkaloids of the isoquinoline-group which are characteristic in the first instance for *Hydrastis, Coptis, Xanthorrhiza, Thalictrum* and the *Aquilegia*-branch of Ranunculaceae, we come to the conclusion that chemistry can be a valuable aid to taxonomy. The Thalictreae and Coptideae of Gregory (Table 2),'also accepted by Tamura, seem to represent natural groups. For the R-type of genera our results combined with those of Egger (1959) and Nijland and Uffelie (1965) suggest that *Helleborus, Anemone, Clematis* and *Ranunculus*, may be more intimately related than is supposed, for instance, by Tamura (Fig. 1).

ACKNOWLEDGEMENTS

We are much indebted for fresh plants to the following: Professor Beal, Columbus (Ohio); Professor Böcher, Copenhagen; Dr. Cronquist, New York; Professor Emberger, Montpellier; Professor Flück, Zürich; Professor Gloor, Leiden; Professor Widder, Graz.

Part of this investigation was supported by a grant of the Netherlands Organisation for the Advancement of Pure Research (Z.W.O.).

REFERENCES

Asahina, Y. and Fujita, A. (1922). *Acta Phytochim. (Tokyo)* **1**, 1.
Bate-Smith, E. C. (1962). *J. Linn. Soc. (Bot.)* **58**, 95.
Beckurts, H. (1892). *Arch. Pharm.* **230**, 182.
Bergmann, M. (1944). *Ber. Schweiz. Bot. Ges.* **54**, 399.
Bredenberg, J. B. (1961). *Suomen Kemistilehti* B34, 80.
Egger, K. (1959). *Z. Naturf.* **14**B, 401.
Egger, K. (1961). *Z. Naturfh.* **16**B, 430.
Gilg, E. and Schürhoff, P. N. (1932). *Arch. Pharm.* **270**, 217.
Gregory, W. C. (1939–1941). *Trans. Amer. phil. Soc.* **31**, 443.
Greshoff, M. (1909). *Kew Bull.* 397.
Harborne, J. B. (1965). *Phytochemistry* **4**, 107.
Harborne, J. B. and Janet Corner, J. (1961). *Biochem. J.* **81**, 242.
Hegnauer, R. (1959). *III. Pharm. Weekbl.* **94**, 248.
Hegnauer, R. (1961). *IV. Pharm. Weekbl.* **96**, 577.
Hellström, N. (1959). *Kgl. Lantbrukshögsk. Ann.* **25**, 171.
Hill, R. and van Heyningen, R. (1951). *Biochem. J.* **49**, 342.
Hutchinson, J. (1959). "The Families of Flowering Plants." University Press, Oxford.
Janchen, E. (1949). *Denkschr. Akad. Wiss. Wien, Math.-naturw. Kl.*, 108, 4. Abhandl. 1.
Langlet, O. (1927). *Svensk bot. Tidskr.* **21**, 1.
Langlet, O. (1932). *Svensk bot. Tidskr.* **26**, 381.
Müller, J. (1850). *Arch. Pharm.* **113**, 1.
Nijland, M. N. and Uffelie, O. F. (1965). *Pharm. Weekbl.* **100**, 49.
Prantl, K. (1891). Ranunculaceae. *In* "Die natürlichen Pflanzenfamilien." Bd. III/2, Engelmann, Leipzig.

Ruijgrok, W. L. (1963). *Planta Medica* **11**, 338.

Schneider, G. (1930). Ueber die Berücksichtigung des Saponins für die Pflanzen-systematik innerhalb der Ranunculaceentribus der Anemoneae, Diss. Berlin.

Shearer, G. D. (1938). *Vet. J.* **94**, 22.

Tamura, M. (1962). *Acta Phytotax. Geobot.* **20**, 71.

Tamura, M. (1963). *Sci. Rep. Osaka Univ.* **4**, 115.

CHAPTER 11

The Distribution of Sulphur Compounds

ANDERS KJÆR

*Department of Organic Chemistry, Royal Veterinary and
Agricultural College, Copenhagen, Denmark*

I. Introduction

A large number of organic sulphur compounds have been recorded as constituents of plant species ranging from micro-organisms and algae to higher plants. Great variation occurs in the chemical nature of these compounds, but little is known about the metabolic pathways by which they are formed or degraded in living cells. Several sulphur-containing constituents, such as cysteine, methionine, certain vitamins and coenzymes are of decisive importance for the maintenance of the steady state in living cells and for cell division and growth and are, therefore, ubiquitously distributed and hence of limited interest for chemotaxonomic consideration. Other compounds, however, are of much more restricted occurrence and hence of greater interest in the present connection. In the following discussion emphasis will be placed on selected groups of sulphur compounds, mainly from higher plants, which are believed to be of special importance from a chemotaxonomic point of view.

II. Various Important Groups of Sulphur-Containing Plant Constituents

The great majority of the organic sulphur compounds encountered in the plant kingdom contain sulphur in its lowest oxidation level, i.e. in sulphidic linkage. This applies to the metabolic key compounds such as the protein amino acids cysteine and methionine, the peptide glutathione, the widely distributed thiol ergothioneine, and certain coenzymes (thiamine pyrophosphate, coenzyme A, thioctic acid and biotin). As mentioned above most of

187

these are universally distributed in the plant kingdom. There are, however, a large number of other sulphur compounds which are of much more restricted occurrence and these will be surveyed according to their chemical character.

A. THIOLS

Many references can be found in the literature to the presence or formation of evil-smelling principles in plant material which are generally attributed to thiols or sulphides (cf. Kjær, 1963). It seems very doubtful, however, if the low molecular weight, volatile thiols are actually present as genuine plant constituents. It is more probable that they are formed by enzymic or chemical degradations of mainly unknown precursors. A distinct odour of mercaptans is associated with several members of the crucifer family, and in these cases it seems likely that cleavage of certain thioglucosides or amino acids is responsible for the formation of the volatiles. Likewise, several species of Rubiaceae (e.g. of the genera *Lasianthus* and *Coprosma*) have been reported as sources of methanethiol produced from unknown progenitors (cf. Kjær, 1963). In *Allium cepa* L. (onion) 1-propanethiol has been identified (Challenger and Greenwood, 1949), apparently arising from the enzymic fission of *S*-propylcysteine or its sulphoxide, the latter a well-established constituent of onions (Virtanen and Matikkala, 1959). A rather remarkable thiol, 3,3'-dimercaptoisobutyric acid, has been reported in *Asparagus* (Jansen, 1948), and it appears likely that this compound also arises as a product of an unknown enzymic reaction. Again, several species of the family Mimosaceae have long been reputed for their ability to produce obnoxiously smelling volatile products on being cut or bruised. This phenomenon was explained by Gmelin *et al.* (1957) who found in the seeds of one species of this family, *Albizzia lophanta* Benth, an enzyme which would degrade djenkolic acid (I), a long-known constituent of many Mimosaceae, to ammonia, pyruvic acid, and methane dithiol (II), the latter being responsible for the unpleasant odour.

$$
\begin{array}{ccc}
\ce{S-CH2-CH(NH2)COOH} & & \ce{SH} \\
\diagup & \xrightarrow{\text{Enzyme}} & \diagup \\
\ce{H2C} & \ce{H2C} & + 2\ce{CH3COCOOH} + 2\ce{NH3} \\
\diagdown & & \diagdown \\
\ce{S-CH2-CH(NH2)COOH} & & \ce{SH}
\end{array}
$$

I. Djenkolic acid   II. Methane dithiol

The few and scattered observations on hand, however, do not permit any conclusion to be drawn as to the distribution of thiol-producing compounds in nature. Yet, the fact that the thiols themselves are so readily detectable should make the further search for progenitors an equally interesting and rewarding task.

B. SULPHIDES

Numerous sulphides, including some cyclic compounds, have been recorded as constituents of plant species. Several are formed exclusively by micro-organisms, such as the sulphur-containing antibiotics (e.g. penicillin, gliotoxin, bacitracin, etc.), whereas others seem to be limited to algae or higher plants. The many recently discovered *S*-substituted cysteine-derivatives will be dealt with separately below. Naturally occurring thiophene-derivatives are being reported at an ever-increasing rate in connection with studies of acetylene compounds in higher plants; due to their close biogenetic relationship with the acetylenes they are more profitably discussed in connection with the natural distribution of the latter (Sørensen, 1963; Bu'Lock, Chapter 5).

The numerous reports on the production of simple, volatile sulphides from thetins and other sulphonium precursors in marine algae as well as in higher plants have been recently surveyed (cf. Kjær, 1963). Here again the botanical sources of the observed sulphides are so different that no taxonomically useful conclusions can be drawn. It should be realized that the same sulphide may well arise by two entirely different biosynthetic pathways, and hence in itself provide little information of systematic interest. The identification of the individual progenitors is, therefore, of obvious importance.

Again, the isolation of various di- and poly-sulphides, substituted with identical or different alkyl- and alkenyl-groups, from plants of quite unrelated character (cf. Kjær, 1963) gives little hope that these compounds themselves will provide any immediately useful chemotaxonomic information. On the other hand, a better understanding of the precursors whence they are derived should eventually lead to some useful correlations.

There seems to be a great preponderance in nature of methyl sulphides, probably a reflection of their biological derivation from methionine or the facile *in vivo* methylation of thiol-groupings. Thus the isomeric *cis*- and *trans*-3-methylthioacrylic acids occur as esterifying acids in the *S*-petasitolides A and B from *Petasites officinalis* Moench (Novotný *et al.*, 1964), and an extensive series of thioglucosides containing ω-methylthioalkyl side chains has been encountered in cruciferous species (cf. Kjær, 1963).

Recent reports of the occurrence of sulphur-containing alkaloids are of considerable interest. Thus, Nelson and Price (1952) isolated, *inter alia*, 4-methylthiocanthin-6-one (IIIa) from bark of *Pentaceras australis* Hook.f. (Rutaceae), and Achmatowicz and Bellen (1962) obtained four crystalline sulphur-containing alkaloids from *Nuphar luteum* (L.) SM. (Nymphaeaceae). One of these, thiobinupharidine, is a stereoisomeride of neothiobinupharidine, more recently isolated from the same source (Achmatowicz and Wróbel, 1964).

IIIa. 4-Methylthiocanthin-6-one   IIIb. Neothiobinupharidine

A structure was proposed for the latter (Achmatowicz *et al.*, 1964), slightly modified to (III*b*) by X-ray analysis (Birnbaum, 1965). Two other sulphur-containing alkaloids have recently been reported from an unidentified plant source (Döpke, 1965). Information on the further distribution of sulphur-containing alkaloids must be awaited with considerable interest.

C. NON-PROTEIN SULPHUR-CONTAINING AMINO ACIDS

Recent years have witnessed a considerable increase in our knowledge of the occurrence of sulphur-containing, non-protein amino acids in plants. Nearly all of these are formally derived from cysteine by *S*-substitution, and it appears likely that they are synthesized in the plant from cysteine and other suitable precursors. The individual amino acids of this type have been discussed recently (Kjær, 1963), and it suffices here to outline the main types of compound which have so far been encountered.

S-Methylcysteine and the corresponding sulphoxide are substances of rather well-defined botanical distribution. The former is found in the seeds of *Phaseolus vulgaris*, and its γ-glutamyl derivative in *Phaseolus lunatus*, whereas the sulphoxide has been repeatedly observed in Cruciferae and *Allium* species, probably formed from *S*-methylcysteine by enzymic oxidation of the sulphide precursor (Arnold and Thompson, 1962). Generally, there seems to be a striking resemblance between these two systematically remote groups of plants in their ability to produce optically active sulphoxides by stereospecific, enzymic oxidation. In this connection it may be significant that sulphoxides of both enantiomeric series now are known from the same or closely related species. By X-ray crystallography it has been established that (+)-*S*-methyl-L-cysteine sulphoxide, occurring in e.g. turnip root or cabbage juice, possesses the absolute configuration (IV) (Hine and Rogers, 1956), whereas all ω-methyl-sulphinylalkyl isothiocyanates, produced by enzymic fission of thioglucosidic

$$CH_3-\overset{O}{\underset{\overset{\displaystyle\parallel}{..}}{S}}-CH_2-\overset{\overset{\displaystyle\oplus}{NH_3}}{\underset{H}{C}}-COO^{\ominus} \qquad CH_3-\overset{O}{\underset{..}{S}}-(CH_2)_n-NCS$$

IV. (+) *S*-Methyl-L-cysteine sulphoxide   V. ω-Methylsulphinylalkyl isothiocyanates

progenitors in cruciferous species, have recently been shown to belong to the enantiomeric series (V) (Cheung, Kjær and Sim, 1965). The possible chemo-taxonomic value of this enzymic stereospecificity, suggesting the action of two different oxidative enzymes, remains to be established.

Other important *S*-substituted cysteines are reported from various species of the genus *Allium*. They comprise *S*-allyl-, *S*-propenyl, and *S*-propyl derivatives, as well as the corresponding sulphoxides, and they often appear together with their γ-glutamyl derivatives. These C₃-substituted derivatives seem to be strictly limited to the genus *Allium* with significant differences in the chemical patterns of the important species *A. cepa*, *A. sativum*, and *A. schoenoprasum*. In the former two, the cysteine-derivatives function as substrates for enzymic disproportionations leading to the characteristic lachrymatory and odoriferous principles of onion and garlic, respectively (cf. Kjær, 1963). The various compounds of this class have recently been reviewed by Virtanen (1965).

As mentioned earlier another center of sulphur-containing amino acids is the Mimosaceae. Several species of various genera have been found to contain *S*-substituted cysteines, such as djenkolic acid (I), its *N*-monoacetyl-derivative (Gmelin *et al.*, 1962), *S*-(2-carboxyethyl)-L-cysteine (VI) (Gmelin *et al.*, 1958), *S*-(2-carboxyisopropyl)-L-cysteine (VII) (Gmelin and Hietala, 1960), and dichrostachic acid (VIII) (Gmelin, 1962).

$$HOOC-CH_2CH_2-S-CH_2-CH(NH_2)COOH$$

VI. *S*-(2-Carboxyethyl)-L-cysteine

$$HOOC-CH_2-CH-S-CH_2-CH(NH_2)COOH$$
$$|$$
$$CH_3$$

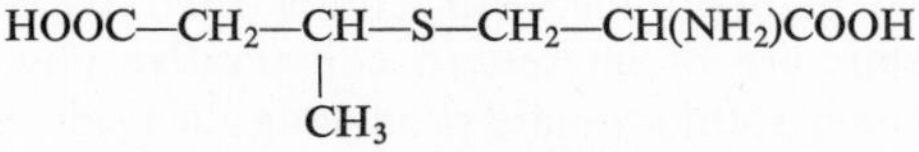

VII. *S*-(2-Carboxyisopropyl)-L-cysteine

$$S-CH_2CH(NH_2)COOH$$
$$\diagup$$
$$H_2C$$
$$\diagdown$$
$$SO_2CH_2CH(OH)COOH$$

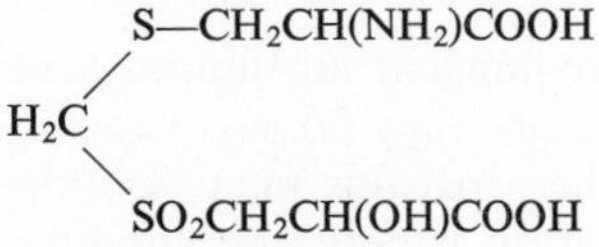

VIII. Dichrostachic acid

Again, it appears as if these amino acids are biogenetically derived from the condensation of cysteine and a hydroxy- or oxo-fragment from basal metabolism. Too few data are on hand to permit any conclusions as to the botanical distribution of these compounds. Thus far, however, none of them have been encountered in species outside Mimosaceae.

In general, species of higher plants seem to possess little ability to elaborate sulphur-containing amino acids other than those derivable from cysteine. This can be taken as a reflection of the key role of cysteine in sulphur metabolism. The records of the natural distribution of sulphur-containing amino acids in higher plants are still far too limited and scattered to permit any conclusions as to the chemotaxonomic value of the observed, rather specific occurrence patterns of the various *S*-substituted cysteines. It may well be, however, that the observed consistencies can be extrapolated to other families, thus gradually providing a valuable chemical character of taxonomic significance. All attempts to broaden and intensify the search for old and new non-protein sulphur amino acids in higher plants should therefore be encouraged. More importantly, however, our fragmentary knowledge of the biogenesis of these amino acids and their unique enzymic degradations should also be supplemented and experiments undertaken in the hope of providing data of sufficient specificity to make the compounds even more significant as chemotaxonomic characters.

D. ISOTHIOCYANATES AND THEIR GLUCOSIDIC PRECURSORS

Isothiocyanate-producing thioglucosides, possessing the general structure (IX), readily undergo enzymic hydrolysis followed by intramolecular rearrangement to give mustard oils, (X), glucose, and sulphate. They constitute a

$$R-C \begin{matrix} S-Glucose \\ \\ N-OSO_2O^{\ominus} \end{matrix} \quad \xrightarrow[H_2O]{Myrosinase} \quad R-N{=}C{=}S + Glucose + HSO_4^{\ominus}$$

IX. Thioglucosides   X. Mustard oils

rather unique class of plant products with a limited and characteristic botanical distribution and thus are of interest in comparative phytochemistry. The chemistry of the known isothiocyanate-producing glucosides has been reviewed repeatedly within recent years (cf. e.g. Kjær, 1960, 1963), and the following discussion will be mainly centered on specific data of particular interest in taxonomy.

Most isothiocyanates are pungent substances, easily detected by their biting taste. In higher concentrations they possess vesicant and, frequently, lachrymatory properties. For these reasons many isothiocyanate-producing plant species have found application as potherbs, condiments, and remedies in folk medicine. Normally, the parent thioglucosides (IX) are accompanied by the corresponding hydrolysing enzyme, myrosinase, which is contained in particular cells (idioblasts) and liberated on disintegration of the plant tissues. Histochemical methods are available for the detection of myrosinase, and these might be advantageously used for the rapid screening of a larger number of species.

Comprehensive studies during the last ten years have greatly increased our knowledge as to the chemical character of the individual glucosides. Glucose appears to be a constant sugar moiety of all naturally occurring compounds of this class and they invariably seem to conform to the general structure (IX), differing only in the chemical nature of the side-chain, R. Today, the number of structurally known isothiocyanate-producing glucosides runs close to fifty. They comprise compounds with simple straight and branched alkyl side-chains, variously hydroxylated derivatives of these, alkenyl groups, straight-chain monoketo-alkyls, ω-methylthioalkyls, the corresponding sulphoxides and sulphones, aralkyl groups and hydroxylated derivatives, and a few heterocyclic substituents. Thus, the structural variation is considerable, although many of the observed groupings are reminiscent of molecular fragments known from the metabolic pathways of the common α-amino and α-keto-acids. Important work is already in progress in several research groups, aimed at understanding the biosynthesis of the natural thioglucosides (Benn, 1962; Underhill *et al.*, 1962), and these efforts should indicate the significance of the biosynthetic pathways in the phytochemical classification of individual botanical taxa.

The most prominent source of isothiocyanate-producing glucosides is undoubtedly the family Cruciferae. On basis of the available evidence, including a systematic investigation of some 300 of the about 1500 species usually considered as belonging to this family, it can be concluded that the presence of one or more of the thioglucosides is a family-characteristic: no cruciferous species has been rigorously proved to be devoid of these compounds. Equally important is the fact that virtually all investigated species of the related families Capparidaceae, Resedaceae and Moringaceae are sources of the same general type of compound. It seems, therefore, that these families, traditionally grouped together in the order Rhoeadales, also constitute a close alliance, when judged on this chemical character. Rhoeadales *sensu* Wettstein, however, includes the large family Papaveraceae. Careful analyses of a series of species of this latter family have failed to reveal even the smallest trace of thioglucosides. This fact, in conjunction with other conspicuous differences between the phytochemical composition of Papaveraceae and the remaining families of Rhoeadales *sensu* Wettstein, lends strong support to the idea that Papaveraceae may be of entirely different ancestry. It is interesting that Tákhtajan (1959), on these and other grounds, reached the same conclusion. Hutchinson's formulation (1959) of Capparidaceae and Moringaceae as allied families, differing in origin from Cruciferae, Resedaceae and Papaveraceae, is certainly not supported by the chemical patterns of these families.

Other families, with no obvious systematic relationship to Rhoeadales *sensu stricto*, also seem to be consistent sources of thioglucosides. This is true for Tropaeolaceae, Salvadoraceae, Limnanthaceae, and Caricaceae, with the reservation that only a limited number of species have yet been studied. There is much interest attached to further chemical studies of these families for thioglucosides and other types of constituents. Apart from these last-named families, a number of isolated species from widely differing families have been listed as sources of thioglucosides, such as *Jatropha multifida* L. and *Putranjiva roxburghii* Wall. (Euphorbiaceae), *Codonocarpus cotinifolius* (Phytolaccaceae), and *Plantago major* L. (Plantaginaceae). In these cases it has been reported that quite closely related species are often devoid of glucosides and no explanation, other than parallelism or convergence, can be offered for these sporadic occurrences. In general, many more species must be examined before we can properly evaluate the more detailed distribution patterns of the isothiocyanate-producing glucosides.

III. CONCLUDING REMARKS

In conclusion it can be stated that the organic sulphur-compounds produced in the plant kingdom are numerous and diverse in their chemical structure. Continued efforts are being made to locate and identify new sulphur compounds and partly due to the rapid development of new and efficient analytical tools, important new discoveries will undoubtedly be forthcoming at an ever-increasing rate. To what extent mere chemical knowledge, even of a greatly increased number of sulphur compounds, will provide help for taxonomic problems remains to be seen. It appears likely that, here, as in practically all provinces of phytochemistry, a far higher return may result from efforts to increase our insight into the biosynthetic pathways, including the highly specific enzyme systems, of the individual sulphur compounds in living cells. It is to be expected that these synthetic sequences will prove to be more characteristic for limited groups of taxa than just the chemical end-products, and hence provide a far more secure basis for the assistance which natural product chemists are anxious to offer to the taxonomists. In the field of sulphur-containing plant products, therefore, nothing can be more important in the future than to increase the efforts to augment our understanding of the general, as well as the more specialized, pathways utilized by the plant from its uptake of sulphur as sulphate ions all way through to the final steps in its production of this very large, important, and diverse group of organic compounds.

REFERENCES

Achmatowicz, O. and Bellen, Z. (1962). *Tetrahedron Letters* 1121.
Achmatowicz, O. and Wróbel, J. T. (1964). *Tetrahedron Letters* 129.
Achmatowicz, O., Banaszek, H., Spiteller, G. and Wróbel, J. T. (1964). *Tetrahedron Letters* 927.
Arnold, W. N. and Thompson, J. F. (1962). *Biochim. biophys. Acta* **57**, 604.
Benn, M. H. (1962). *Chem. & Ind.* 1907.
Birnbaum, G. I. (1965). *Tetrahedron Letters* 4149.
Challenger, F. and Greenwood, D. (1949). *Biochem. J.* **44**, 87.
Cheung, K. K., Kjær, A. and Sim, G. A. (1965). *Chem. Comm.* 100.
Döpke, W. (1965). *Naturwissenschaften* 133.
Gmelin, R. (1962). *Hoppe-Seyl. Z.* **327**, 186.
Gmelin, R., Hasenmaier, G. and Strauss, G. (1957). *Z. Naturf.* **12b**, 687.
Gmelin, R. and Hietala, P. K. (1960). *Hoppe-Seyl. Z.* **322**, 278.
Gmelin, R., Kjær, A. and Olesen Larsen, P. (1962). *Phytochemistry* **1**, 233.
Gmelin, R., Strauss, G. and Hasenmaier, G. (1958). *Z. Naturf.* **13b**, 252.
Hine, R. and Rogers, D. (1956). *Chem. & Ind.* 1428.
Hutchinson, J. (1959). "The Families of Flowering Plants", Vol. 2. Oxford University Press, London.
Jansen, E. F. (1948). *J. biol. Chem.* **176**, 657.
Kjær, A. (1960). *Fortschr. Chem. org. Naturst.* **18**, 122.
Kjær, A. (1963). *In* "Chemical Plant Taxonomy" (T. Swain, ed.), p. 453. Academic Press, London and New York.
Nelson, E. R. and Price, J. R. (1952). *Aust. J. sci. Res.* **5A**, 768.
Novotný, L., Herout, V. and Šorm, F. (1964). *Coll. Czech. Chem. Commun.* **29**, 2182.
Sørensen, N. A. (1963). *In* "Chemical Plant Taxonomy" (T. Swain, ed.). Academic Press, London and New York.
Tákhtajan, A. (1959). "Die Evolution der Angiospermen". G. Fischer, Jena.
Underhill, E. W., Chisholm, M. D. and Wetter, L. R. (1962). *Canad. J. biochem. Physiol.* **40**, 1505.
Virtanen, A. I. (1965). *Phytochemistry* **4**, 207.
Virtanen, A. I. and Matikkala, E. J. (1959). *Acta chem. scand.* **13**, 1898.

CHAPTER 12

Amino Acids and Related Compounds

E. A. BELL

Department of Biochemistry, King's College, London

I. INTRODUCTION

The first naturally occurring amino acid to be described, asparagine, was originally obtained from the juice of asparagus (Delaville, 1802; Vauquelin and Robiquet, 1806), and several others, such as glutamine (Schulze, 1877), phenylalanine (Schulze and Barbieri, 1879) and arginine (Schulze and Steiger, 1886a, b), were also first discovered in the free state in plant extracts.

When the relationship between the amino acids and the proteins became recognized towards the end of the last century however, interest in the amino acids was largely directed towards the isolation and characterization of those present in protein hydrolysates, and to the determination of the amino acid composition of proteins from different sources.

The methods used in this work were those of classical organic chemistry which involved the preliminary isolation and purification, not only of unknown amino acids for characterization, but also of known amino acids for identification. To isolate a sufficient quantity of a pure amino acid for such purposes was frequently a difficult and tedious procedure. Nevertheless all the amino acids normally found in proteins were isolated and characterized by such methods. The last major ones, asparagine and glutamine, being isolated from hydrolysates of plant proteins in 1932 (Damodaran, 1932; Damodaran *et al.*, 1932). Comparative studies suggested that these amino acids (about twenty in number) were present in the free or bound form in all living organisms.

Information about other amino acids which are not usually found in protein hydrolysates was obtained much more slowly, and at the time when asparagine

195

and glutamine were isolated from protein hydrolysates only about six "non-protein" amino acids and an equal number of betaines had been found in plants (Greenstein and Winitz, 1961). Moreover, these were known only as rare constituents of isolated species.

It is scarcely surprising then that early workers in the field of chemical taxonomy should have largely disregarded the amino acids. Even had they been readily identifiable, the majority were apparently so common and the minority so rare that studies of their distribution could not have been expected to yield information of much systematic value.

A. CHROMATOGRAPHY AND IONOPHORESIS

The development of paper chromatography, and more recently of complimentary methods such as paper ionophoresis and thin-layer chromatography, altered this situation in several fundamental respects. They provide simple, sensitive techniques for detecting different amino acids simultaneously; they can be used for the rapid analysis of small volumes of extracts and thus enable large numbers of species to be examined in a short period of time; and finally their use reveals that "non-protein" amino acids are far more numerous than had been supposed. Over one hundred of these have now been isolated and characterized (Fowden, 1964), many of them from higher plants.

The first use of a chromatographic survey of amino acids for taxonomic purposes was made by Buzzati-Traverso and Rechnitzer (1953) who determined the amino acid composition of fish muscle protein from various species. These workers reported a general correlation between their findings and the accepted classification of the species, although geographical variations between different members of the same species were found in certain instances.

Reuter (1957) applied this technique to botanical material when he studied the distribution of the soluble nitrogen compounds in various organs of 166 species representing 48 plant families. He noted the presence of high concentrations of particular free amino acids (both "protein" and "non-protein") in certain families and suggested that these "principal amino acids" (*hauptaminosäuren*) were of systematic significance. Of the "protein" amino acids he reported arginine as being the "principal amino acid" of the Saxifragaceae, Hamamelidaceae and Rosaceae and proline as that of the Papilionatae.

B. TAXONOMIC SIGNIFICANCE OF DIFFERENT AMINO ACIDS

While the occurrence of high concentrations of "protein" amino acids in plants may be of systematic significance, this significance is not always easy to assess because all species are able to synthesize such amino acids and variations in their concentration may be due either to genetically controlled differences or to differing environmental conditions.

Studies of the distribution of the less commonly occurring "non-protein" amino acids would seem therefore to be potentially more valuable for taxonomic purposes, as the presence of such compounds, irrespective of their concentration, provides unequivocal evidence of inherited differences between the species which do and do not contain them. The actual systematic value of such studies cannot, however, be assessed until they have been completed and the distribution patterns of the particular compounds established. An amino acid may, like azetidine-2-carboxylic acid (I) [reported by Reuter (1957) as the "principal amino acid" of the Liliaceae and found by Fowden and Steward (1957) to occur only in that family and the closely related Agavaceae and Amaryllidaceae] be restricted to one or a few closely related families. Another may be restricted, like lathyrine (II), to a single genus (Alston and Turner, 1963) or to some small sub-group of related species. In such instances the presence of the particular amino acid may only serve to emphasize a phylogenetic relationship which clearly exists already. It may also, however, be of value in assigning a species to a particular genus or family, or in confirming or establishing a relationship between different genera or families. Alternatively a "non-protein" amino acid may occur in taxonomically remote species. For example, δ-acetylornithine (III) the "principal amino acid" of the Fumariaceae

$$CH_2$$
$$H_2C \qquad CH \cdot COOH$$
$$N$$
$$H$$

I. Azetidine-2-carboxylic acid

$$H_2N \quad N \quad CH_2{-}CH{-}COOH$$
$$N \qquad NH_2$$

II. Lathyrine

$$CH_3CO \cdot NH(CH_2)_3 \cdot CH \cdot COOH$$
$$NH_2$$

III. δ-Acetylornithine

$$H_2N \cdot C \cdot NH \cdot O \cdot (CH_2)_2 \cdot CH \cdot COOH$$
$$\| \qquad\qquad\qquad\quad$$
$$NH \qquad\qquad NH_2$$

IV. Canavanine

has been identified in such different plants as the fern *Asplenium nidum* (Virtanen and Linko, 1955) and the grass *Brachypodium sylvaticum* (Fowden, 1958).

Other types of distribution have been discussed previously (Bell and Fowden, 1964), but these present examples will suffice to show that the occurrence of an unusual amino acid in different plant species may be due either to a relatively recent evolutionary divergence of these species, a divergence likely to be reflected in morphological similarities, or to the independent evolution of similar biosynthetic pathways in species which may have little else in common.

Even where an amino acid is restricted to a single genus or family, the fact that it does not occur in all species of that genus or family may throw doubt on the systematic significance of its distribution. Tschiersch (1959) doubted whether the sporadic distribution of canavanine (IV) in the Papilionatae was of systematic significance while Birdsong *et al.* (1960) took the opposite view and suggested that its absence from various species and indeed genera of that sub-family might indicate only a minor biochemical difference, possibly the lack of a single enzyme in those species without the amino acid.

II. Variation in Amino Acid Biosynthesis in the Papilionatae

In an attempt to elucidate biochemical differences between species of the same genus and to discover if comparable differences occurred between closely related genera, surveys have been made in my laboratory of the free amino acids and related compounds present in high concentration in the seeds of 52 species of *Lathyrus* and 42 species of *Vicia*. These two genera of the Papilionatae were chosen after earlier surveys had failed to reveal any canavanine-containing species in *Lathyrus* and had shown that *Vicia* contained some species with, and some without that amino acid. Seeds were used since they were readily obtainable, contained high concentrations of free amino acids and provided, as nearly as possible, a comparable stage in the life-cycles of the different species.

The results of these surveys (summarized in Tables 1 and 2) showed that both genera could be sub-divided on the basis of the free amino acids and related compounds occurring in high concentrations in the seeds of the different species. The sub-genera so formed were not characterized by the presence of a single ninhydrin-reacting compound in arbitrary concentration (usually the least amount detectable by the analytical procedure used), but rather by the presence of associations of such compounds which showed themselves as characteristic patterns of spots after chromatography or ionophoresis on paper.

Without knowing the identity of the compounds involved, the systematic value of these characteristic patterns was as limited as that of any other single superficial variable character. By identifying the unknowns and establishing their biosynthetic relationships, however, it should be possible to obtain a fuller "phylogenetic" understanding of the relationships between the different groups of species. Although the elucidation of these problems is far from complete, much progress has been made during the last three years which allows us now to see the relationships more clearly.

A. *Lathyrus* SPECIES

Of the eleven "non-protein" amino acids or related compounds found in high concentrations in various species of *Lathyrus* (Bell, 1962a), β-(γ-glutamyl-amino)propionitrile (V) had been characterized previously (Schilling and Strong, 1955), lathyrine (II) had been isolated (Bell, 1961) but not characterized, while homoarginine (VI) had been tentatively identified from R_f values (Bell, 1962a) but not isolated.

The structure of lathyrine (II) has since been established as β-(2-amino-pyrimidin-4-yl)alanine (Bell and Foster, 1962), the presence of L-homoarginine (VI) confirmed by isolation (Bell, 1962b) three of the unknowns identified unequivocally as α,γ-diaminobutyric acid (VII) (Ressler *et al.*, 1961), α-amino-β-oxalylaminopropionic acid (VIII) (Adiga *et al.*, 1963; Murti *et al.*, 1964) and γ-hydroxyhomoarginine (IX) (Bell, 1964a) and a fourth tentatively as α-amino-γ-oxalylaminobutyric acid (X) (Bell, 1964b). The distribution of

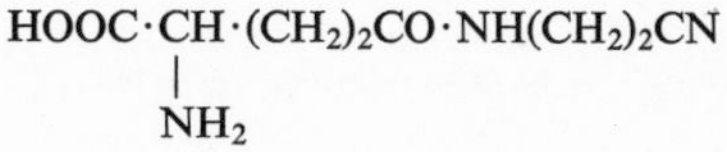

$$HOOC \cdot CH \cdot (CH_2)_2CO \cdot NH(CH_2)_2CN$$
$$| $$
$$NH_2$$

V. β-(γ-Glutamylamino)propionitrile

$$H_2N \cdot C \cdot NH \cdot (CH_2)_4 \cdot CH \cdot COOH$$
$$\| \qquad\qquad\qquad | $$
$$NH \qquad\qquad\qquad NH_2$$

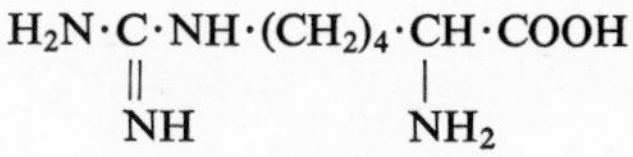

VI. Homoarginine

$$H_2N \cdot (CH_2)_2 \cdot CH \cdot COOH$$
$$| $$
$$NH_2$$

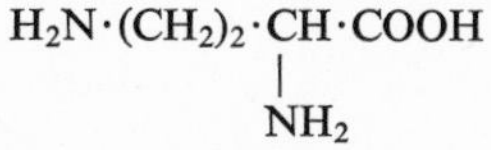

VII. α, γ-Diaminobutyric acid

$$HOOC \cdot CO \cdot NH \cdot CH_2 \cdot CH \cdot COOH$$
$$| $$
$$NH_2$$

VIII. α-Amino-β-oxalylaminopropionic acid

$$H_2N \cdot C \cdot NH \cdot (CH_2)_2 \cdot CHOH \cdot CH_2 \cdot CH \cdot COOH$$
$$\| \qquad\qquad\qquad\qquad\qquad\qquad | $$
$$NH \qquad\qquad\qquad\qquad\qquad\qquad NH_2$$

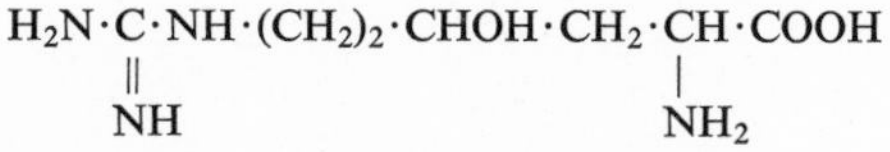

IX. γ-Hydroxyhomoarginine

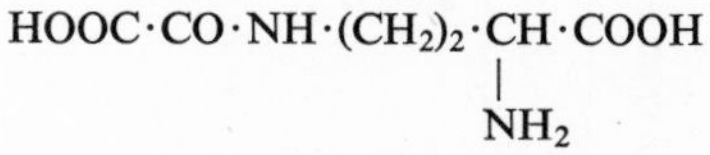

$$HOOC \cdot CO \cdot NH \cdot (CH_2)_2 \cdot CH \cdot COOH$$
$$| $$
$$NH_2$$

X. α-Amino-γ-oxalylaminobutyric acid

these and the remaining four unknowns is set out in Table 1. A schematic representation of the genus based on the distribution of these compounds, and what is known of their biosynthesis, is given in Fig. 1. This figure shows a primary subdivision of *Lathyrus* into species containing α, γ-diaminobutyric acid (VII) and species containing L-homoarginine (VI); although some species of each group contain α-amino-β-oxalylaminopropionic acid (VIII). Those species of the α, γ-diaminobutyric acid-containing group which contain (VIII) also contain an oxalyl derivative of α, γ-diaminobutyric acid (X) suggesting that the oxalylating enzyme(s) involved may be non-specific.

The group of species containing β-(γ-glutamylamino)propionitrile (V) is shown tentatively as an offshoot of the α, γ-diaminobutyric acid-containing group. That these two groups may be related is suggested by recent work of Butler and Conn (1964) indicating that α-amino carboxylic acids can undergo decarboxylation and dehydrogenation to give nitriles in higher plants. Alternatively β-aminopropionitrile may be formed by the decarboxylation of β-cyano-alanine (XI) and the labelling of the glutamyl derivative in *L. odoratus* after supplying the plant with $H^{14}CN$ (Tschiersch, 1964) could be explained in terms of such a mechanism.

The species containing L-homoarginine (VI) are again divided into two groups. Those containing high concentrations of the free amino acid and those in which a lower concentration of homoarginine is accompanied by varying concentrations of γ-hydroxyhomoarginine (IX) and lathyrine (II). Using ^{14}C labelled L-homoarginine and γ-hydroxyhomoarginine (Bell, 1964a; Bell and Przybylska, 1965), it has been shown that γ-hydroxyhomoarginine is an intermediate in the biosynthesis of lathyrine (II) from homoarginine (VI) in *L. tingitanus* and the biochemical difference between the two groups of species, with respect to these amino acids, is apparently the presence or absence of the enzyme systems required for the hydroxylation of homoarginine and subsequent cyclization to give lathyrine. The eleven species of Group 3 (Table 1)

TABLE 1. Amino-acids and related compounds in seeds of *Lathyrus* species

Group	Species	Compound present[1]											
		B₁	B₂	B₃	B₄	N₁	N₂	A₁	A₂	A₃	A₄	Lat.	Arg.
1	*aurantius*		+										T
	luteus		+										T
	laevigatus sp. *aureus*		+										T
	sylvestris		+					+	+	T			+ +
	latifolius		+					+	+	T			+ +
	heterophyllus		+				T	+	+	T			+ +
	gorgoni		+					+	+	T			T
	grandiflorus		+					+	+	T			+
	cirrhosus		+					+	+	T			+
	rotundifolius		+				T	T	+	T			+
	tuberosus		+				T	T	+	T			+
	multiflora		+						+				+
	undulatus		+				+		+				
2	*alatus*	+					T	+					T
	articulatus	+						+					T
	arvense	+						+					+
	setifolius	+						+					T
	pannonicus	+						+					T
	ochrus	+						+					T
	clymenum	+						+					T
	sativus	+						T					T
	megallanicus	+						T					T
	quadrimarginatus	+						T					T
	cicera	+											
3	*pratensis*	+										T	
	laevigatus sp. *occidentalis*	T										T	+

varius	+										T
niger	+									+	T
machrostachys	+									+	T
maritimus	+			+					+	+	T
aphaca	+	+		+						+	T
sphaerious	+	+		+						+	T
tingitanus	+	+		+					+	+	
cyanus	+	+							+	+	
alpestris	+			+					+	+	
variegatus	+			+					+	+	
venetus	+								+	T	
inconspicuous	+								+	T	
incurvus	++								+		+
4 *vernus*	+++	+++									T
montanus	+++	++									T
palustris	++	+									
aureus	+++	+									
neurobolus	++	++									
nissolia	++	++									
5 *roseus*		T									
hirsutus		+									
odoratus	+	+									+
pisiformis²	T										+
annuus²	T										+
angulatus²	T										+
laetiflorus	T									+	+
venosus	+										T

¹ T = trace; B₁ = homoarginine (VI); B₂ = α,γ-diaminobutyric acid (VII); B₃ = γ-hydroxyhomoarginine (IX); B₄ = unidentified; N₁ = β-(γ-glutamylamino)propionitrile (V); N₂ = unidentified; A₁ = α-amino-β-oxalylaminopropionic acid (VIII); A₂ = α-amino-γ-oxalylamino-butyric acid (X) (tentative); A₃ = second acidic derivative of α,γ-diaminobutyric acid; A₄ = unidentified; Lat. = lathyrine (II); Arg. = arginine.
² These three species contained low concentrations of amino-acids and were not assigned to the previous groups; *L. laetiflorus* contained a neutral compound and *L. venosus* a basic compound not seen in other species. All species contained traces of glutamic acid and aspartic acid.

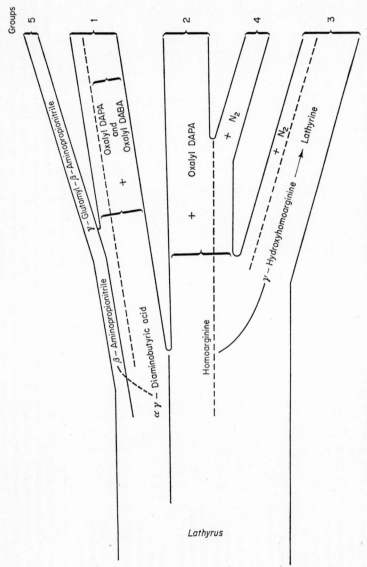

FIG. 1. Schematic representation of the distribution of the amino acids in *Lathyrus*.

in which no γ-hydroxyhomoarginine was found in the original survey, have all since been shown to contain it (Bell, 1964a), but at concentrations below those detectable by the original analytical procedure.

B. *Vicia* SPECIES

The thirteen "non-protein" amino acids and related compounds found in high concentration in the seeds of *Vicia* (Bell and Tirimanna, 1965) include canavanine (IV) γ-hydroxyarginine (XII) and γ-hydroxyornithine (XIII) (Bell and Tirimanna, 1964), β-cyanoalanine (XI) (Ressler, 1962), γ-glutamyl-β-cyanoalanine (XIV) (Ressler *et al.*, 1963), a ninhydrin-reacting ureido compound tentatively identified as γ-hydroxycitrulline (XV) (Bell and Tirimanna, 1965) and seven unknowns. The distribution of these compounds is given in Table 2 and a schematic representation of the genus in Fig. 2. This figure shows

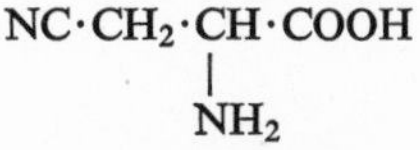

$$NC \cdot CH_2 \cdot \underset{\underset{NH_2}{|}}{CH} \cdot COOH$$

XI. β-Cyanoalanine

$$H_2N \cdot \underset{\underset{NH}{\|}}{C} \cdot NH \cdot CH_2 \cdot CHOH \cdot CH_2 \cdot \underset{\underset{NH}{|}}{CH} \cdot COOH$$

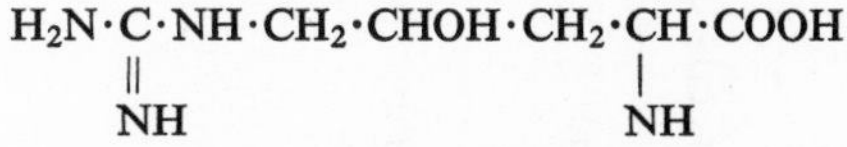

XII. γ-Hydroxyarginine

$$H_2N \cdot CH_2 \cdot CHOH \cdot CH_2 \cdot \underset{\underset{NH_2}{|}}{CH} \cdot COOH$$

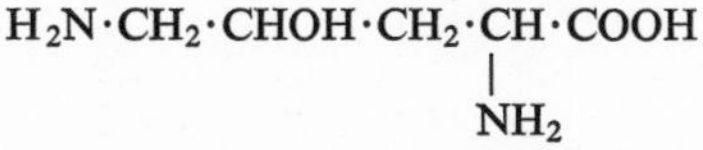

XIII. γ-Hydroxyornithine

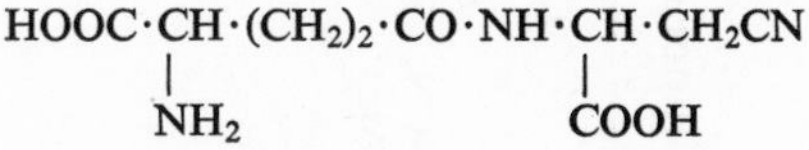

$$HOOC \cdot \underset{\underset{NH_2}{|}}{CH} \cdot (CH_2)_2 \cdot CO \cdot NH \cdot \underset{\underset{COOH}{|}}{CH} \cdot CH_2CN$$

XIV. γ-Glutamyl-β-cyanoalanine

$$H_2N \cdot CO \cdot NH \cdot CH_2 \cdot CHOH \cdot CH_2 \cdot \underset{\underset{NH_2}{|}}{CH} \cdot COOH$$

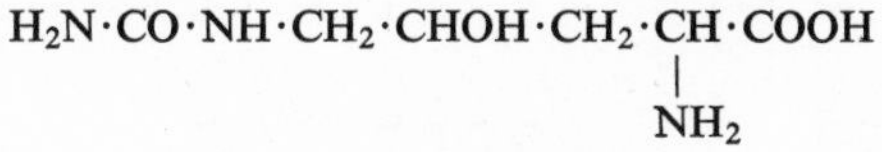

XV. γ-Hydroxycitrulline

a primary division of the genus into those species with β-cyanoalanine (XI) and its γ-glutamyl derivative (XIV) and those without. Tracer studies have shown (Fowden and Bell, 1965) that the apparent absence of β-cyanoalanine from representative species of the second group is due to the high activity of a nitrilase in these species which hydrolyses the β-cyanoalanine to asparagine preventing a build up of the nitrile or its γ-glutamyl derivative. Slight labelling of asparagine in some species accumulating the cyano compounds and of the cyanocompounds in species accumulating asparagine suggests that both enzyme systems exist together, but that the nitrilase is dominant in some species and the γ-glutamyl transferase in the rest.

γ-Hydroxyarginine (XII), an amino acid which has not been found in other plant genera, occurs in species of *Vicia* with, and species without, high concentrations of cyano compounds. In three of the second group it is accompanied by γ-hydroxyornithine (XIII) and γ-hydroxycitrulline (XV) which are possibly derived from it.

TABLE 2. Amino-acids and related compounds in seeds of *Vicia* species

Group	Species	Arg	γ-OH Arg	Asp (NH₂)	Homo-ser	Can	β-CN, Ala	Glu-β-CN, Ala	Pip	V.A₁	V.A₃	V.A₄	V.N	V.B₁	V.B₂
1	*articulata.* Hormen.	T		T		++									
	silvatica L.	+		T		++								++	
	ervilia Wild.	T		T		+								+	
	tennuisima Bieb.	++		+		++								++	+
	cassubica L.	T			T	+++				+	+		++	+	
	disperma D.C.	+		T	T	+++				++	++		T	+	
	benghalensis L.	T		T	T	+++				++	+		++	+	
	dalmatica Kern.	T		++	T	+++				+	++		++	T	
	selloi Vog.	+		++	T	++					++		++	T	
	villosa Roth	T		++	T	++				++	+		T		
	cracca L.	T		++	T	+++				+	++		++		
	tennuifolia Roth.	++		++	T	++					+		+		
	hirsuta Gray	T		T		++					+				
	tetrasperma Moench.	+		T							T		+		
2	*pannonica* Crantz	++								+	++				
	lutea L.	+									+				
	orobus D.C.	++		++						++	++				
	bithinica L.	++		T							+		T		
	hyrcanica Fisch.	+++								+	++		T		
	fulgens Barr.	++									+		T		
	onobrychoides L.	+	+	T							+				

Species									
unijuga Braun.	T	+							+
narbonensis L.	++	+						T	+
dumetorum L.	T	+			T			++	+
baicalensis Fedtsch.	+	+	++					++	+
3									
peregrina L.	+	+	+	+	++	T	+	+++	++
sepium L.	+	T	+	+	+			+++	+
picta Fisch.	T	+	+++	+++	+++			+++	T
lathyroides L.	+	T	+	+	+++			+++	+
cornigera Chaub.	+	+	+	+	++			+++	T
cordata L.	+	+	+	+	+++	+		+++	+
grandiflora Scop.	T	+	T	+	+++	+	+	+++	+
sativa L.	+	+	+++	+++	+++	+++		+++	+
ludoviciana L.	+	+	+	+	++	+		+++	++
michauxii L.	T	+	+	+	+			+++	++
sicula Guss.	+	+	+		++	+		+	
hybrida L.	+	+	+	+	+			+	
amphicarpa L.	+	++	+						
augustifolia L.	+	+	T						
aurantica Boiss.[3]	T			T				++	
cylindrica Skeals.[3]	+	+				++	T	++	
faba L.[3]	++	T					T		

[1] The letters V.A$_1$, etc. represent unidentified compounds; Arg = Arginine; γ-OH Arg = γ-Hydroxyarginine (XII); Asp(NH$_2$) = Asparagine; Homo-ser = Homoserine; Can = Canavanine (IV); β-CN, Ala = β-Cyanoalanine (XI); Glu-β-CN, Ala = γ-Glutamyl-β-cyanoalanine (XIV) (formerly V.A$_2$); Pip = Pipecolic acid. T = trace.

[2] Glutamic acid and aspartic acids were found in all species: high concentrations of proline were found in *V. articulate*, *V. sativa* and *V. hybrida*, of serine in *V. tennuisima* and *V. villosa*, of an amino acid tentatively identified as γ-hydroxycitrulline (γ-OH Cit) in *V. fulgens* and *V. unijuga*, of γ-hydroxyornithine (γ-OH Orn) in *V. onobrychoides* and *V. unijuga*, of α, β-diaminopropionic acid (DAPA) in *V. baicalensis*, of α, γ-diaminobutyric acid (DABA) in *V. aurantica* and an unidentified basic compound (VB$_3$) in *V. peregrina*, *V. sepium* and *V. picta*.

[3] These species could not be assigned to the previous groups.

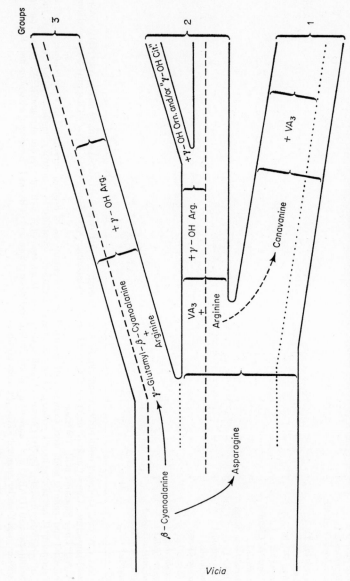

Fig. 2. Schematic representation of the distribution of the amino acids in *Vicia*.

The species in which high concentrations of cyano compounds do not accumulate may be further sub-divided into those which accumulate canavanine and those which do not. The biosynthesis of canavanine (IV) in these species is at present under investigation, but the relative concentrations of free arginine in Groups 1 and 2 (Table 2) suggest that the synthesis of canavanine may involve the enzymatic transfer of an amidine group from arginine.

From these results, summarized in Tables 1 and 2, it will be seen that the two genera (*Lathyrus* and *Vicia*) may each be subdivided on the basis of the biosynthetic variations, with respect to the amino acids, shown by their species. Within each genus however the different sub-genera are clearly related in terms of overall amino acid distribution pattern even as they are in terms of morphology, and the most dissimilar members of each genus can be visualized as the end-product of a gradual step-wise divergence of biochemical characteristics traceable through intermediate forms, rather than as unrelated species.

No such relationship is apparent between the two genera themselves however. Arginine is the only amino acid found in high concentration in the seeds of species of both genera, and if the anomalous *V. aurantica* is disregarded, no "non-protein" amino acids or related compound has been found common to species of both *Lathyrus* and *Vicia*. The accumulation of C_6 guanidino compounds or canavanine by *Vicia* and C_7 guanidino compounds (including lathyrine) or α, γ-diaminobutyric acid by *Lathyrus* being the most striking difference.

III. Non-Protein Amino Acids in Other Genera

This account is concerned primarily with two genera which have been examined in some detail in my laboratory. Other groups of structurally related "non-protein" amino acids are known in other groups of species however.

The occurrence in the Mimosaceae of α, β-diaminopropionic acid (XVI), albizziine (L-2-amino-3-ureidopropionic acid) (XVII) and willardiine [β-(uracil-3-yl)alanine] (XVIII), all discovered by Gmelin and co-workers (Gmelin *et al.*, 1958 and 1959; Gmelin, 1959) and of mimosine [β-(1,4 dehydro-3-hydroxy-4-oxo-pyrid-1-yl)-L-alanine] (XIX) which was found earlier by Renz (1936), provides one example of such a structurally related group of compounds, while in the same family a second group of sulphur containing compounds has also been found by Gmelin and his associates (Gmelin, 1962, Gmelin *et al.*, 1962); these include djenkolic acid (2-amino-2-carboxyethane-thiomethyl-L-cysteine) (XX), S-(2-carboxyethyl)-L-cysteine (XXI), dichrostachic acid [S-(2-hydroxy-2-carboxyethanesulphonylmethyl)-L-cysteine] (XXII) and N-acetyl-L-djenkolic acid.

Another group of sulphur-containing amino acids and derivatives are those of *Allium* which have been studied by Virtanen and co-workers (Virtanen, 1962; Matikkala and Virtanen, 1964), while in the Rosaceae attention has been drawn previously (Fowden, 1964) to the possible phylogenetic significance of

$$H_2N \cdot CH_2 \cdot CH \cdot COOH$$
$$|$$
$$NH_2$$

XVI. α, β-Diaminopropionic acid

$$H_2N \cdot CO \cdot NH \cdot CH_2 \cdot CH \cdot COOH$$
$$|$$
$$NH_2$$

XVII. Albizziine

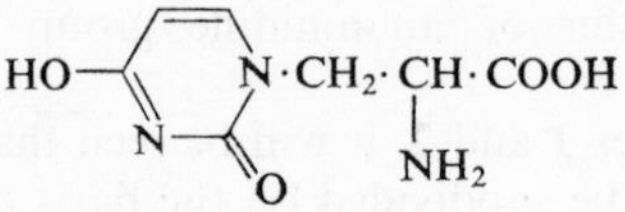

XVIII. Willardiine

and

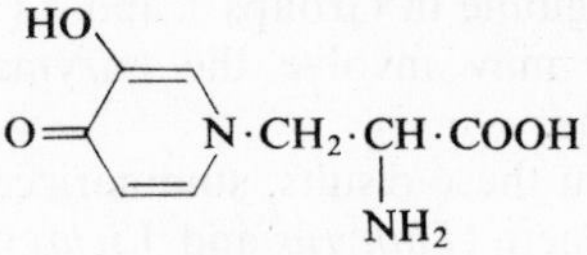

XIX. Mimosine

$$HOOC \cdot CH \cdot CH_2 \cdot S \cdot CH_2 \cdot S \cdot CH_2 \cdot CH \cdot COOH$$
$$| \qquad\qquad\qquad | $$
$$NH_2 \qquad\qquad\qquad NH_2$$

XX. Djenkolic acid

$$HOOC \cdot (CH_2)_2 \cdot S \cdot CH_2 \cdot CH \cdot COOH$$
$$|$$
$$NH_2$$

XXI. S-(2-Carboxyethyl)-L-cysteine

$$HOOC \cdot CHOH \cdot CH_2 \cdot SO_2 \cdot CH_2 \cdot S \cdot CH_2 \cdot CH \cdot COOH$$
$$|$$
$$NH_2$$

XXII. Dichrostachic acid

three structurally related amino acids (4-methylproline, 4-hydroxymethyl-proline and 4-methyleneproline) which occur in various species.

Although the number of genera for which detailed information of amino acid distribution and metabolism is available is small, similarities and differences which are apparent within these encourage the belief that the study of metabolic pathways involved in the biosynthesis of plant amino acids may provide results of phylogenetic significance in other parts of the plant kingdom.

IV. CONCLUSIONS

The genera *Lathyrus* and *Vicia* may be distinguished from one another on the basis of variations in amino acid metabolism shown by their species. Variations in this respect also occur within each genus enabling both *Lathyrus* and *Vicia* to be sub-divided into sub-genera of biochemically related species.

The occurrence of other associated groups of "non-protein" amino acids elsewhere in the plant kingdom suggests that detailed study of the distribution and metabolism of such compounds may contribute to the understanding of the relationships and differences between certain plant species and the evolutionary processes which have produced them.

ACKNOWLEDGEMENTS

Acknowledgement is made to the Editors of *Nature* and the *Biochemical Journal* for permission to reproduce (with modification) Tables 1 and 2.

REFERENCES

Adiga, P. R., Rao, S. L. N. and Sarma, P. S. (1963). *Curr. Sci.* **32**, 153.
Alston, R. E. and Turner, B. L. (1963). "Biochemical Systematics", Prentice-Hall, Englewood Cliffs, N.J.
Bell, E. A. (1961). *Biochim. biophys. Acta* **47**, 602.
Bell, E. A. (1962a). *Biochem. J.* **83**, 225.
Bell, E. A. (1962b). *Biochem. J.* **85**, 91.
Bell, E. A. (1964a). *Biochem. J.* **91**, 358.
Bell, E. A. (1964b). *Fed. Europ. Biochem. Socs. Abs.* **1**, 53.
Bell, E. A. and Foster, R. G. (1962). *Nature, Lond.* **194**, 91.
Bell, E. A. and Fowden, L. (1964). *In* "Taxonomic Biochemistry and Serology" (C. A. Leone, ed.) Chap. 10. Ronald Press, New York.
Bell, E. A. and Przybylska, J. (1965). *Biochem. J.* **94**, 35P.
Bell, E. A. and Tirimanna, A. S. L. (1964). *Biochem. J.* **91**, 356.
Bell, E. A. and Tirimanna, A. S. L. (1965). *Biochem. J.* **97**, 104.
Birdsong, B. A., Alston, R. E. and Turner, B. L. (1960). *Canad. J. Bot.* **38**, 499.
Butler, G. W. and Conn, E. E. (1964). *J. biol. Chem.* **239**, 1674.
Buzzati-Traverso, A. A. and Rechnitzer, A. B. (1953). *Science* **117**, 58.
Damodaran, M. (1932). *Biochem. J.* **26**, 235.
Damodaran, M., Jaaback, G. and Chibnall, A. C. (1932). *Biochem. J.* **26**, 1704.
Delaville (1802). *Ann. Chim.* **41**, 298.
Fowden, L. (1958). *Nature, Lond.* **182**, 406.
Fowden, L. (1964). *Ann. Rev. Biochem.* **33**, 173.
Fowden, L. and Bell, E. A. (1965). *Nature, Lond.* **206**, 110.
Fowden, L. and Steward, F. C. (1957). *Ann. Bot.* **21**, 53.
Gmelin, R. (1959). *Z. phys. Chem.* **316**, 164.
Gmelin, R. (1962). *Z. phys. Chem.* **327**, 186.
Gmelin, R., Kjaer, A. and Olesen Larsen, P. (1962). *Phytochemistry* **1**, 233.
Gmelin, R., Strauss, G. and Hasenmaier, G. (1958). *Z. Naturf.* **13b**, 252.
Gmelin, R., Strauss, G. and Hasenmaier, G. (1959). *Z. phys. Chem.* **314**, 28.
Greenstein, J. P. and Winitz, M. (1961). "Chemistry of the Amino Acids", John Wiley and Sons, New York.
Matikkala, E. J. and Virtanen, A. I. (1964). *Acta chem. scand.* **18**, 2009.
Murti, V. V. S., Seshadri, T. R. and Venkitasubramanian, T. A. (1964). *Phytochemistry* **3**, 73.
Renz, J. (1936). *Z. phys. Chem.* **244**, 153.
Ressler, C. (1962). *J. biol. Chem.* **237**, 733.
Ressler, C., Nigam, S. N., Giza, Y.-H. and Nelson, J. (1963). *J. Amer. chem. Soc.* **85**, 3311.
Ressler, C., Redstone, P. A. and Erenberg, R. H. (1961). *Science* **134**, 188.
Reuter, G. (1957). *Flora* **145**, 326.
Schilling, E. D. and Strong, F. M. (1955). *J. Amer. chem. Soc.* **77**, 2843.
Schulze, E. (1877). *Ber.* **10**, 85.
Schulze, E. and Barbieri, J. (1879). *Ber.* **12**, 1924.
Schulze, E. and Steiger, E. (1886[a]). *Ber.* **19**, 1177.
Schulze, E. and Steiger, E. (1886[b]). *Z. phys. Chem.* **11**, 43.
Tschiersch, B. (1959). *Flora* **147**, 405.
Tschiersch, B. (1964). *Phytochemistry* **3**, 365.
Vauquelin, L. N. and Robiquet, P. J. (1806). *Ann. Chim.* **57**, 88.
Virtanen, A. I. (1962). *Angew. Chem. (Int.)* **1**, 299.
Virtanen, A. I. and Linko, P. (1955). *Acta chem. scand.* **9**, 531.

CHAPTER 13

Comparative Phytochemistry of Alkaloids

R. HEGNAUER

Laboratorium voor Experimentele Plantensystematiek, Leiden, Netherlands

I. DEFINITION AND CLASSIFICATION

Alkaloids are basic, nitrogen-containing plant constituents which have been known to man for a long time because of their toxic and medicinal properties (e.g. aconitine, colchicine, hyoscyamine, quinine, etc.). Medical scientists, pharmacists and organic chemists have all been interested in them; those in the first two disciplines naturally classified alkaloids according to their pharmacological action, whereas chemists divided them mainly by structure, especially on the basis of the type of carbon skeleton and heterocyclic systems which they contained.

In the last decade classification based according to the biogenetic pathways by which alkaloids are elaborated in plants has come more and more into use. In using plant constituents in the study of problems of taxonomy, a biogenetic classification seems to be most appropriate. For the following survey, alkaloids are grouped in three main categories based both on knowledge and speculation about their biogenesis (compare Leete, 1963).

1. *True alkaloids.* The true alkaloids are compounds in which the nitrogen-containing heterocyclic system is derived from a biogenetic amine, formed by decarboxylation from an amino acid. Amino acids which are known with certainty to be involved in this type of alkaloid biogenesis are phenylalanine (isoquinolines), tryptophan (most indolic alkaloids), ornithine (hygrine, tropane, ecgonine and necine group of alkaloids), lysine (the papilionaceous group of alkaloids; some α-substituted piperidines), and histidine (pilocarpine group of alkaloids). Nicotinic acid and anthranilic acid (acridine and

furanoquinoline groups of alkaloids) may also be mentioned; however these two widespread nitrogen-containing carboxylic acids are not amino acids in the ordinary sense and are not generally decarboxylated when they are used as alkaloid precursors.

2. *Alkaloid-like compounds* (=*protoalkaloids*). These compounds, like the true alkaloids, are derived from amino acids or biogenetic amines, but contain no heterocyclic system (if not derived from tryptophan). They are represented in nature by the biogenetic amines themselves and their methylated derivatives, betaines and so on.

3. *Alkaloids, which are apparently unrelated to amino acids* (=*pseudo-alkaloids*). These compounds are nitrogen-containing, basic substances which, however, have a carbon skeleton derived from monoterpenes, diterpenes, sterols, triterpenes and acetate-derived, aliphatic polyketo acids.

It appears logical when comparing plants and the constituents they synthesize that we use in the first instance biogenetic pathways and biogenetic groups of compounds, and not categories of compounds grouped together by man for his own convenience (e.g. alkaloids as a whole). If we do this, the great biological difference between the true alkaloids and the so-called pseudoalkaloids is immediately apparent.

II. DISTRIBUTION OF ALKALOIDS IN CORMOPHYTES

Tables 1 and 2 show the present-day picture of our knowledge about the distribution of alkaloids. Members of all three main categories are included. The botanical system used is that of Engler as modified in the 12th Edition of his *Syllabus der Pflanzenfamilien* (Engler, 1964).

The study of Tables 1 and 2 reveals several aspects of the distribution of alkaloids in plants and some of these will now be discussed in more detail.

True alkaloids seem to be very rare in the lower plants (Table 1). In fact, based on the knowledge acquired with higher plants (species of *Nicotiana*), the only compound of known structure which may be interpreted as a true alkaloid is nicotine, but the latter does occur only in trace amounts in *Lycopodium* and *Equisetum*, and nobody has proved that its biogenesis proceeds in the same way as it does in *Nicotiana*. The characteristic alkaloids of *Lycopodium* and the α,α_1-disubstituted piperidine (=pinidine) of pines are believed to be built up from acetate-derived polyketo acids (Conroy, 1960; Hill *et al.*, 1965). Nothing can be said about the *Cephalotaxus*-alkaloids until their structure is fully elucidated.

In Angiosperms (Table 2) alkaloids are very unevenly distributed. Besides taxa which contain alkaloids and alkaloid-like substances throughout, there are many families and orders in which these constituents occur sporadically or not at all. Especially poor in alkaloids are the amentiferous group (Orders 1–7), Proteales, Cucurbitales, the first four Orders of Sympetalae and most Orders of Monocotyledones except the Liliiflorae. The main centres of alkaloid synthesis and accumulation are to be found in the Centrospermae, Magnoliales, Ranunculales, Papaveraceae, Leguminosae-Papilionatae, Rutaceae,

TABLE 1. Alkaloids in Pteridophytes and Gymnosperms

Classes	Number of recent families	Approximate number of recent genera and species	Families in which alkaloids and alkaloid-like compounds are known	Approximate number of compounds identified
		Pteridophyta		
1. Lycopsida	3	4 (1160)	Lycopodiaceae	60[1]
2. Psilotopsida	2	2 (5)	0	0
3. Articulatae	1	1 (30)	Equisetaceae	4[2]
4. Filices	23	245 (9120)	0	0
		Gymnospermae		
1. Cycadopsida	2	10 (90)	0	0[3]
2. Coniferopsida	6	42 (490)	Cephalotaxaceae	5[4]
			Pinaceae	2[5]
3. Taxopsida	1	2 (8)	Taxaceae	10[6]
4. Chlamydospermae	3	3 (80)	Ephedraceae	6[7]

[1] All typical *Lycopodium*-alkaloids can theoretically be derived from two molecules of 3,5,7-triketoöctanoic acid; traces of nicotine in many species.

[2] Spermidine is the basic moiety of *Equisetum*-alkaloids; traces of nicotine and 3-methoxypyridine were found in several species.

[3] Nitrogen-containing glycosides of macrozamine-type, which are not alkaloids.

[4] Cephalotaxine, $C_{18} H_{21} O_4 N$; structure still unknown.

[5] Piperidine-type alkaloids, pinidine and α-pipecoline.

[6] Diterpenes with taxane skeleton.

[7] Ephedrine and related protoalkaloids.

TABLE 2. Alkaloids in Angiosperms*†

Order	Number of families	Approximate number of genera and species	Families in which alkaloids and alkaloid-like compounds are known[1]	Approximate number of compounds identified[1]
		Dicotyledoneae—Archichlamydeae		
1. Casuarinales	1	1 (45)	0	0
2. Juglandales	2	11 (115)	0	0
3. Balanopales	1	1 (9)	0	0
4. Leitneriales	2	2 (3)	0	0
5. Salicales	1	2 (350)	0	0
6. Fagales	2	13 (700)	0	0
7. Urticales	5	120 (2400)	Moraceae	5[2]
8. Proteales	1	62 (1400)	0	0
9. Santalales	7	112 (2110)	Loranthaceae	2[3]
			Santalaceae	4[4]
10. Balanophorales	1	18 (100)	0	0

* For footnotes see pp. 216–217.

† See also Addendum, p. 230.

TABLE 2—*continued**

Order	Number of families	Approximate number of genera and species	Families in which alkaloids and alkaloid-like compounds are known[1]	Approximate number of compounds identified[1]
11. Medusandrales	1	2 (6)	0	0
12. Polygonales	1	40 (800)	Polygonaceae	5[5]
13. Centrospermae[6] (incl. Didiereaceae)	13	450 (8000)	Aizoaceaeae	5[7]
			Amaranthaceae	1[8]
			Chenopodiaceae	20[9]
			Nyctaginaceae	1[10]
14. Cactales	1	200 (2000)	Cactaceae	12[11]
15. Magnoliales	22	240 (5600)	Annonaceae	16[12]
			Calycanthaceae	4[13]
			Hernandiaceae	4[14]
			Himantandraceae	10[15]
			Lauraceae	30[16]
			Magnoliaceae	30[17]
			Monimiaceae	20[18]
16. Ranunculales	7	150 (3190)	Berberidaceae	40[19]
			Menispermaceae	50[20]
			Nymphaeaceae	15[21]
			Ranunculaceae	150[22]
17. Piperales	4	20 (1475)	Piperaceae	5[23]
18. Aristolochiales	3	18 (675)	Aristolochiaceae	10[24]
19. Guttiferales	16	165 (3200)	Actinidiaceae	1[25]
			Ancistrocladaceae	1[26]
20. Sarraceniales	3	8 (190)	0	0
21. Papaverales	6	450 (4500)	Capparidaceae	5[27]
			Cruciferae	8[28]
			Moringaceae	1[29]
			Papaveraceae	100[30]
22. Batales	1	1 (2)	0	0
23. Rosales	19	930 (20300)	Crassulaceae	8[31]
			Krameriaceae	1[32]
			Leguminosae	150[33]
			Rosaceae	2[34]
			Saxifragaceae	3[35]
24. Hydrostachyales	1	1 (30)	0	0
25. Podostemales	1	43 (200)	0	0
26. Geraniales	9	370 (10250)	Daphniphyllaceae	5[36]
			Erythroxylaceae	12[37]
			Euphorbiaceae	25[38]
			Zygophyllaceae	6[39]
27. Rutales	12	340 (5570)	Akaniaceae	1[40]
			Malpighiaceae	3[41]
			Meliaceae	0[42]
			Rutaceae	75[43]
			Simarubaceae	2[44]
28. Sapindales	10	235 (2950)	Aceraceae	1[45]
29. Julianiales	1	2 (5)	0	0

* For footnotes see pp. 216–217.

TABLE 2—continued*

Order	Number of families	Approximate number of genera and species	Families in which alkaloids and alkaloid-like compounds are known[1]	Approximate number of compounds identified[1]
30. Celastrales	13	150 (2170)	Buxaceae	20[46]
			Celastraceae	8[47]
31. Rhamnales	3	71 (1670)	Rhamnaceae	6[48]
32. Malvales	7	250 (3565)	Malvaceae	1[49]
33. Thymelaeales	5	61 (990)	Elaeagnaceae	4[50]
34. Violales	20	180 (4430)	Caricaceae	2[51]
			Flacourtiaceae	1[52]
			Passifloraceae	1[53]
35. Cucurbitales	1	100 (850)	0	0
36. Myrtiflorae	17	420 (9410)	Combretaceae	1[54]
			Lythraceae	13[55]
			Punicaceae	4[56]
37. Umbelliflorae	7	390 (3830)	Alangiaceae	4[57]
			Umbelliferae	5[58]

Dicotyledoneae—Sympetalae

Order	Number of families	Approximate number of genera and species	Families in which alkaloids and alkaloid-like compounds are known[1]	Approximate number of compounds identified[1]
1. Diapensiales	1	6 (18)	0	0
2. Ericales	5	130 (3010)	0	0
3. Primulales	3	65 (1910)	0	0
4. Plumbaginales	1	10 (350)	0	0
5. Ebenales	7	70 (1710)	Sapotaceae	1[59]
			Symplocaceae	5[60]
6. Oleales	1	27 (600)	0	0
7. Gentianales	7	1050 (12650)	Apocynaceae	250[61]
			Asclepiadaceae	5[62]
			Gentianaceae	2[63]
			Loganiaceae	100[64]
			Menyanthaceae	1[65]
			Rubiaceae	60[66]
8. Tubiflorae	26	1360 (21500)	Acanthaceae	3[67]
			Bignoniaceae	3[68]
			Boraginaceae	35[69]
			Convolvulaceae	15[70]
			Labiatae	6[71]
			Orobanchaceae	1[72]
			Scrophulariaceae	5[73]
			Solanaceae	65[74]
9. Plantaginales	1	3 (265)	Plantaginaceae	3[75]
10. Dipsacales	4	39 (1030)	Dipsacaceae	1[76]
			Valerianaceae	3[77]
11. Campanulales	8	1020 (21600)	Campanulaceae	30[78]
			Compositae	60[79]

Monocotyledoneae

Order	Number of families	Approximate number of genera and species	Families in which alkaloids and alkaloid-like compounds are known[1]	Approximate number of compounds identified[1]
1. Helobiae	9	36 (410)	0	0
2. Triuridales	1	7 (80)	0	0

* For footnotes see pp. 216–217.

R. Hegnauer

TABLE 2—continued

Order	Number of families	Approximate number of genera and species		Families in which alkaloids and alkaloid-like compounds are known[1]	Approximate number of compounds identified[1]
3. Liliiflorae	17	465	(7910)	Amaryllidaceae	150[80]
				Dioscoreaceae	2[81]
				Liliaceae	130[82]
				Stemonaceae	14[83]
4. Juncales	2	9	(300)	0	0
5. Bromeliales	1	46	(1700)	0	0
6. Commelinales	8	110	(2600)	0	0
7. Graminales	1	700	(8000)	Gramineae	20[84]
8. Principes	1	236	(3400)	Palmae	4[85]
9. Synanthae	1	11	(180)	0	0
10. Spathiflorae	2	115	(1825)	0	0
11. Pandanales	3	5	(915)	0	0
12. Cyperales	1	70	(3700)	Cyperaceae	3[86]
13. Scitamineae	5	90	(2125)	0	0
14. Microspermae	1	600	(20000)	Orchidaceae	4[87]

[1] Purines (e.g. caffeine) and trace amounts of alkaloids (e.g. nicotine or biogenic amines in many species) not included. [2] Tylophorine and related alkaloids; pseudopelletierine. [3] Biogenic amines only; if growing on alkaloid-containing hosts alkaloids may pass into the parasite. [4] Necine-type bases (thesinine, etc.); possibly originating in the host. [5] α-Picoline from *Rumex obtusifolius* L.; carbolines from *Calligonum minimum* Lipski. [6] All members except Caryophyllaceae and Molluginaceae produce betacyanins or betaxanthins (see Chapter 14). [7] Mesembrine and related alkaloids; piperidine; trianthemine. [8] Betaine. [9] Betaine; different types of protoalkaloids; simple isoquinolines (salsolidine, etc.); carbolines (leptocladine); piperidine derivatives; anabasine; papilionaceous alkaloids. [10] Dopamine from *Hermidium alipes* Watson. [11] Simple isoquinolines and related biogenic amines; furthermore betacyanins and betaxanthins are of common occurrence. [12] Benzylisoquinolines from many members; quinidine and hydroquinidine from *Enantia polycarpa* Engl. et Diels. [13] Indolic alkaloids closely related to tryptamine. [14] Benzylisoquinolines. [15] Complex α, α₁-disubstituted piperidines (e.g. himbacine, himbosine). [16] Benzylisoquinolines probably present in most members. [17] Benzylisoquinolines found in most species investigated. [18] Benzylisoquinolines detected in several species; sparteine probably found in one species (see Table 3). [19] Isoquinolines present in *Berberis, Caulophyllum, Epimedium, Leontice, Mahonia* and *Nandina*. Papilionaceous alkaloids found in two genera (see Table 3). [20] Benzylisoquinolines in all members; protostephanine and dihydroerysodine from one species each. [21] Benzylisoquinolines in *Nelumbo*; sesquiterpene-type alkaloids (nupharidine, etc.) in *Nuphar*. [22] Benzylisoquinolines in *Aquilegia, Aconitum, Caltha, Coptis, Hydrastis, Isopyrum* (?), *Thalictrum* and *Zanthorrhiza* (compare also Tables 4 and 5). Damascenine in *Nigella*. Diterpene alkaloids in *Aconitum* and *Delphinium*. [23] Piperidine amides (e.g. piperine); jaborandine and piperovatine of unknown structure. [24] Benzylisoquinolines; aristolochic acid-type nitrophenanthrenes; aristolactame. [25] Monoterpenoid, skytanthine-like actinidine. [26] Structure not known. [27] Tetramine; stachydrine-type betaines. [28] Sinapine; macrocyclic alkaloids like lunarine; hygrines and the tropane-type alkaloid cochlearine from *Cochlearia arctica* Schlecht. (compare Fig. 1). [29] Benzylamine from *Moringa pterygosperma* Gaertn. [30] Benzylisoquinolines in all members; ephedrine in one genus; traces of sparteine in one species (compare Table 3). [31] α-Substituted piperidines (e.g. sedridine, sedamine, etc.) from several species of *Sedum*. Nicotine demonstrated to be

TABLE 2—*continued*

present in species of *Echeveria, Sedum* and *Sempervivum*. [32] *N*-Methyltyrosine (=ratanhine) from *Krameria triandra* R. et P. [33] Biogenic amines and related protoalkaloids in many Mimosoideae. Different types of pseudoalkaloids (chaksine, cassiine, *Erythrophleum*-alkaloids) in Caesalpinioideae. Papilionaceous alkaloids wide spread in three tribes (Sophoreae, Podalyrieae, Genisteae) of Papilionatae; rarely these alkaloids are replaced by simple isoquinolines (e.g. calycotomine); in the genus *Crotalaria* they are replaced by necine-type bases. Erythrinaceous alkaloids in all species of *Erythrina* (Phaseoleae); eserine-type alkaloids in *Physostigma* (Phaseoleae); guanidine-type bases in Galegeae. [34] Probably diterpenoid alkaloids in *Spiraea japonica* L. [35] Quinazolone-type alkaloids (febrifugine, etc.). [36] Isolated but not yet chemically characterized. [37] Hygrines, ecgonines, tropeines. [38] Compare Table 6 and Fig. 5. [39] Harmane-type alkaloids; vasicine (=peganine). [40] Structure unknown. [41] Harmane-type alkaloids. [42] Alkaloids undoubtedly present but not yet investigated. [43] Very different types: Several types of benzylisoquinolines (compare also Tables 4 and 5); furanoquinoline- and acridine-type alkaloids present in most members; pilocarpine-type alkaloids in *Pilocarpus*; indolic alkaloids (e.g. rutaecarpine, canthinone) in some genera. [44] Canthinone-type indolic alkaloid. [45] Gramine. [46] Steroidal and triterpenic alkaloids probably in all members. [47] Ephedrine-type protoalkaloids; wilforine-type polyesters. [48] Benzylisoquinolines (compare Table 4); ceanothine of unknown structure from several species of *Ceanothus*. [49] Ephedrine in several species of *Sida*. [50] Harmane-type alkaloids probably in all members. [51] α,α_1-Disubstituted piperidines (carpaine). [52] Ryanodine, an ester of pyrrole-α-carboxylic acid with a C_{20}-polyol. [53] Harmane in several species of *Passiflora*. [54] Pyridine in *Quisqualis indica* L. [55] Structure not yet elucidated. [56] Isopelletierine-type alkaloids; pseudopelletierine (Fig. 1). [57] Emetine and related alkaloids. [58] Coniine-type alkaloids (i.e. α-substituted piperidines). [59] Possibly yohimbine. [60] Harmane; benzylisoquinolines (compare Table 4). [61] Skytanthine-type monoterpenes; pregnane-type sterols; complex indolic alkaloids. [62] Tylophorine-type alkaloids (Fig. 1); cryptolepine. [63] Gentianine and related bases. [64] Complex indolic alkaloids; monoterpene-type bases (bakankosine, gentianine); quinine in *Strychnos pseudoquina* St.Hil. [65] Gentianine. [66] Harmane-type alkaloids; complex indolic alkaloids; emetine-type alkaloids; tryptophan-derived quinolines (quinine, etc.); tubulosine from *Pogonopus tubulosus* (DC.) Schum. was recently described as a new emetine-type base (one of the two phenylethylamine residues replaced by tryptamine) (Brauchli *et al.*, 1964). [67] Vasicine-type quinazolines. [68] Skytanthine-type monoterpenes. [69] Necine-type alkaloids. [70] Ergot-type indolic alkaloids; hygrines; esters of tropanol (compare Fig. 1). [71] Stachydrine-type betaines; leonurine of still uncertain structure; diterpenoid alkaloids (?). [72] Orobanchamine (the hosts of these parasitic plants should also be investigated). [73] Vasicine from species of *Linaria*; nicotine in *Herpestis monniera* L.; papilionaceous alkaloids from two members of Rhinanthoideae (compare Table 3) (the hosts of the semiparasitic Rhinanthoideae should be also investigated). [74] C_{27}-sterols (=solanines); nicotine and anabasine; hygrines and tropane-group of alkaloids; isopelletierine (compare Fig. 1); monoterpene-derived alkaloids like fabianine in *Fabiana imbricata* R. et P. [75] Isolated; structure not yet elucidated. [76] Gentianine. [77] Isolated; structure still unknown. [78] α,α_1-Disubstituted piperidines (lobeline-type alkaloids) and lobinaline-type dimeric alkaloids. [79] Many different types: Necine-type bases in *Senecio, Petasites* and *Eupatorium*; echinopsine-type quinoline alkaloids in *Echinops*; nicotine in a few genera (*Eclipta, Zinnia*); Betaines in *Achillea*; lycoctonine-type diterpenoid alkaloids in *Inula*; alkaloid-like compounds (acanthoine, acanthoidine) in *Carduus acanthoides* L. Many alkaloids of this family have not yet been investigated chemically. [80] Amaryllidaceous alkaloids. [81] Dioscorine and related alkaloids. [82] C_{27}-Sterols in *Veratreae* and *Fritillaria*; hordenine in *Eremurus*; colchicines in Wurmbaeoideae; aporphine-type isoquinolines in one species (compare Table 4); many alkaloids of still unknown structure. [83] Tuberostemonine-type alkaloids. [84] Very different types: Hordenine, gramine, tryptamine-derivatives, loline, perloline, annuloline, thelepogine. [85] Arecoline-type alkaloids from *Areca catechu* L. [86] Harmane and related alkaloids from *Carex brevicollis* DC. [87] Picrotoxine-type alkaloids (dendrobine, dendrobamine); shihunine. Alkaloids demonstrated to be present in 10–20% of the species investigated (Lüning 1964).

Buxaceae, Gentianales, some families of Tubiflorae, Campanulaceae-Lobelioi-deae, Compositae-Senecioneae and some of the families of Liliiflorae.

The same is true if we look at the distribution of the different classes of alkaloids or of individual alkaloids. Again we find main centres of distribution besides sporadic occurrences in botanically quite unrelated groups. These latter observations which indicate the frequent convergence of biochemical characters will be discussed further in Section III.

In many orders and families different types of alkaloids occur which are chemically quite unrelated. This indicates that in some taxa alkaloid bio-genesis may be very versatile. Some aspects of this divergence will also be discussed in Section III.

The climate- and habitat-dependent "clines" of alkaloids which were discussed by McNair (1931) do not now appear to exist in the light of the vast amount of knowledge acquired since he wrote his paper. At that time the picture was undoubtedly incorrect because thorough investigations were made of relatively more species of temperate regions than of tropical regions. This is one of the most serious drawbacks, even today, when we wish to make use of plant constituents in taxonomic work. We are still far away from a representative knowledge of the distribution of individual compounds and biogenetic classes of compounds in the plant kingdom as a whole.

III. PARALLELISM AND DIVERSIFICATION

If plants, which on other grounds cannot be regarded as closely related by descent, have certain characters in common, one usually speaks of parallelism or convergence with regard to such characters. If on the other hand plants clearly related by descent are very unlike in some aspects, we explain this difference by divergence of the character in question. With regard to bio-chemical characters, I define the terms parallelism, convergence, analogy, diversification, divergence and homology as follows:

Parallelism: Plants, clearly not related by descent, synthesize the same compound or type of compound but the biosynthetic pathways used for their elaboration is not yet known, at least not in both taxa being com-pared (e.g. *N*-methylcytisine which occurs in Berberidaceae, Leguminosae and Scrophulariaceae).

Convergence: The same compound or type of compound occurs in taxa which seem not to be very closely related by descent, and both taxa use the same biogenetic pathway to elaborate these substances (e.g. sparteine in *Lupinus* and *Chelidonium majus* (Schütte and Hindorf, 1964)).

Analogy: The same type of constituents is found in taxa which, on other grounds cannot be regarded as closely related by descent. The taxa con-cerned, however, use different pathways for synthesis (e.g. anabasine occurs in Chenopodiaceae and Solanaceae but the species concerned possibly use different pathways to synthesize this alkaloid).

Diversification: Taxa clearly related to each other accumulate chemically dissimilar compounds of still unknown biogenetic origin (e.g. *Conium* produces coniine and related alkaloids, while other members of umbelliferae accumulate monoterpenes, sesquiterpenes or acetylenic compounds).

Divergence: Taxa which must be regarded on other grounds as related by descent accumulate chemically different representatives of a class of substances and these are formed from different biogenetic pathways (e.g. *Calycotome spinosa* Link and *Genista purgans* L. accumulate isoquinoline-type alkaloids (calycotomine, salsoline, salsolidine) while related species and genera synthesize papilionaceous alkaloids, which clearly must have a biogenetic origin different from the isoquinoline bases).

Homology: The constituents of related taxa are chemically very unlike but they issue from the same biogenetic pathway (e.g. the alkaloid-like substance bakankosine of *Strychnos vacacoua* Baill. is, doubtless, homologous with nitrogen-free iridoids (see Chapter 9) like swertiamarin (Loganiaceae, Gentianaceae) and gentiopicrin (Gentianaceae)).

At present, however, there are two main difficulties in applying strictly these terms as defined above.

Firstly we generally do not yet know the biogenetic pathways used by all taxa accumulating a distinct class of compounds. Therefore, in most instances, we can only distinguish between parallelism and diversification.

Secondly in many instances morphology, anatomy, palynology and cytology have not yet produced sufficient evidence for an unequivocal estimation of true phylogenetic relationships of taxa. Alangiaceae, for example, are usually regarded as rather closely related to Cornaceae, Araliaceae and Umbelliferae (i.e. are included in the same order, Umbelliflorae). The alkaloids of the Alangiaceae are emetine-type bases which suggests a rather close affinity with the Rubiaceae. If the former concept is true, however, we have to speak of parallelism with regard to alkaloid metabolism. On the other hand if the Rubiaceae evolved from cornaceous-alangiaceous ancestors, both the Cornaceae and Alangiaceae would better be placed in the Rubiales and then the chemical character would be in line with phylogeny. This demonstrates clearly the difficulties which are encountered when we wish to interpret more thoroughly the parallelism and diversification of chemical characters, and to make use of them as guides for classification. In each case all available points of evidence should be carefully weighed before taxonomic conclusions are drawn.

Doubtless a thorough knowledge of the chemistry and biochemistry of plant constituents and of the distribution of individual compounds and biogenetic groups of compounds will prove highly valuable to plant taxonomy. The taxonomist will then be able to rely on still another category of characters when he attempts to discern the true relationships of taxa. Very often the examination of chemical characters will stimulate new taxonomic research and suggest relationships which have not as yet been seriously considered. In all instances, however, it will not be the phytochemist himself, who will come nearest to the truth in elaborating a phylogenetic system of plants, but rather the taxonomist, who will combine facts from morphology, anatomy, palynology and cytology with the new information furnished by phytochemistry.

For taxonomy it is essential that the distinction between convergence, divergence, homology and analogy of characters can be made. With regard to alkaloids the following seven examples will illustrate clearly that both parallelism and diversification commonly occur, but that in most instances a thorough interpretation of these phenomena is not yet possible.

A. ISOPELLETIERINE AND PSEUDOPELLETIERINE (FIG. 1)

Isopelletierine has been detected hitherto in three plant families:

Crassulaceae: *Sedum acre* L. (Franck, 1958; Franck and Hartmann, 1963); Punicaceae: *Punica granatum* L. (original source); Solanaceae: *Duboisia myoporoides* R.Br. (Mortimer, 1957; Mortimer and Wilkinson, 1957); *Salpiglossis sinuata* R. et P. (Schröter, 1963).

The closely related pseudopelletierine occurs in *Punica granatum* L. and is present also, according to Altman (1958), in the latex of *Ficus anthelmintica* Martius.

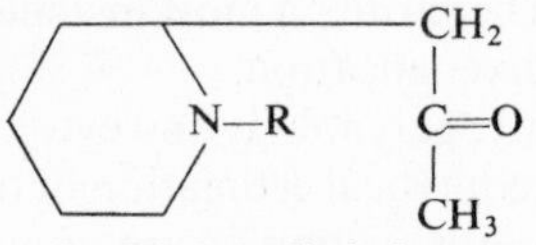

R=H: Isopelletierine
R=CH₃: N-Methyl-isopelletierine

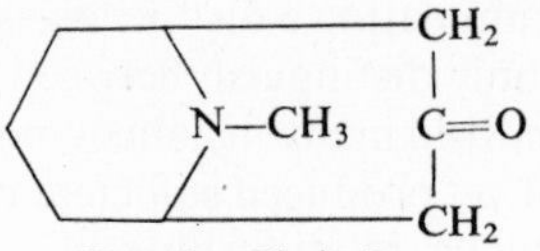

Pseudopelletierine

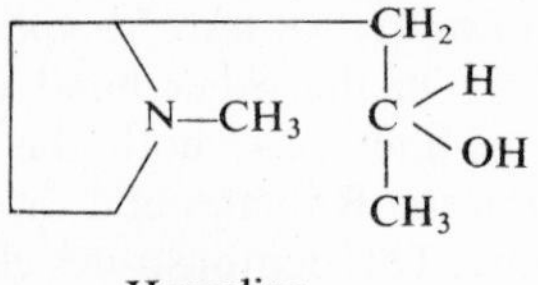

Hygroline

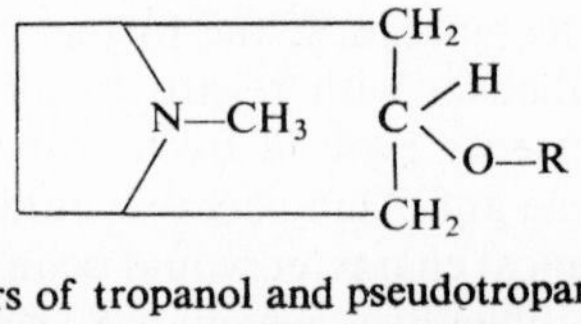

Esters of tropanol and pseudotropanol;
R = Acyl for the following acids:

Tropic	Hyoscyamine
Veratric	Convolamine
Vanillic	Phyllalbine
Benzoic	Tropacocaine
m-Hydroxybenzoic	Cochlearine

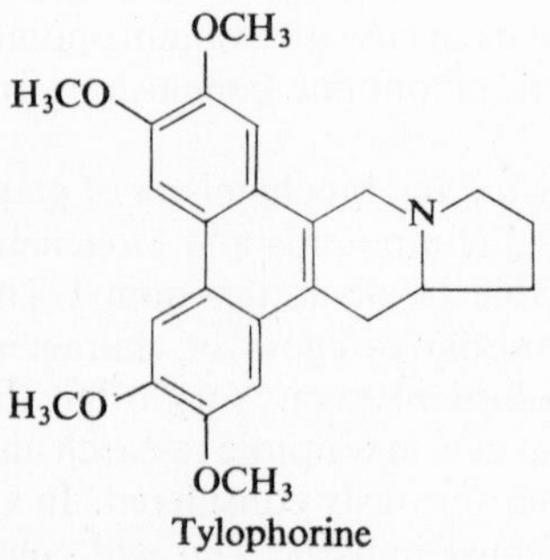

Tylophorine

FIG. 1. Different types of alkaloids for which parallelism is mentioned in Sections III A, B and C.

It is highly probable that species of Solanaceae synthesize isopelletierine along the same lines as hygrine and the tropeines, i.e. starting with lysine instead of ornithine. For the Punicaceae and Crassulaceae the mode of bio-synthesis of this compound may be different; e.g. derivation from acetate (compare Hill *et al.*, 1965). Until biogenetic experiments have been carried out with all the taxa mentioned this question, which is most important from a taxonomic point of view, cannot be answered.

B. TROPANE GROUP OF ALKALOIDS (FIG. 1)

Esters of tropanol or pseudotropanol occur in at least five families of plants:

Convolvulaceae: Several species of *Convolvulus* (convolvine and convol-amine; hygrine and cuscohygrine occur too). Cruciferae: *Cochlearia arctica* Schlecht. (cochlearine, together with hygrine and hygroline: Platonova and Kuzovkov, 1964). Erythroxylaceae: *Erythroxylum coca* Lamk. (tropacocaine, together with ecgonines and hygrines). Euphorbiaceae: *Phyllanthus discoideus* Muell. Arg. (phyllalbine: see Table 6). Solanaceae: Several genera (hyoscine, hyoscyamine and related alkaloids; often together with cuscohygrine).

The co-occurrence of hygrines suggests that at least four of the five families mentioned use the same pathway to elaborate tropeines. If this is true, con-vergence is a frequent feature in the production of this type of alkaloid.

C. TYLOPHORINE (FIG. 1)

Tylophorine and related alkaloids represent a significant character of the asclepiadaceous genus *Tylophora*. Recently tylophorine and related alkaloids were isolated from *Ficus septica* Burm.f. (Russel, 1963). As nothing is known

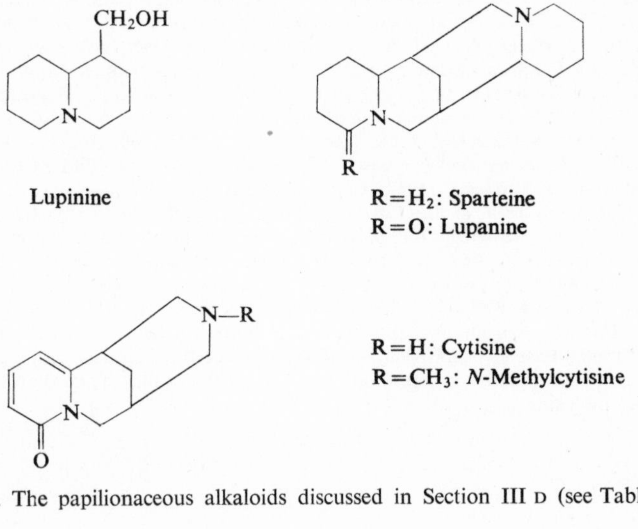

Lupinine

R=H₂: Sparteine
R=O: Lupanine

R=H: Cytisine
R=CH₃: *N*-Methylcytisine

FIG. 2. The papilionaceous alkaloids discussed in Section III D (see Table 3).

TABLE 3. Distribution of lupinine, sparteine, lupanine and N-methylcytisine (compare Fig. 2)

Alkaloid	Order[1]	Family	Genus (species)[2]
Lupinine	13 Centrospermae	Chenopodiaceae	*Anabasis aphylla* L.
	23 Rosales	Leguminosae	*Lupinus.*
Sparteine	15 Magnoliales	Monimiaceae	*Peumus boldo* Molina[3]
	16 Ranunculales	Berberidaceae	*Leontice ewersmanii* Bunge
		Ranunculaceae	*Aconitum napellus* L.[4]
	21 Papaverales	Papaveraceae	*Chelidonium majus* L.[5]
	23 Rosales	Leguminosae	In 3 tribes and 15 genera of Papilionatae
	8 Tubiflorae	Scrophulariaceae	*Leptorhabdos parviflora* Benth.[6]
Lupanine	16 Ranunculales	Berberidaceae	*Caulophyllum robustum* Maxim.[7]
			Leontice ewersmanii Bunge[8]
	23 Rosales	Leguminosae	In 3 tribes and 8 genera of Papilionatae.
	8 Tubiflorae	Solanaceae	*Solanum lycocarpum* St. Hil.[9]
N-Methyl-cytisine	16 Ranunculales	Berberidaceae	*Caulophyllum robustum* Maxim. *C. thalictroides* Michx.[10]
			Leontice albertii Regel[11]
		Ranunculaceae	*Cimicifuga racemosa* (L.) Nutt.[12]
	23 Rosales	Leguminosae	In 3 tribes and 10 genera of Papilionatae.
	8 Tubiflorae	Scrophulariaceae	*Pedicularis olgae* Regel[13]

[1] For numbers of Orders see Table 2. [2] If no references are given, compare Boit (1961). [3] Schindler, H. (1957). *Arzneimittelforsch.* 7, 747. Valonzuela, M. and Rebolledo, L. (1961). *Chem. Abstr.* 55, 6784. [4] Freudenberg, W. and Rogers, E. F. (1937). *J. Amer. chem. Soc.* 59, 2572. In fact traces of sparteine, together with ephedrine, have been isolated from commercial amorphous aconitine extracted from tuberous roots of *Aconitum napellus*. The proof that sparteine does in fact occur in the plant is still lacking. [5] Schütte, H. R. and Hindorf, H. (1964). *Naturwissenschaften* 51, 463. Traces in leaves, not in roots. [6] Bocharnikova, A. V. and Massagetov, P. S. (1964). *Chem. Abstr.* 60, 16214. (+)—sparteine (=pachycarpine) and matrine-type bases. [7] Safronich. L. N. *et al.* (1961). *Chem. Abstr.* 55, 18892. [8] Platonova, T. F. *et al.* (1954). *Chem. Abstr.* 48, 3987; Platanova, T. F. *et al.* (1956). 50, 378; Platanova, T. F. *et al.* (1957). 51, 5102. (+)—Sparteine (=pachycarpine); also matrine-type bases. [9] Ribeiro, O. *et al.* (1952). *Chem. Abstr.* 46, 3221. [10] Power, F. B. and Salway, A. H. (1913). *J. chem. Soc.* 103, 191. Davy, E. D. and Chu, H. P. (1927). *J. Amer. Pharm. Assoc.* 16, 302. [10] Yunusov, S. Yu. and Sorokina, cit. ex McShefferty, J. *et al.* (1956). *J. Pharm. Pharmacol.* 8, 1117. [12] Gemeinhardt, K., Sadé, H. and Schenck, G. (1956). *Pharm. Z.* 101, 238; cytisine or methylcytisine or both. [13] Ubaev, Kh. *et al.* (1963). *Chem. Abstr.* 59, 15602. Base $C_{12}H_{16}ON_2$, M.p. 134–135°, probably identical with methylcytisine; together with pedicularine (structure unknown) and the two plantaginaceous bases plantagonine and indicaine.

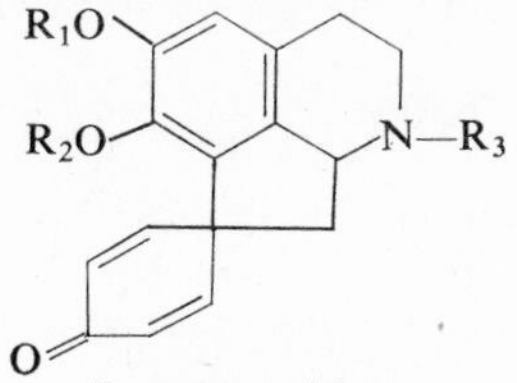

Protoaporphines

Alkaloid	R_1	R_2	R_3
Crotonsine	H	CH_3	H
Homolinearisine	H	CH_3	CH_3
Pronuciferine	CH_3	CH_3	CH_3
Glaziovine	CH_3	H	CH_3

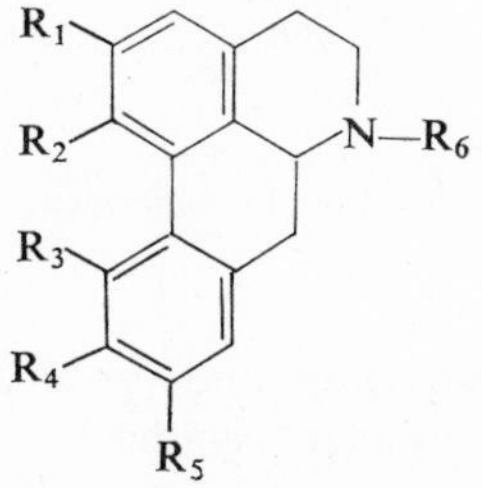

Aporphines

Alkaloid	R_1	R_2	R_3	R_4	R_5	R_6
Boldine	OH	OCH_3	H	OCH_3	OH	CH
Caaverine	OCH_3	OH	H	H	H	H
Domesticine	OCH_3	OH	H	$O—CH_2—O$		CH_3
Glaucine	OCH_3	OCH_3	H	OCH_3	OCH_3	CH_3
Isocorydine	OCH_3	OCH_3	OH	OCH_3	H	CH_3
Laurotetanine	OCH_3	OCH_3	H	OCH_3	OH	H
N-Methyllaurotetanine	OCH_3	OCH_3	H	OCH_3	OH	CH_3
Isoboldine	OCH_3	OH	H	OCH_3	OH	CH_3
Nornuciferine	OH	OCH_3	H	H	H	CH_3
Nuciferine	OCH_3	OCH_3	H	H	H	CH_3
Nantenine	OCH_3	OCH_3	H	$O—CH_2—O$		CH_3
Rogersine	OCH_3	OCH_3	H	OH	OCH_3	CH_3
Roemerine	$O—CH_2—O$		H	H	H	CH_3
Thalicmidine	OCH_3	OH	H	OCH_3	OCH_3	CH_3

FIG. 3. Proaporphines and aporphines mentioned in Table 4.

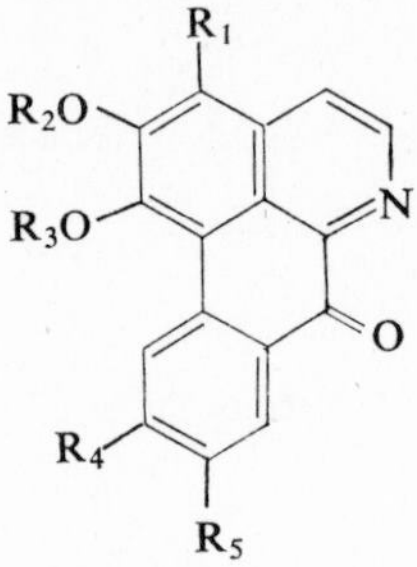

R. *Hegnauer*

Alkaloid	R_1	R_2	R_3	R_4	R_5
Liriodenine (=Oxoushinsonine= Spermatheridine)	H	—CH_2—		H	H
Liriodendron—Base II	H	CH_3	CH_3	OCH_3	OCH_3
Atherospermidine	OCH_3	—CH_2—		H	H

Fig. 3—*continued*

about the biogenesis of these compounds in both taxa, we can only speak of parallelism. Of course the character " synthesis of tylophorine-type alkaloids " remains a good one for *Tylophora* but as a guide for classification it is meaningless at present.

D. PAPILIONACEOUS ALKALOIDS (FIG. 2)

In Table 3 our present-day knowledge about distribution of three representatives of this group of alkaloids has been summarized. It has been demonstrated that in papilionaceous plants these alkaloids are mainly derived from lysine. For most other taxa which contain cytisine, sparteine or lupanine such knowledge is still lacking.

E. PROTOAPORPHINES AND APORPHINES (FIG. 3)

Taxa accumulating protoaporphine- and aporphine-type isoquinoline alkaloids are given in Table 4. Biogenetic experiments with members of the Berberidaceae, Ranunculaceae and Papaveraceae have demonstrated that they all produce these alkaloids by the same pathway. Experiments with Euphorbiaceae, Rhamnaceae and Symplocaceae would be highly desirable.

TABLE 4. Distribution of some proaporphine- and aporphine-type isoquinoline alkaloids (compare Fig. 3)*

Order[1]	Family	Genera (species)[2]	Individual alkaloids[2]
		Monocotyledoneae	
3 Liliiflorae	Liliaceae	*Camptorrhiza strumosa* (Bak.) Oberm.	Isocorydine[3].
		Dicotyledoneae—Archichlamydeae	
15 Magnoliales	Annonaceae	Several	Several.
	Hernandiaceae	*Illigera pulchra* Bl.	Laurotetanine?[4]
	Lauraceae	Many	More than 10 aporphines; also proaporphines (glaziovine).[5]
	Magnoliaceae	Several	Glaucine, magnoflorine, liriodenine, base II.[6]
	Monimiaceae	Several	Several; also liriodenine-type bases.[7]
16 Ranunculales	Berberidaceae	5 Genera	Magnoflorine, domesticine, nantenine.
	Menispermaceae	Several	Several.
	Nymphaeaceae	*Nelumbium*	Pronuciferine, nuciferine, nornuciferine, roemerine.[8]
	Ranunculaceae	Several	Magnoflorine, thalicmine, thalicmidine.
18 Aristolochiales	Aristolochiaceae	*Aristolochia, Bragantia*	Magnoflorine, menisperine.[9]
21 Papaverales	Papaveraceae	Many	Many.
26 Geraniales	Euphorbiaceae	*Croton linearis* Jacq.	See Table 6.
27 Rutales	Rutaceae	*Fagara, Phellodendron, Zanthoxylum*	Magnoflorine, menisperine, laurifoline, xanthoplanine.[10]
31 Rhamnales	Rhamnaceae	*Phylica rogersii* Pillans	Isocorydine, *N*-methyl-laurotetanine, rogersine.[11]
		Dicotyledoneae—Sympetalae	
5 Ebenales	Symplocaceae	*Symplocos celastrinea* Mart.	Caaverine, isoboldine(= *N*-methyl-laurelliptine).[12]

* See footnotes to Table 4, at foot of page 226.

F. THE QUATERNARY ALKALOIDS MAGNOFLORINE AND BERBERINE
(FIG. 4)

Table 5 demonstrates that even individual alkaloids may be rather wide-spread in nature. Quaternary bases have often been overlooked in the past, and it is to be expected that magnoflorine will be found in many other taxa in future. For the present I am inclined to consider this character as biogenetically identical in all the taxa mentioned in Table 5, and, at the same time, as indicating a true phylogenetic relationship between them.

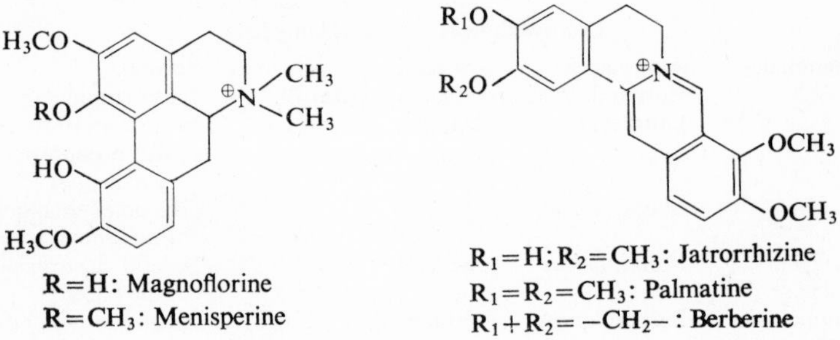

R=H: Magnoflorine
R=CH₃: Menisperine

$R_1=H; R_2=CH_3$: Jatrorrhizine
$R_1=R_2=CH_3$: Palmatine
$R_1+R_2= -CH_2-$: Berberine

FIG. 4. Magnoflorine and berberine-type quaternary alkaloids discussed in Section III F (see Table 5).

Footnotes to Table 4, p. 225.

[1] For numbers of orders see Table 2. [2] If no references are given, compare Boit (1961). [3] Kaul, J. L. et al. (1964). Coll. Czech. Chem. Commun. 29, 1689. [4] Greshoff, M. (1898). Meded. PlTuin., Batavia 25. [5] Gilbert, B. et al. (1964). J. Amer. chem. Soc. 86, 694. [6] Majumder, P. L. and Chatterjee, A. (1963). J. Indian Chem. Soc. 40, 929. Taylor, W. I. (1961). Tetrahedron 14, 42. Tomita, M. and Furukawa, H. (1962). J. Pharm. Soc. Japan 82, 616, 1199. Tsang-Hsiung Yang et al. (1962). J. Pharm. Soc. Japan 82, 816. Tsang-Hsiung Yang et al. (1963). 83, 216. [7] Bick, I. R. C. and Douglas, C. K. (1964) Tetrahedron Letters 1629. [8] Bernauer, K. (1963). Helv. chim. acta 46, 1783. Kupchan, S. M. et al. (1963). Tetrahedron 19, 227. [9] Katritzky, A. R. et al. (1960). J. Chem. Soc. 1950; Pailer, M. and Pruckmayr, G. (1959). M. Chemie 90, 145. Tomita, M. and Kura, S. (1957). J. Pharm. Soc. Japan 77, 812. [10] Cannon, J. R. et al. (1953). Aust. J. Chem. 6, 86. Comin, J. and Deulofeu, V. (1954). J. org. Chem. 19, 1774. Ishii, H. (1961). J. Pharm. Soc. Japan 81, 243. Ishii, H. and Harada, K. (1961). J. Pharm. Soc. Japan 81, 238. Jun-ichi Kunitomo (1962). J. Pharm. Soc. Japan 82, 611. Riggs, N. V. et al. (1961). Canad. J. Chem. 39, 1330. Tomita, M. and Ishii, H. (1957). J. Pharm. Soc. Japan 77, 810. Tomita, M. and Ishii, H. (1958). J. Pharm. Soc. Japan 78, 1180, 1141. Tomita, M. and Ishii, H. (1959). J. Pharm. Soc. Japan 79, 1228. [11] Arndt, R. R. (1963). J. chem. Soc. 2547. Arndt, R. R. and Baarschers, W. H. (1964). J. chem. Soc. 2244. Tschesche, R. et al. (1964). Tetrahedron 20, 1435. Tschesche, R. et al. (1965). Tetrahedron Letters 445.

TABLE 5. The distribution of magnoflorine, menisperine and berberine in Dicotyledoneae (see Fig. 4)

Order[1]	Family	Genera (species)[2]
	Magnoflorine (I) and menisperine (II)	
15 Magnoliales	Lauraceae	*Cryptocarya* (II).
	Magnoliaceae	*Magnolia* (I), *Michelia* (I).
16 Ranunculales	Berberidaceae	*Berberis* (I), *Epimedium* (I), *Mahonia* (I), *Nandina* (I), (II), *Caulophyllum* (I).
	Menispermaceae	*Cissampelos* (I), *Cocculus* (I), *Legnephora* (II), *Menispermum* (II), *Sinomenium* (I).
	Ranunculaceae	*Aconitum* (I),[4] *Aquilegia* (I),[5] *Caltha* (I),[6] *Coptis* (I), *Thalictrum* (I), *Zanthorrhiza* (I).[7]
18 Aristolochiales	Aristolochiaceae	*Aristolochia* (I), *Bragantia* (II).[8]
21 Papaverales	Papaveraceae	(I) isolated from opium, which is derived from *Papaver somniferum* L.[9]
27 Rutales	Rutaceae	*Fagara* (I), (II), *Phellodendron* (I), (II), *Zanthoxylum* (II).[10]
	Berberine[3]	
15 Magnoliales	Annonaceae	*Enantia*.[11]
16 Ranunculales	Berberidaceae	*Berberis, Mahonia, Nandina.*
	Menispermaceae	8 Genera.
	Ranunculaceae	*Aquilegia*,[5] *Coptis, Hydrastis, Thalictrum, Zanthorrhiza.*
21 Papaverales	Papaveraceae	8 Genera.
27 Rutales	Rutaceae[12]	*Orixa japonica* Thunb.[13], *Phellodendron*,[14] *Zanthoxylum*.[13]

[1] For numbers of orders see Table 2. [2] If no reference is given compare Boit (1961). [3] Including the closely related alkaloids palmatine and jatrorrhizine. [4] Nijland, M. M. (1963). *Pharm. Weekbl.* **98**, 623. [5] Winek, Ch.L. *et al.* (1963). *Lloydia* **26**, 205; Winek, Ch.L. *et al.* (1964). *Lloydia* **27**, 111; Winek, Ch.L. *et al.* (1964). *J. Pharm. Sci.* **53**, 734. [6] Nijland, M. M. (1963). *Pharm. Weekbl.* **98**, 261; M. M. Nijland and Uffelie, O. F., also demonstrated magnoflorine to be present in the subterranean organs ($\geqq 0.05\%$ of dry weight) of species of the following genera: *Adonis, Anemonella, Delphinium, Isopyrum, Nigella, Trautvetteria* and *Trollius* (1965). *Pharm. Weekbl.* **100**, 49. [7] Hussein, F. T. *et al.* (1963). *Lloydia* **26**, 254. [8] For references see table 4, number (7). [9] Nijland, M. M. (1965). *Pharm. Weekbl.* **100**, 88. [10] For references see table 4, number (8). [11] Buzas, A. *et al.* (1959). *C.R. Acad. Sci., Paris* **248**, 1397, 2791. Seitz, G. (1959). *Naturwissenschaften* **46**, 263. [12] Berberine has often been confused with other coloured quaternary alkaloids (e.g. chelerythrine, sanguinarine; compare Widera (1902)[13]; Govindachari, T. R. and Thyagaranjan, B. J. (1956). *J. chem. Soc.* 769). Only occurrences which seem to be firmly established are mentioned. [13] Widera, R., *Pharmakognostisch-chemische Studien über die Verbreitung des Berberins*, Thesis, Strassburg 1902. [14] Tomita, M. and Kunitomo, J. (1958). *J. Pharm. Soc. Japan* **78**, 1444.

G. THE ALKALOIDS OF EUPHORBIACEAE (FIG. 5)

Euphorbiaceous species with alkaloids of known structure are collected together in Table 6. This Table clearly demonstrates that alkaloid metabolism in plants of one family (admittedly a rather heterogeneous one!) can diverge greatly.

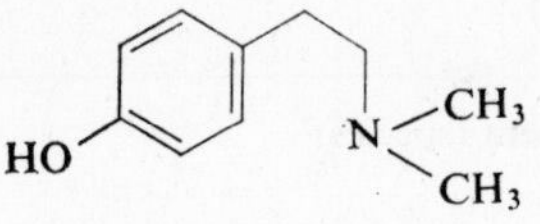

228   R. Hegnauer

Hordenine (= Flueggeine)

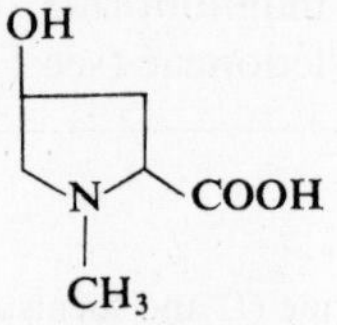

4-Hydroxy-
hygrinic acid

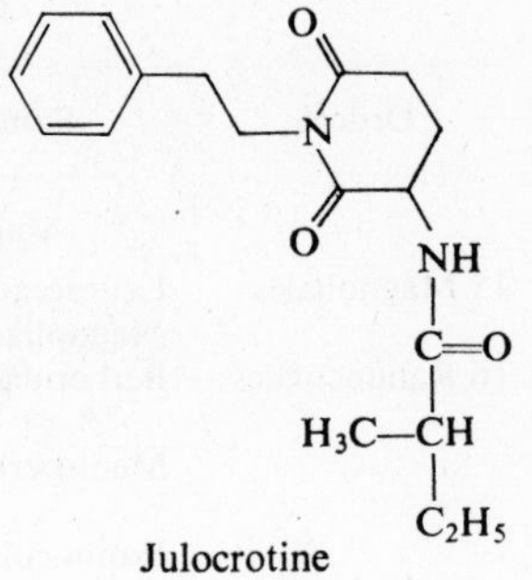

Julocrotine

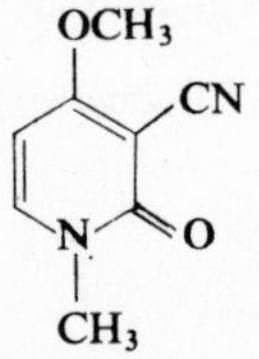

Ricinine

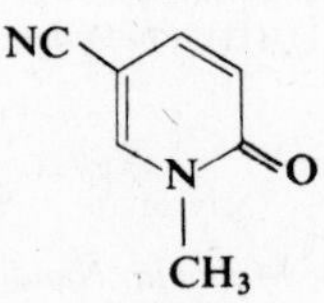

Nudiflorine

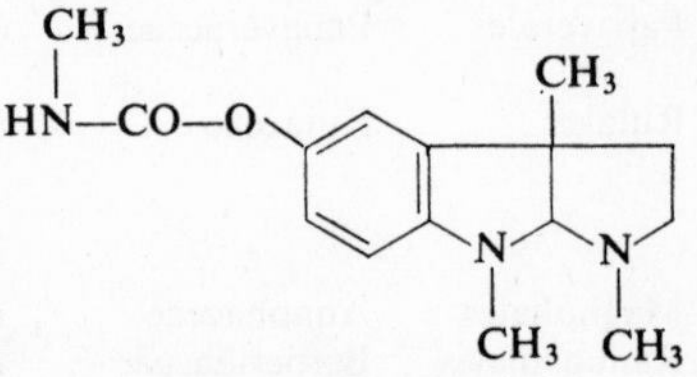

Physostigmine

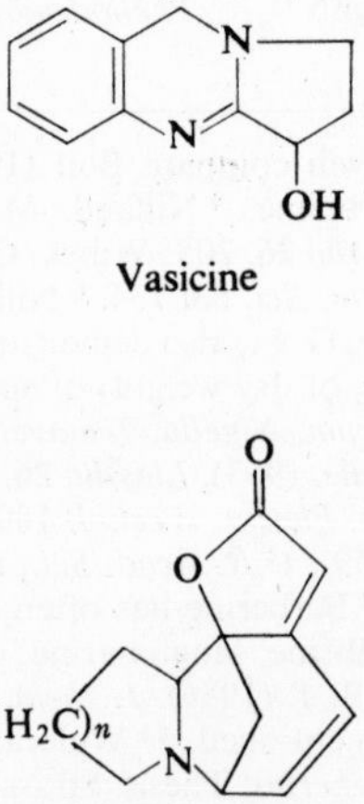

Vasicine

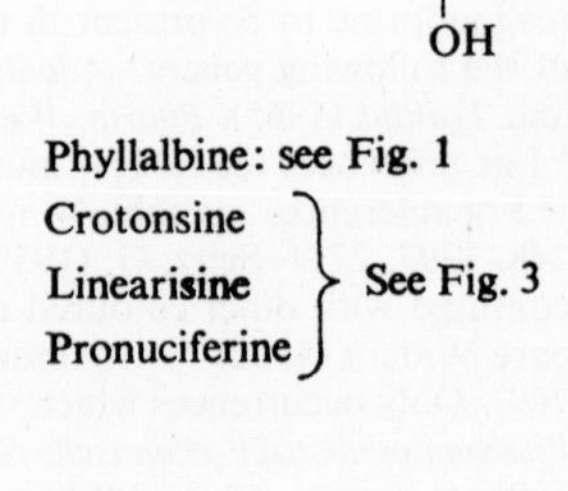

Yohimbine

Phyllalbine: see Fig. 1
Crotonsine
Linearisine   } See Fig. 3
Pronuciferine

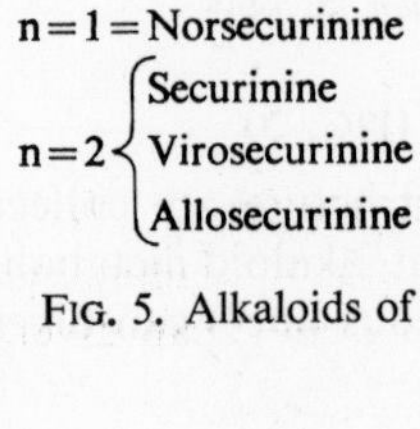

n = 1 = Norsecurinine

n = 2 { Securinine
        Virosecurinine
        Allosecurinine (= Phyllochrysine)

FIG. 5. Alkaloids of Euphorbiaceae discussed in Section III G (compare Table 6).

TABLE 6. Alkaloids and Alkaloidal Types found in the Euphorbiaceae (compare Fig. 5)*

Sub-family	Species	Alkaloids	Alkaloidal Types
Phyllanthoideae (=Antidesmataceae)	*Phyllanthus discoideus* Muell. Arg.	Securinine Phyllochrysine	} Securinine-type[3]
		Phyllalbine[4]	Tropane-type
	Securinega suffruticosa (Pall.) Rehd.	Securinine Suffruticodine Dihydrosecurinine Allosecurinine (= Phyllochrysine)	} Securinine-type.[5]
	S. virosa (Willd.) Pax et Hoffm. (= *Flueggea virosa* Baill.)	Virosecurinine Norsecurinine	} Securinine-type[6, 7]
		Hordenine (= Flueggeine)[7]	} Tyrosine derived biogenic amine.
Euphorbioideae (=Crotonoideae)[1]	*Croton*[2] *linearis* Jacq.	Pronuciferine Linearisine Homolinearisine Crotonsine	} Proaporphines.[8]
	C. sparsiflorum Morong	Sparsiflorine[9]	?Proaporphine.
	C. gabouga S. Moore	4-Hydroxyhygrinic acid[10]	Stachydrine-type protoalkaloid.
	C. draco Schlecht.?	Ester of vasicine[11]	Vasicine-type.
	C. turumiquirensis Stayerm.	Turumiquirensine	Bisbenzylisoquinoline-type?[12]
	Julocroton[2] *montevidensis* Klotzsch	Julocrotine	Phenylalanine derived protoalkaloid.[13]
	Alchornea[2] *floribunda* Muell. Arg.	Yohimbine[14]	Complex indolic base-type.
	Hippomane mancinella L.	Physostigmine	Simple indolic base.[15]
	Ricinus communis L.	Ricinine	Nicotinic acid-type.
	Trewia nudiflora L.	Nudiflorine[16]	Nicotinic acid-type.

* See also Addendum p. 230.
[1] Alkaloids are present in several other genera of this sub-family but their structure is not yet known. [2] Other species of these genera are known to contain alkaloids. The latter however have not yet been identified with certainty. [3] Parello, J. *et al.* (1963) *Bull. Soc. Chim. Fr.* 898. Bevan, C. W. L. *et al.* (1964). *Chem. & Ind.* 838. [4] Parello, J. *et al.* (1963). *Bull. Soc. Chim. Fr.* 2787. [5] Horii, Z. *et al.* (1964). *Chem. Pharm. Bull.* (*Tokyo*) **12**, 1118. Saito, S. *et al.* (1963). *Chem. & Ind.* 689; Saito, S. *et al.* (1963). *J. Pharm. Soc. Japan* **83**, 800. Satoda, I. *et al.* (1962). *Tetrahedron Letters* 1199. [6] Nakano, T. *et al.* (1963). *Tetrahedron Letters* 665; Nakano, T. *et al.* (1963) *Chem. & Ind.* 1034, 1763. Iketobosin, G. O. and Mathieson, D. W. (1963). *J. Pharm. Pharmacol.* **15**, 810. [7] Iketobosin-Mathieson, *l.c. sub* 6. [8] Haynes, L. J. and Stuart, K. L. (1963). *J. chem. Soc.* 1784, 1789; Haynes, L. J. *et al.* (1963). *Proc. chem. Soc.* 1887; Haynes, L. J. *et al.* (1964). *Proc. chem. Soc.* 261. [9] Srisha Kumar, Saha (1959). *Sci. Culture* (*Calcutta*) **24**, 572. [10] Goodson, J. A. and Clewer, H. W. B. (1919). *J. chem. Soc.* **115**, 923. [11] Sodi Pallares, E. (1946). *Biochem. Arch.* **10**, 235. Botanical identity of the bark investigated uncertain. [12] Burnell, R. H. and Della Casa, D. (1964). *Nature, Lond.* **203**, 296. [13] Nakano, T. *et al.* (1959). *Tetrahedron Letters* No. 14 8–12; Nakano, T. *et al.* (1961). *J. Org. Chem.* **26**, 1184. [14] Paris, R. R. and Goutarel, R. (1958). *Ann. Pharm. Fr.* **16**, 15. [15] Lauter, W. M. and Foote, P. A. (1955). *J. Amer. Pharm. Assoc.* **44**, 361. [16] Mukherjee, R. and Chatterjee, A. (1964). *Chem. & Ind.* 1524.

IV. CONCLUSIONS

Alkaloids, in some instances, may be highly valuable guides for classification. There are, however, several difficulties which temper optimism: their distribution in plants is still very incompletely known and parallelism and diversification of alkaloid metabolism occur frequently.

The fragmentary present-day knowledge of distribution may cause premature taxonomic conclusions to be drawn and parallelism and diversification prevent a sound taxonomic use of these compounds as long as they cannot be interpreted in terms of convergence, divergence, analogy and homology.

Notes added in proof: Since this manuscript was written several new alkaloidal types and alkaloid-containing taxa have been described.

Addenda to Table 2 (numbers = numbers of orders):

Archichlamydeae—7. Moraceae: *Ficus pantoniana* King: New, flavonoidal alkaloids (ficine etc.). 13. Amaranthaceae: Canthinone-type alkaloid from *Charpentiera obovata* Gaudich. 30. Icacinaceae: Tubulosine and deoxytubulosine from *Cassinopsis ilicifolia* Kuntze. Pandaceae: Peptide alkaloids from *Panda oleosa* Pierre. 36. Lythraceae: Alkaloids of this family represent a new type of quinolizidine alkaloids. Rhizophoraceae: Hygroline from leaves of *Carallia brachiata* (Lour.) Merr. 37. Araliaceae: *Mackinleya* species produce a new type of quinazoline alkaloids.

Sympetalae—8. Acanthaceae: *Macrorungia longistrobus* C. B. Clarke, synthesizes a new type of quinoline alkaloids (macrorine etc.). Bignoniaceae: Indolic alkaloids from *Newbouldia laevis* Seem.

Monocotyledoneae—3. Liliaceae: Androcymbine and melianthioidine (bis-benzylisoquinoline group) represent new types of alkaloids. 7. Gramineae: Festucin is a pyrrolizidine-type base. 14. Orchidaceae: Chysin A and B from *Chysis bractescens* Lindl. are pyrrolizidine-type alkaloids.

Addenda to Table 6. Phyllanthoideae: *Astrocasia phyllanthoides* Robinson et Millsp. contains astrocasin, which may be biogenetically related to the papilionaceous alkaloid adenocarpine. Crotonoideae: *Croton cumingii* Muell. Arg. contains magnoflorine!. Sparsiflorine is *N*-norapoglaziovine.

REFERENCES

Altman, R. F. A. (1958). Inst. Nacional de Pesquisas da Amazonia, Quimica Publ. No. 3, Rio de Janeiro.

Boit, H.-G. (1961). "Ergebnisse der Alkaloidchemie bis 1960, Akademie-Verlag, Berlin.

Brauchli, P., Deulofeu, V., Budzikiewicz, H. and Djerassi, C. (1964). *J. Amer. chem. Soc.* **86**, 1895.

Conroy, H. (1960). *Tetrahedron Letters* **10**, 34.

Franck, B. (1958). *Chem. Ber.* **91**, 2803.

Franck, B. and Hartmann, W. (1963). *Abhandl. dtsch. Akad. Wiss. Berl.* **4**, 111.

Hill, R. K., Chan, T. H. and Joule, J. A. (1965). *Tetrahedron* **21**, 147.

Leete, E. (1963). *In* "Biogenesis of Natural Compounds" (P. Bernfeld, ed.), Chapter 13. Pergamon Press, Oxford.

Lüning, B. (1964). *Acta chem. scand.* **18**, 1507.

McNair, J. B. (1931). *Amer. J. Bot.* **18**, 416.

Mortimer, P. I. (1957). *Aust. J. Sci.* **20**, 87.

Mortimer, P. I. and Wilkinson, S. (1957). *J. chem. Soc.* 3967.

Russel, J. H. (1963). *Naturwissenschaften* **50**, 444.

Schröter, H.-B. (1963). *Abhandl. dtsch. Akad. Wiss., Berl.* **4**, 99.

Schütte, H. R. and Hindorf, H. (1964). *Naturwissenschaften* **51**, 463.

CHAPTER 14

The Betacyanins and Betaxanthins

TOM J. MABRY

Department of Botany, The University of Texas, Austin, Texas, U.S.A.

I. INTRODUCTION

Notable advances in the chemistry of the betacyanin-betaxanthin pigments since the preparation of the first crystalline samples of "nitrogenous anthocyanins" in 1957 (Wyler and Dreiding, 1957; Schmidt and Schönleben, 1957), have been paralleled by significant new insights into the possible phylogenetic implications of these substances. The recent reviews by Dreiding (1961) and Mabry (1964) cover comprehensively the older publication pertaining to the betacyanins, and no attempt will be made to recount these early observations here.

The betacyanins and betaxanthins are the red-violet and yellow pigments, respectively, which are only found, so far as is known, in ten families of flowering plants belonging to the order Centrospermae. The expressions betacyanin and betaxanthin were derived from the generic name of the red beet, *Beta vulgaris* and from anthocyanin and anthoxanthin (flavones, etc.), respectively, reflecting the erroneous but long prevailing inference of a direct chemical relationship between the betacyanin-betaxanthin class of pigments and the flavonoids. A correlation between the occurrence of betacyanins and the Centrospermae has been recognized for many years, and the recent discovery that neither betacyanins nor betaxanthins are related chemically to the anthocyanins, as was previously believed, has stimulated a re-evaluation of their possible phylogenetic significance.

The taxonomic reliability of the betacyanins and betaxanthins rests not only on their chemically unique structures and the observation that their distribution is limited to the Centrospermae, but also on the fact that they and the much more widely distributed anthocyanin pigments are mutually exclusive (although other classes of flavonoid pigments occur in the Centrospermae). For these reasons, together with morphological considerations, the order Centrospermae has been recognized as containing the ten betacyanin families: Chenopodiaceae, Portulacaceae, Amaranthaceae, Nyctaginaceae, Phytolaccaceae, Stegnospermaceae (often treated as a subfamily of the Phytolaccaceae), Aizoaceae, Basellaceae, Cactaceae and Didieraceae (Mabry, 1964; Mabry *et al.*, 1963). Furthermore, it has been proposed that the anthocyanin-containing families formerly included by many taxonomists in this order be separated. Thus the families Caryophyllaceae and Illecebraceae are recognized as belonging to a related but distinct order, the Caryophyllales.

II. Structure of the Betacyanins

Prior to 1957, when both Dreiding (Wyler and Dreiding, 1957) and Schmidt (Schmidt and Schönleben, 1957) and their co-workers succeeded almost simultaneously in preparing for the first time authentic samples of crystalline betanin (I) (the betacyanin from *Beta vulgaris*), little structural evidence had been accumulated regarding the true nature of the betacyanins, which were still generally described at that time as "nitrogenous anthocyanins". This name acknowledged the nitrogen content of the pigments but incorrectly presumed a direct structural relationship to the anthocyanins.

Much of the recent degradative work was carried out with betanidin, the aglycone obtained from the acidic hydrolysis of the monoglucoside betanin (Pucher *et al.*, 1938). In the period 1959–62, Wyler and Dreiding (1959, 1962) identified three important products from the alkaline degradation of betanidin (III): 5,6-dihydroxy-2,3-dihydroindole-2-carboxylic acid (IV), formic acid, and 4-methylpyridine-2,6-dicarboxylic acid (V) (Fig. 1).

Mabry *et al.* (1962) further clarified the structural features of this class of pigments when the structures of neobetanidin, an oxidized form of betanidin, and a number of neobetanidin derivatives including VII and VIII were established (Fig. 2). These results suggested the provisional formulation of betanidin and betanin as III and I, respectively, and in addition, demonstrated that the structures were unrelated to the anthocyanins and indeed were members of a new class of plant pigments. The misleading expression "nitrogenous anthocyanins" was therefore finally discarded. The detailed structure for betanidin (III), which emerged from the analysis of 5,6-di-*O*-acetylbetanidin (Wyler *et al.*, 1963), was in some respects quite remarkable; it demonstrated the presence of an unusual combination of structural features including dihydroindole and dihydropyridine rings and a polymethylene cyanine chromophore. An alternative resonance structure for betanidin would have a positive charge on the dihydropyridine nitrogen.

The absolute stereochemical configuration at C-2 in betanidin was shown to be *S* by the isolation of 5,6-dihydroxy-2,3-dihydro-2*S*-indole-2-carboxylic acid (IV) from the alkaline degradation of betanidin (Wyler and Dreiding, 1962; Wyler *et al.*, 1963). More work is needed to clarify the configuration at the only other asymmetric centre in betanidin, C-15, but the results of Schmidt

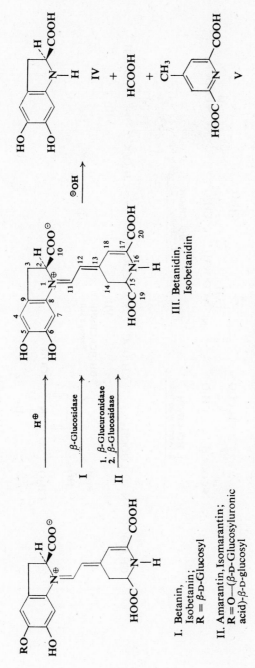

FIG. 1. Hydrolytic and degradative reactions given for several betacyanin pigments.

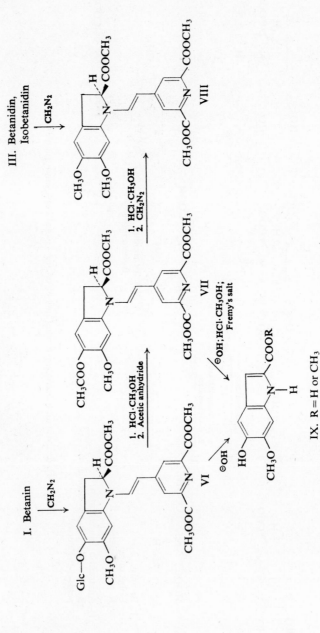

FIG. 2. Synthesis and degradation of neobetanin and neobetanidin derivatives.

et al. (1960) for the oxidative cleavage of betanidin indicate that this centre may also have an *S*-configuration. Schmidt and co-workers detected L-aspartic acid from the ozonolysis and peracetic acid oxidation of betanidin. Although L-aspartic acid might theoretically arise from either the dihydroindole or dihydropyridine rings of (III), it is likely that it was derived from the latter, analogously with the oxidative results observed for indicaxanthin (X) (Fig. 3).

Acidic treatment of betanidin or betanin transforms the pigments into isomeric compounds, isobetanidin and isobetanin, respectively. The difference between the members of these isomeric pairs of pigments presumably arises from inversion of the configuration at C-15 since both betanidin and isobetanidin could be transformed by diazomethane treatment into the same product, 5,6-di-*O*-methyl-2*S*-neobetanidin trimethyl ester (VIII) Wyler *et al.*, 1963.

The remaining major problems in the structural analysis of betanin, in addition to the still unresolved configuration at C-15, were the detailed structure of the glucosyl moiety and the manner and position of its linkage to betanidin. Hydrolytic studies with β-glucosidase (Piattelli *et al.*, 1964a; Wilcox *et al.*, 1965) and the N.M.R. spectral data for betanin in trifluoroacetic acid (Wilcox *et al.*, 1965) demonstrated that (I) was an *O*-β-D-glucopyranoside of (III) (Fig. 1). The position of the linkage of the glucose in betanin was only recently clarified by Piattelli *et al.* (1964) and Wilcox *et al.* (1965). Using different procedures, both groups cleaved the neobetanin derivative, (VI), to 5-hydroxy-6-methoxy-indole-2-carboxylic acid or its methyl ester, (IX) (Fig. 2). This product required that betanin be a 5-*O*-glucoside of betanidin. Thus, betanin and isobetanin can be formulated as 5-*O*-β-D-glucopyranosyl-betanidin and isobetanidin, respectively.

Another pair of isomeric betacyanins, amarantin and isoamarantin (II), were recently described (Piattelli *et al.*, 1964b) from the leaves of *Amaranthus tricolor*. Careful hydrolysis of amarantin and isoamarantin with β-glucuronidase followed by β-glucosidase transformed the glycosides into betanidin and isobetanidin, respectively. As with betanin, amarantin could be fragmented after methylation to (IX), thereby establishing the position of the glycosyl moiety. These betacyanins could, therefore, be formulated as *O*-(β-D-glucopyranosyluronic acid)-5-*O*-β-D-glucopyranosides of betanidin and isobetanidin (II), respectively.

Piattelli and Minale (1964) reported the occurrence of at least 44 different betacyanins. This report is of considerable interest, particularly since 29 of the substances were hydrolysed and yielded either betanidin or isobetanidin. No data were reported for the sugar components. The latter information is pertinent to any arguments regarding structural variations among the betacyanins. The available evidence suggests, of course, that the differences in these betacyanins involve only modifications in the glycosidic patterns, the configuration at C-15 and perhaps esterification of the carboxyl groups.

III. STRUCTURE OF THE BETAXANTHINS

Although long suspected (Spragg, 1960; Dreiding, 1961; Mabry, 1964), a direct structural relationship between the two classes of Centrospermae

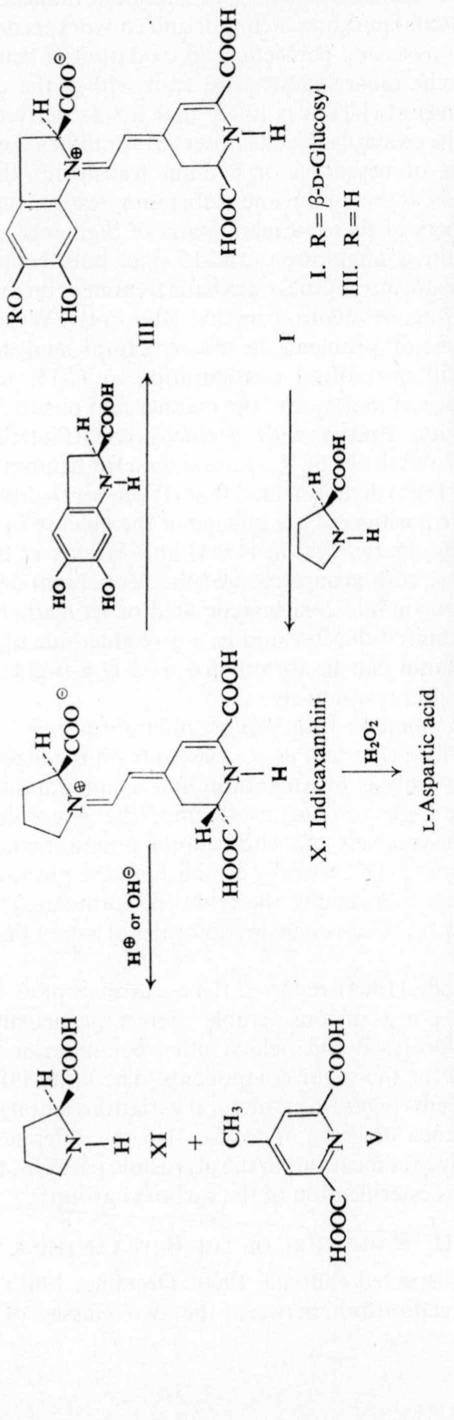

FIG. 3. Degradative reactions given by indicaxanthin; interconversion of betacyanins and indicaxanthin.

pigments, the red-violet betacyanins and the yellow betaxanthins, was only recently confirmed with the isolation and structural analysis of indicaxanthin (X), the betaxanthin from *Opuntia ficus-indica* fruits (Piattelli *et al.*, 1964c). The structure of indicaxanthin was established by acidic and alkaline degradations of indicaxanthin to L-proline and 4-methylpyridine-2,6-dicarboxylic acid (Fig. 3). Moreover, the detection of L-aspartic acid from the hydrogen peroxide oxidation of this betaxanthin, combined with the other data, allowed the formulation of the absolute structure (X) for indicaxanthin. A partial synthesis of (X) was effected by a novel reaction: indicaxanthin was formed by the careful treatment of betanin (I) with L-proline (Wyler *et al.*, 1965). Furthermore indicaxanthin could be transformed into betanidin by allowing it to react with 5,6-dihydroxy-2,3-dihydroindole-2-carboxylic acid. These imino acid exchange reactions with the dihydropyridine component of the betacyanin-betaxanthin pigments are of particular interest relevant to the possible biosynthetic relationships among these substances. Several betaxanthin pigments have been reported in the literature, all having visible absorption maxima in the region 474–486 mμ (Reznik, 1955, 1957; Piattelli *et al.*, 1964c; Mabry and Turner, 1964; Wyler *et al.*, 1965). It seems reasonable, on the basis of the imino acid exchange experiments, that modifications of the proline component or substitution of alternative amino acids for proline in (X) would account for still other members of this class of yellow pigments. Piattelli *et al.* (1965) have reported two additional betaxanthins from *Beta vulgaris*: vulgaxanthin—I and vulgaxanthin—II. They have glutamine and glutamic acid, respectively, combined with the dihydropyridine moiety found in all betacyanin-betaxanthin pigments.

IV. BIOSYNTHESIS OF THE BETACYANINS AND BETAXANTHINS

A tentative biogenetic scheme is offered for the betacyanins and betaxanthins (Fig. 4). It is known that beet seedlings grown on a medium containing L-β-(3,4-dihydroxyphenyl)alanine-2-[14]C (DOPA) incorporate ^{14}C into betanin (Hörhammer *et al.*, 1964; Mabry and Wohlpart, 1965). As yet, however, no degradative results for labelled betanin have been reported. Therefore, important questions such as the position of the label and the number of molecules of DOPA incorporated into betanin cannot be resolved.

Betalamic acid (XII), though not yet isolated, was proposed as an intermediate in the interconversion of betacyanins and betaxanthins (Wyler *et al.*, 1965). This acid might be formed from DOPA by an oxidative cleavage of the aromatic ring and subsequent closure to a dihydropyridine system which, on condensation with an appropriate amino acid, could yield directly betanidin, indicaxanthin, or other members of the betacyanin-betaxanthin class of pigments.

V. TAXONOMIC SIGNIFICANCE OF THE BETACYANINS AND BETAXANTHINS

The betacyanins are restricted to ten families belonging to the order Centrospermae: Chenopodiaceae, Amaranthaceae, Portulacaceae, Nycta-

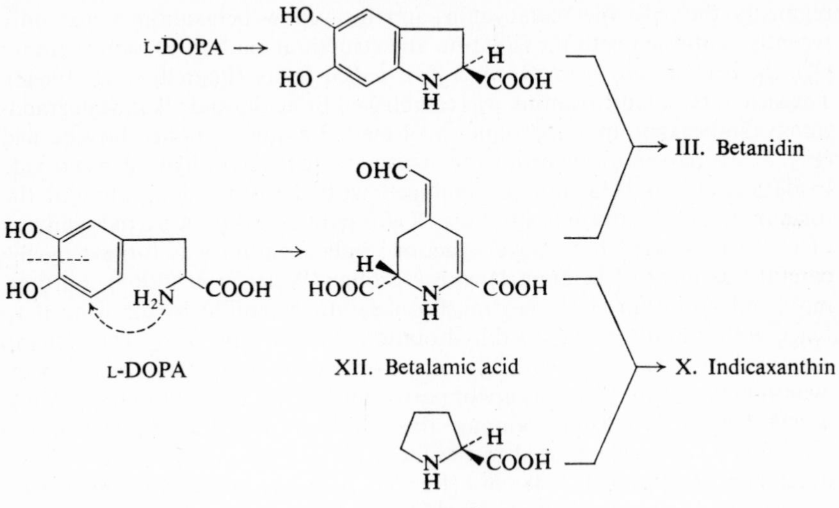

238   *Tom J. Mabry*

FIG. 4. Tentative biosynthetic pathways for betacyanins and betaxanthins.

ginaceae, Phytolaccaceae, Stegnospermaceae, Aizoaceae, Basellaceae, Cactaceae and Didieraceae. All of these families, except the Basellaceae, Didieraceae, and Stegnospermaceae, are known to contain members producing betaxanthins as well. However, most of the genera and species in these families remain uninvestigated. Unfortunately most workers have not published their negative results (i.e., when the pigments were absent or else present in quantities below reliable analytical detection); hence it is not possible to give an accurate estimate of the number of taxa that have actually been investigated. For example, of the more than 650 genera and 8,000 species in these ten families (Eckhardt, 1964) only about 80 genera and 170 species have been described as containing betacyanins (Table 1). Our knowledge of the distribution of the betazanthins is even more limited. About 30 genera from seven families are known to contain these yellow pigments (Table 1) (Reznik, 1955, 1957; Mabry and Turner, 1964; Mabry and Wohlpart, 1965). The expression "betacyanin families" is, however, now frequently used to designate these ten families, implying the expectation that more extensive surveys will further substantiate the present view that plants containing the betacyanin-betaxanthin series of pigments are phylogenetically closely related.

As noted above, it was suggested recently that the order Centrospermae might best be recognized as containing the ten betacyanin families. While major taxonomic importance would not ordinarily be accorded a single character or even several chemical characters, the totally different structures of the two classes of pigments, the betacyanins and anthocyanins, which indicate different biosynthetic pathways, their mutual exclusion, and the limited distribution of the betacyanins make the presence of these latter substances of

TABLE 1. Betacyanin-betaxanthin containing genera[1,2]

Chenopodicaceae: (over 100 genera, 1500 species)		**Portulacaceae: (19 genera; 500 species)**		**Cactaceae: (about 200 genera, 2000 species)**	
Atriplex	(150: 5)	*Anacampseros*	(70: 1)	*Ariocarpus*	(8: 5)
Beta[3]	(12: 1)	*Calandrinia*	(150: 1)	*Aylostera*	(10: 1)
Chenopodium	(200: 7)	*Claytonia*	(20: 3)	*Cereus*[3]	(40: 3)
Corispermum	(50: 2)	*Montia*	(50: 1)	*Chamaecereus*[3]	(11: 1)
Cycloloma	(1: 1)	*Portulaca*[3]	(100: 3)	*Cleistocactus*[3]	(35: 1)
Kochia	(80: 2)	*Spraguea*	(5: 1)	*Gymnocalycium*	(70: 3)
Salicornia	(30: 1)	*Talinum*[3]	(50: –)	*Hariota*[3]	(2: 1)
Salsola	(100: 1)			*Hylocereus*	(20: 1)
Spinacia	(3: 1)	**Basellaceae: (4 to 5 genera; 20 species)**		*Lobivia*[3]	(75: 2)
Suaeda	(100: 2)	*Anredera*	(2: 1)	*Mammillaria*[3]	(300: 7)
		Basella	(5: 2)	*Melocactus*	(35: 1)
Amaranthaceae: (over 60 genera, 900 species)				*Monvillea*	(5: 1)
Achyranthes	(5: 1)	**Stegnospermaceae: (one genus; 3 species)**		*Neoporteria*[3]	(40: 1)
Aerva	(10: 1)	*Stegnosperma*	(3: 1)	*Nopalxochia*	(2: 1)
Alternanthera[3]	(170: 6)			*Notocactus*	(20: 1)
Amaranthus[3]	(50: 13)	**Phytolaccaceae: (about 17 genera, 120 species)**		*Opuntia*[3]	(200: 6)
Celosia[3]	(60: 4)	*Phytolacca*	(35: 3)	*Parodia*[3]	(35: 3)
Froelichia	(10: 1)	*Rivina*[3]	(3: 2)	*Pereskia*[3]	(20: 1)
Gomphrena	(100: 4)	*Trichostigma*	(3: 1)	*Phyllocactus*	(20: 1)
Iresine	(70: 2)			*Rebutia*[3]	(40: 3)
Mogiphanes	(12: 1)	**Aizoaceae (Ficoidaceae); 130 genera; 2500 species**		*Selinocereus*	(23: 1)
Tidestromia	(6: 1)	*Aptenia*[3]	(2: 1)	*Thelocactus*	(20: 1)
		Bergeranthus[3]	(15: –)	*Zygocactus*[3]	(2: 1)
Didieraceae: (4 genera; 11 species)		*Conophytum*[3]	(300: 17)		
Alluandia	(6: 3)	*Dorotheanthus*	(6: 1)		
Alluandiopsis	(2: 2)	*Faucaria*[3]	(35: –)		
Decarya	(1: 1)	*Fenestraria*[3]	(2: 1)		
Didierea	(2: 2)	*Gibbaeum*	(30: 2)		
		Glottiphyllum[3]	(60: –)		
Nyctaginaceae: (30 genera; 300 species)		*Lampranthus*[3]	(200: 2)		
Abronia	(20: 1)	*Lithops*[3]	(70: 1)		
Boerhaavia	(20: 1)	*Malephora*	(1: 1)		
Bougainvillea[3]	(14: 2)	*Mesembryanthemum*	(45: 4)		
Cryptocarpus	(2: 1)	*Pleiospilos*[3]	(30: 2)		
Cyphomeris	(2: 1)	*Rabiaea*[3]	(5: –)		
Mirabilis[3]	(60: 3)	*Rhombophyllum*[3]	(3: –)		
Nyctaginia	(1: 1)	*Sesuvium*	(8 :1)		
Oxybaphus	(30: 1)	*Tetragonia*	(60: 1)		
		Trianthema	(15: 1)		
		Trichodiadema	(26: 1)		

[1] References: Dreiding, 1961; Mabry *et al.*, 1963; Piattelli and Minale, 1964a; Mabry and Wohlpart, 1965.

[2] Numbers in parenthesis following generic names represent the estimated number of species in the genus; the number of species known to contain betacyanins.

[3] Contain betaxanthins (Reznik, 1955, 1957; Mabry and Turner, 1964; Mabry and Wohlpart, 1965).

particular taxonomic significance (Mabry, 1964; Mabry *et al.*, 1963). Furthermore, the fact that another group of closely related pigments, the betaxanthins, may also be restricted to these ten families strengthens the case for the treatment of the Centrospermae as proposed here.

TABLE 2. Recent treatments of betacyanin families

Hutchinson (1959)	Eckhardt (1964)	Buxbaum (1961)	Mabry, *et al.* (1963)
Caryophyllales	Centrospermae	Caryophyllales	Centrospermae
Elatinaceae	Phytolaccaceae[1]	Phytolaccaceae[1]	Chenopodiaceae[1]
Molluginaceae	Gyrostemonaceae[2]	Aizoaceae[1]	Amaranthaceae[1]
Caryophyllaceae	Achatocarpaceae[2]	Cactaceae[1]	Nyctaginaceae[1]
Aizoaceae[1]	Nyctaginaceae[1]	Portulacaceae[1]	Portulacaceae[1]
Portulacaceae[1]	Molluginaceae	Nyctaginaceae[1]	Phytolaccaceae[1]
	Aizoaceae[1]	Basellaceae[1]	(including the
Chenopodiales	Portulacaceae[1]	Amaranthaceae[1]	Petiveriaceae)
Barbeuiaceae[2]	Basellaceae[1]	Batidaceae	Stegnospermaceae[1, 2]
Phytolaccaceae[1]	Caryophyllaceae	Chenopodiaceae[1]	Aizoaceae[1]
Gyrostemonaceae[2]	Dysphaniaceae	Caryophyllaceae	Basellaceae[1]
Agdestidaceae[2]	Chenopodiaceae[1]		Cactaceae[1]
Petiveriaceae[1, 2]	Amaranthaceae[1]		Didieraceae[1]
Chenopodiaceae[1]	Didiereaceae[1]		
Amaranthaceae[1]			
Cynocrambaceae	Cactales		
Batidaceae	Cactaceae[1]		
Basellaceae[1]			
Thymelaeales (in part)			
Nyctaginaceae[1]			
Pittosporales (in part)			
Stegnospermaceae[1, 2]			
Cactales			
Cactaceae[1]			
Sapindales (in part)			
Didieraceae[1]			

[1] Betacyanin families.
[2] These families are often treated as sub-families of the Phytolaccaceae and, except for the betacyanin-containing Petiveriaceae and Stegnospermaceae, to the author's knowledge, they have not been investigated for their pigments.

It is of interest at this point to compare certain recent systematic treatments of the betacyanin families which were based primarily on morphological evidence (Table 2) with the treatment of Mabry *et al.* (1963) and Mabry (1964), who used both morphological and chemical criteria in arriving at their ten-family Centrospermae.

According to Buxbaum (1961), who has presented the most recent comprehensive discussion on the taxonomic constitution and systematic position of the Centrospermae (his Caryophyllales), the order corresponds closely to the natural relationship as deduced for these families by the presence of betacyanins (Table 2). In addition to most of the betacyanin families, he included in this order the Batidaceae, which contain neither anthocyanins nor betacyanin-betaxanthin pigments (Mabry and Turner, 1964) and the Caryophyllaceae. He also excluded the Didieraceae. However, Rauh and Reznik have presented both morphological and chemical evidence for the inclusion of the latter family within the Centrospermae; an interpretation later affirmed by Schölch (1963) who positioned the Didieraceae near the Portulacaceae and Cactaceae purely on morphological grounds. Buxbaum was aware that anthocyanins rather than betacyanins occurred in the Caryophyllaceae but discounted this chemical evidence and suggested the rather unlikely hypothesis that certain members of the Centrospermae had lost the ability to produce betacyanins and subsequently developed a biosynthetic pathway to the anthocyanins from the already present anthoxanthins. He felt there were sufficient morphological reasons for separating from his Caryophyllales the small family Cynocrambaceae, which has been frequently treated as a member of the Centrospermae. Since there are no known pigmented members of this family, its phyletic position cannot be interpreted on the betacyanin-betaxanthin criterion.

Eckhardt (1964), whose treatment of the Centrospermae is most recent, included eight betacyanin families along with the anthocyanin-containing families, Molluginaceae (Beck *et al.*, 1962) and Caryophyllaceae (Table 2). He realized that betacyanins did not occur in the Caryophyllaceae and Molluginaceae but apparently was not aware of the strikingly distinct chemical nature of these compounds. The Gyrostemonaceae and Achatocarpaceae, both previously aligned with the Phytolaccaceae, as well as the small group, Dysphaniaceae, formerly of the Caryophyllaceae, were treated as Centrospermae families; all three, however, remain to be examined for their pigments. The Cactaceae were treated as a closely allied but separate order, the Cactales.

If the betacyanin-betaxanthin criterion for delimiting the Centrospermae is valid, this would tend to discount, in a large measure, the system proposed by Hutchinson (1959). His division of the Dicotyledoneae into two fundamental groups, woody and herbaceous, results in the separation of the betacyanin families into several remote orders (Table 2). For example, the betacyanin families Nyctaginaceae, Didieraceae, Cactaceae, Stegnospermaceae were placed in the predominantly woody orders, Thymelaeales, Sapindales, Cactales and Pittosporales, respectively. Furthermore these orders are widely separated from his herbaceous orders Chenopodiales and Caryophyllales, which contain most of the betacyanin families. Such an arrangement would require the unlikely hypothesis that the betacyanins evolved independently in all these orders.

VI. Conclusions

Betacyanins and betaxanthins apparently developed solely in the Centrospermae at a very early time, perhaps even before the anthocyanins appeared generally in the angiosperms. Several classes of flavonoids other than the anthocyanins do occur in the betacyanin families. Although detailed knowledge of the biosynthetic relationships among the flavonoids is not available,

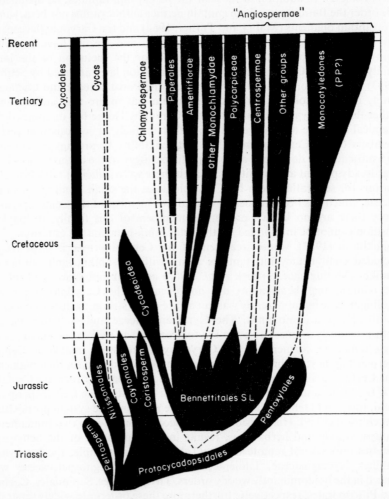

FIG. 5. Phylogeny of the angiosperms as proposed by Meeuse. The Centrospermae is shown as one of several pre-Cretaceous parallel lines. Reproduced by the kind permission of *Acta Biotheoretica*.

it is known, however, that most, if not all, flavonoids including the anthocyanins are formed from similar precursors, presumably chalcones. Furthermore, the *in vivo* transformation of either chalcones or other more highly oxidized flavonoids to anthocyanins is probably not enzymatically complex.

Therefore it appears that the betacyanin families did not evolve pathways to the anthocyanins because the betacyanin pigments perform the same function in these families that is filled by anthocyanins in other families. That function probably involves, in part, floral colorations to attract animal vectors which aid in reproduction. In addition, betacyanins, when present, may serve as nitrogen reservoirs.

It is interesting to note that Meeuse (1963) (Fig. 5) has proposed a tentative phylogeny which shows the Centrospermae rising from a Bennettitalian complex some 150 million years ago along with several other major parallel lines. Although Meeuse did not list the families in his Centrospermae, he probably includes most, if not all, of the betacyanin families. Meeuse arrived at this conclusion from purely morphological considerations, but his proposal for the Centrospermae, as recognized here, is supported, in part, by the available chemical evidence.

Although 44 betacyanins were recently described (Piattelli and Minale, 1964), their taxonomic distribution does not as yet appear to have any systematic significance at either the species or the generic level. However, this does not reduce the taxonomic value of these substances at the family level. The fact that anthocyanins are present in such families as the Caryophyllaceae, Molluginaceae[1] and Illecebraceae and that neither anthocyanins nor betacyanin-betaxanthin pigments are detected in the Batidaceae and Cynocrambaceae suggests the separation of these families from the Centrospermae. At least the chemical evidence does not support their inclusion in the order, though admittedly the absence of betacyanin-betaxanthin pigments from the Batidaceae and Cynocrambaceae does not make an especially strong argument for excluding these families. The Batidaceae is, however, so distinct morphologically from the betacyanin families that some workers have aligned it with the Amentiferae (Lawrence, 1951, Benson, 1957). Several small groups such as the Barbeuiaceae, Gyrostemonaceae and Achatocarpaceae are often treated as members of the Centrospermae but have not yet been investigated for their pigments. The constitution of the Centrospermae as proposed here (Table 2) does not preclude the addition of these and other families to the order when warranted by the evidence, both chemical and morphological.

In summary, while the betacyanins and betaxanthins in themselves do not provide conclusive evidence for the ten-family treatment of the Centrospermae, the betacyanin-betaxanthin criterion does discount the arrangement proposed for these families by Hutchinson (1959), who placed what appear to be closely related taxa into totally different phyletic lines.

ACKNOWLEDGEMENT

I am grateful to the National Institutes of Health (U.S.A.) for financial support (Grant No. GM-11111-02).

[1] This family is often treated as a sub-family of the Aizoaceae. In this context, Hegnauer (1964) recently discussed the presence of anthocyanins in the Molluginaceous genus *Hypertelis* as a member of the Aizoaceae.

9

REFERENCES

Beck, E., Merxmüller, H. and Wagner, H. (1962). *Planta* **58**, 220.
Benson, L. (1957). "Plant Classification", p. 313, D. C. Heath and Co., Boston.
Buxbaum, F. (1961). *Beitr. Biol. Pfl.* **36**, 1.
Dreiding, A. S. (1961). *In* "Recent Developments in the Chemistry of Natural Phenolic Compounds" (W. D. Ollis, ed.), p. 194, Pergamon Press, Oxford.
Eckhardt, Th. (1964). *In* "Syllabus der Pflanzenfamilien", (A. Engler, ed.), Vol. II, p. 79, Gebrüder Borntraeger, Berlin.
Hegnauer, R. (1964). "Chemotaxonomie der Pflanzen", Vol. III, p. 631, Birkhauser Verlag, Basel.
Hörhammer, L., Wagner, H. and Fritzsche, W. (1964). *Biochem. Z.* **399**, 398.
Hutchinson, J. (1959). "Families of Flowering Piants", Vols. I, II, Clarendon Press, London.
Lawrence, G. H. M. (1951). "Taxonomy of Vascular Plants", p. 455, MacMillan Co., New York.
Mabry, T. J. (1964). *In* "Taxonomic Biochemistry and Serology" (C. A. Leone, ed.), p. 239, The Ronald Press, New York.
Mabry, T. J. and Turner, B. L. (1964). *Taxon* **13**, 197.
Mabry, T. J. and Wohlpart, A. (1965). Unpublished results.
Mabry, T. J., Wyler, H., Sassu, G., Mercier, M., Parikh, I. and Dreiding, A. S. (1962). *Helv. chim. acta* **45**, 640.
Mabry, T. J., Taylor, A. and Turner, B. L. (1963). *Phytochemistry* **2**, 61.
Meeuse, A. D. J. (1963). *Acta biotheor.* **16** (III), 9.
Piattelli, M. and Minale, L. (1964). *Phytochemistry* **3**, 547.
Piattelli, M., Minale, L. and Prota, G. (1964a). *Ann. Chim.* **54**, 955.
Piattelli, M., Minale, L. and Prota, G. (1964b). *Ann. Chim.* **54**, 963.
Piattelli, M., Minale, L. and Prota, G. (1964c). *Tetrahedron* **20**, 2325.
Piattelli, M., Minale, L. and Prota, G. (1965). *Phytochemistry* **4**, 121.
Pucher, G. W., Curtis, L. C. and Vickery, H. B. (1938). *J. biol. Chem.* **123**, 61.
Reznik, H. (1955). *Z. Bot.* **43**, 499.
Reznik, H. (1957). *Planta* **49**, 406.
Schmidt, O. Th. and Schönleben, W. (1957). *Z. Naturf.* **12b**, 262.
Schmidt, O. Th., Becher, P. and Huebner, M. (1960). *Chem. Ber.* **93**, 1296.
Schölch, H. (1963). *Ber. dtsch. Bot. Ges.* **76**, 49.
Spragg, S. P. (1960). *In* "Phenolics in Plants in Health and Disease" (J. B. Pridham, ed.), p. 17, Pergamon Press, Oxford.
Wilcox, M. E., Wyler, H., Mabry, T. J. and Dreiding, A. S. (1965). *Helv. chim. acta* **48**, 252.
Wyler, H. and Dreiding, A. S. (1957). *Helv. chim. acta* **40**, 191.
Wyler, H. and Dreiding, A. S. (1959). *Helv. chim. acta* **42**, 1699.
Wyler, H. and Dreiding, A. S. (1962). *Helv. chim. acta* **45**, 638.
Wyler, H., Mabry, T. J. and Dreiding, A. S. (1963). *Helv. chim. acta* **46**, 1745.
Wyler, H., Wilcox, M. E. and Dreiding, A. S. (1965). *Helv. chim. acta* **48**, 361.

CHAPTER 15

Comparative Biochemistry of Hydroxyquinones

Department of Chemistry, University of Strasbourg, France

I. Introduction

The naturally occurring quinones are pigments which although occurring mainly in fungi and higher plants, are also found in algae, in bacteria and in the animal kingdom. However, the majority of these quinones have a fairly restricted distribution in nature and only a few occur in both the animal and plant kingdoms, or in both fungi and higher plants.

A number of recent reviews have covered most of the earlier biogenetic and chemotaxonomic studies which have been made on this group of substances (see References). It seems useful, however, to outline their probable mode of biogenesis and to discuss more recent studies of their distribution, especially in the angiosperms.

II. Structure and Distribution of the Naturally Occurring Quinones

The naturally occurring compounds of this class are usually *para*-substituted quinones, but occasionally orthoquinones have also been isolated. One or more aromatic rings may be fused to the quinone nucleus and so the quinones can be classified as: benzoquinones (Fig. 1), naphthoquinones (Table 1), anthraquinones (with three rings, Table 2), and naphthacenequinones (with four rings) and so on.

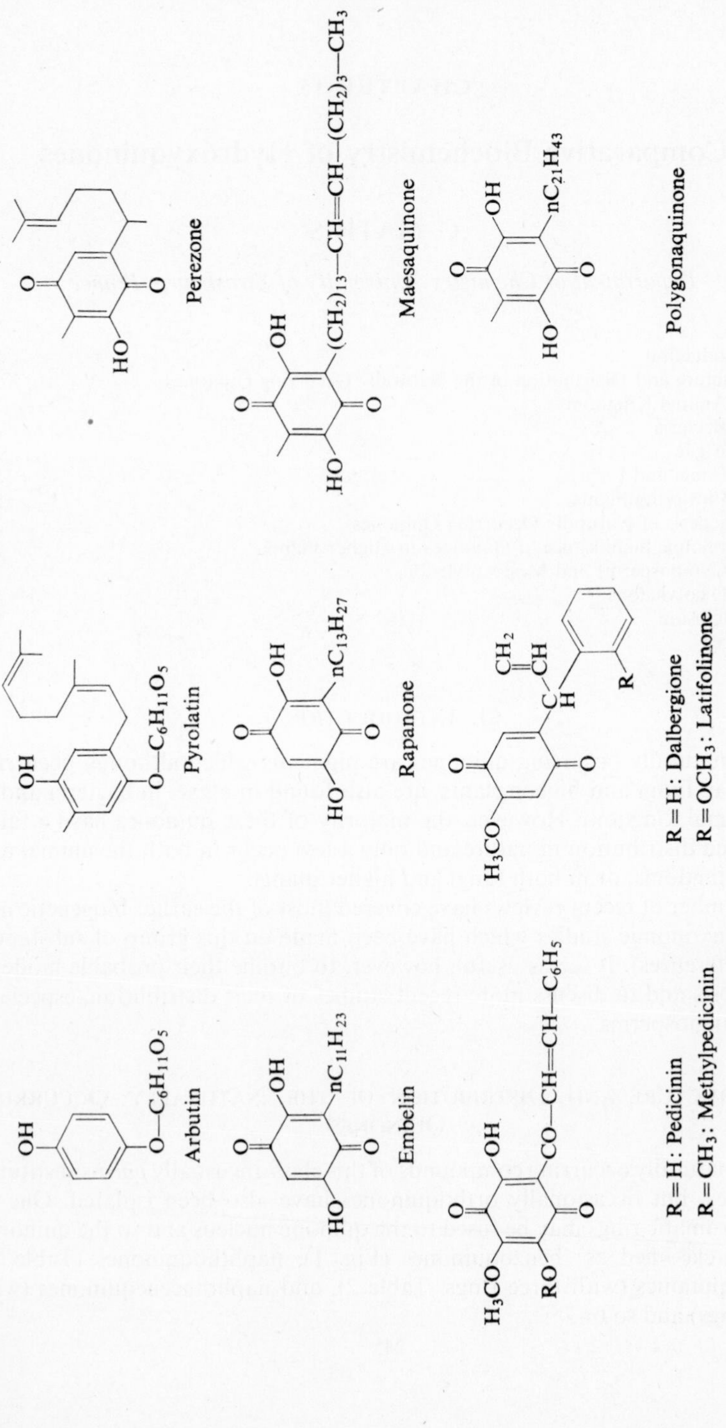

Fig. 1. Benzoquinones occurring in flowering plants.

Relatively few systematic studies have been made on the distribution of quinones in the animal and plant kingdoms, and these are summarized in the remainder of this Section.

A. ANIMAL KINGDOM

A number of closely related compounds have been isolated from a very wide range of animal tissues (mammals, birds, etc.) to which the name, ubiquinones (I)

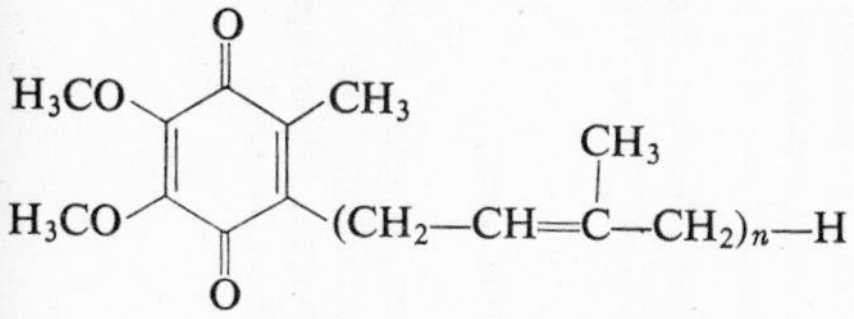

I. Ubiquinones $\begin{cases} n=9: \text{coenzyme } Q_9 \\ n=10: \text{coenzyme } Q_{10} \end{cases}$

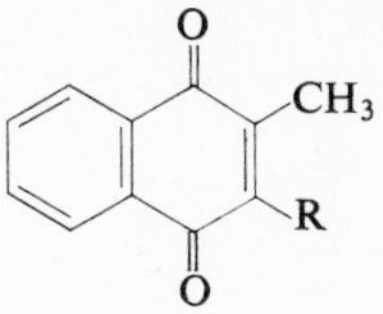

II. K Vitamins $\begin{cases} K_1: R=C_{20}\text{—side chain} \\ K_2: R=C_{30}\text{—side chain} \\ K_3: R=H \end{cases}$

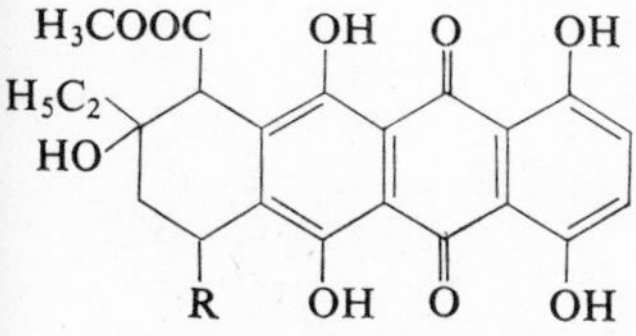

III. Isorhodomycinones   R = H or OH

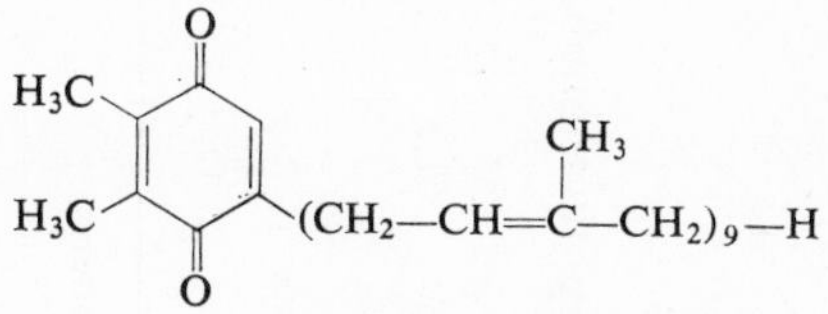

IV. Plastoquinone

has been given because of their wide occurrence in the animal kingdom. These substances are also called coenzymes Q because they are involved in respiration of living cells.

B. BACTERIA

The bacteria are the main natural source of K vitamins (II), which are all naphthoquinones occurring widely among bacteria. Besides the K vitamins, red naphthacenequinones, e.g. isorhodomycinones (III) have been found. These are restricted to bacteria or more exactly to the group of streptomycetes in soil; the presence of these quinones may be a taxonomically very important criterion for this group; the same group of organisms also contain the actino-rhodins which are dinaphthoquinones.

C. ALGAE

Studies so far carried out in the algae have revealed the presence of only the widely occurring quinones such as the ubiquinones (I), K vitamins (II), and plastoquinones (IV), which are of little or no taxonomic interest.

TABLE 1. Distribution of naphthoquinones in flowering plants

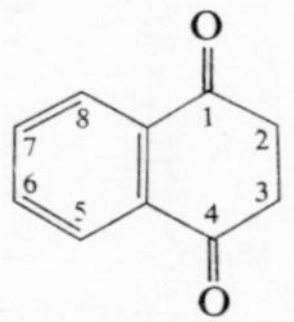

Naphthoquinone	Substitution pattern	Family and genus where found
Chimaphilin	2,7-CH$_3$	Pyrolaceae (*Chimaphila, Pyrola, Moneses*)
Lawsone	2-OH	Lythraceae (*Lawsonia*) Balsaminaceae (*Impatiens*)
2-Methoxynaphthoquinone.	2-OCH$_3$	Balsaminaceae (*Impatiens*)
2,3-Dimethoxynaphthoquinone.	2,3-OCH$_3$	Balsaminaceae (*Impatiens*)
Lapachol	2-CH$_2$—CH=C(CH$_3$)(CH$_3$), 3-OH	Bignoniaceae (*Tecoma, Bignonia*) Verbenaceae (*Tectona, Avicennia*)
Lomatiol	2-CH$_2$—CH=C(CH$_3$)(CH$_2$OH), 3-OH	Proteaceae (*Lomatia*)
Juglone	5-OH	Juglandaceae (*Juglans, Carya, Pterocarya*) Ebenaceae (*Diospyros*)
Plumbagin	2-CH$_3$, 5-OH	Plumbaginaceae (*Plumbago*) Droseraceae (*Drosera*)

7-Methyljuglone	5-OH, 7-CH$_3$	Ebenaceae (*Diospyros*)
Dihydro-7-methyljuglone	5-OH, 7-CH$_3$, dihydro-2,3	Ebenaceae (*Diospyros*)
Droserone	3,5-OH,2-CH$_3$	Droseraceae (*Drosera*)
Hydroxydroserone	3,5,8-OH, 2-CH$_3$	Droseraceae (*Drosera*)
Alkannin	2-CH$_2$—CH=C$\binom{\text{CH}_3}{\text{CH}_3}$, 5,8-OH	Borraginaceae (various genera)
Alkannin	2-CHOH—CH=C$\binom{\text{CH}_3}{\text{CH}_3}$, 5,8-OH	Borraginaceae (various genera)
Dianellinone	5-OH, 6-CO—CH$_3$,7-CH$_3$	Liliaceae (*Dianella*)

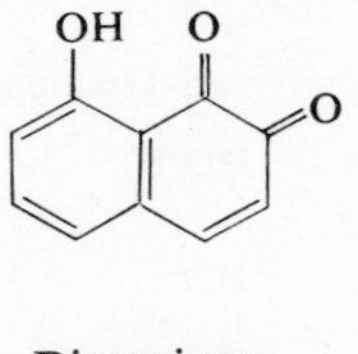

Diosquinone
(Ebenaceae)

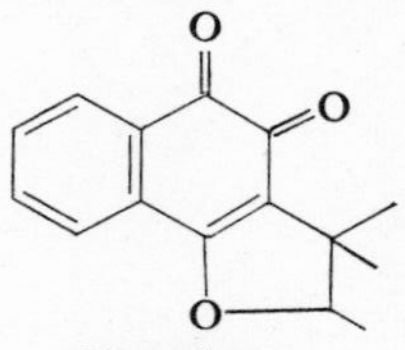

Dunnione
(Gesneriaceae)

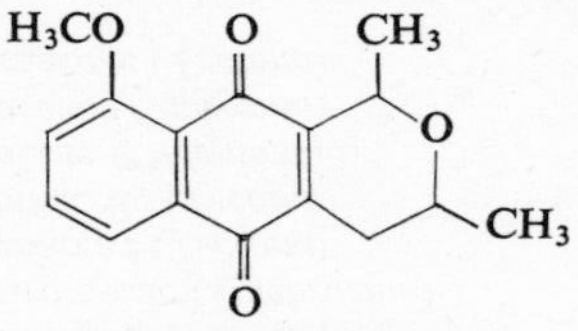

Eleutherin (Eleuthera)

TABLE 2. Distribution of anthraquinones in flowering plants

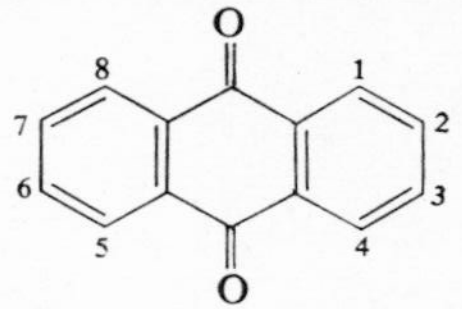

Anthraquinones	Substitution pattern	Family and genus where found
Tectoquinone	3-CH$_3$	Leguminosae (*Cassia*) Anacardiaceae (*Quebrachia*) Verbenaceae (*Tectona*) Bignoniaceae (*Tecoma*)
2-Hydroxyenthraquinone	2-OH	Rubiaceae (*Oldenlandia*)
2-Hydroxy-3-methyl anthraquinone	2-OH, 3-CH$_3$	Rubiaceae (*Coprosma*) Verbenaceae (*Tectona*)
Alizarin	1,2-OH	Rubiaceae (*Oldenlandia, Rubia, Galium, Asperula, Crucianella*)
Alizarin 1-methyl ether	1-OCH$_3$, 2-OH	Rubiaceae (*Oldenlandia, Morinda, Rubia*)
Digitolutein	1-OCH$_3$, 2-OH, 3-CH$_3$	Scrophulariaceae (*Digitalis*)
Hystazarin methyl ether	2-OH, 3-OCH$_3$	Rubiaceae (*Oldenlandia*)
Xanthopurpurin	2,4-OH	Rubiaceae (*Rubia*)
Xanthopurpurin methyl ethers	2-OCH$_3$, 4-OH and 2-OH, 4-OCH$_3$	Rubiaceae (*Rubia*)
Methylxanthopurpurin	2,4-OH, 7-CH$_3$	Rubiaceae (*Morinda*)
Rubiadin	2,4-OH, 3-CH$_3$	Rubiaceae (*Rubia, Galium, Coprosma*)
Rubiadin methyl ether	2-OH, 3-CH$_3$, 4-OCH$_3$	Rubiaceae (*Morinda, Coprosma*)
Lucidin	2,4-OH, 3-CH$_2$OH	Rubiaceae (*Coprosma*)
Damnocanthol	2-OH, 3-CH$_2$OH, 4-OCH$_3$	Rubiaceae (*Damnacanthus, Morinda*)
Nordamnacanthal	2,4-OH, 3-CHO	Rubiaceae (*Damnacanthus*)
Damnacanthal	2-OH, 3-CHO, 4-OCH$_3$	Rubiaceae (*Damnacanthus*)
Munjistin	2,4-OH, 3-CO$_2$H	Rubiaceae (*Rubia*)
Soranjidiol	4,7-OH, 3-CH$_3$	Rubiaceae (*Morinda, Coprosma*)

Compound	Substituents	Occurrence
Chrysophanol	1,8-OH, 3-CH$_3$	Polygonaceae (*Rheum, Rumex, Polygonum*) Rhamnaceae (*Rhamnus*, etc.) Leguminosae (*Cassia*) Euphorbiaceae (*Cluytia*) Saxifragaceae (*Saxifraga*) Ericaceae (*Arbutus*) Lythraceae (*Sonneratia*)
Aloe-emodin	1,8-OH, 3-CH$_2$OH	Polygonaceae (*Rheum*) Rhamnaceae (*Rhamnus*, etc.) Leguminosae (*Cassia*) Liliaceae (*Aloe*)
Rhein	1,8-OH, 3-CO$_2$H	Polygonaceae (*Rheum, Rumex*) Leguminosae (*Cassia*)
Anthragallol	1,2,3,-OH	Rubiaceae (*Coprosma*)
Anthragallol methyl ethers	1,3-OH, 2-OCH$_3$ 1,3-OCH$_3$, 2-OH 1,2-OCH$_3$, 3-OH	Rubiaceae (*Coprosma*) Rubiaceae (*Coprosma*) Rubiaceae (*Coprosma*)
Purpurin	1,2,4-OH	Rubiaceae (*Coprosma*)
Pseudopurpurin	1,2,4-OH, 3-COOH	Rubiaceae (*Rubia, Relbunium*)
Juzunal	2,8-OH, 3-CHO, 4-OCH$_3$	Rubiaceae (*Rubia, Relbunium, Galium, Crucianella*)
Morindone	4,7,8-OH, 3-CH$_3$	Rubiaceae (*Damnacanthus*)
Coelulatin	2,4,5-OH, 3-CH$_2$OH	Rubiaceae (*Morinda, Coprosma*)
Chrysarone	2,7,8-OH, 3-CH$_3$ (?)	Rubiaceae (*Coelospermum*)
Obtusifolin	1-OCH$_3$, 2,8-OH, 3-CH$_3$	Polygonaceae (*Rheum*) Leguminosae (*Cassia*)
Emodin	1,6,8-OH, 3-CH$_3$	Polygonaceae (*Rheum, Rumex, Polygonum*) Rhamnaceae (*Rhamnus*, etc.) Leguminosae (*Cassia*) Lythraceae (*Sonneratia*)
Physcion	1,8-OH, 3-CH$_3$, 6-OCH$_3$	Polygonaceae (*Rheum, Rumex, Polygonum*) Rhamnaceae (*Rhamnus*, etc.)
Copareolatin	4,6,7,8-OH, 3-CH$_3$	Rubiaceae (*Coprosma*)
Alaternin	1,2,6,8-OH, 3-CH$_3$	Rhamnaceae (*Rhamnus*)
Aurantio-obtusine	1,7-OCH$_3$, 2,6,8-OH, 3-CH$_3$	Leguminosae (*Cassia*)
Obtusine	1,6-OCH$_3$, 2,8-OH, 3-CH$_3$	Leguminosae (*Cassia*)
Chryso-obtusine	1,6,7,8-OCH$_3$, 2-OH, 3-CH$_3$	Leguminosae (*Cassia*)

Table 3. Fungal and lichen benzoquinones

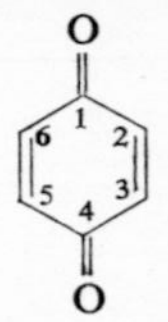

Compound	Substituents	Family (genus)
Gentisylquinone	2-CH_2OH	Aspergillaceae (*Penicillium*)
2-Methyl-5-methoxy q.	2-CH_3, 5-OCH_3	Agaricaceae (*Coprinus, Lentinus*)
2,5-Dimethoxy q.	2,5-OCH_3	Polyporaceae (*Polyporus*)
Fumigatin	2-CH_3, 5-OCH_3, 6-OH	Aspergillaceae (*Aspergillus*)
Aurantiogliocladin	2,3-CH_3, 5,6-OCH_3	Moniliales (*Gliocladium*)
Spinulosin	2-CH_3, 3,6-OH, 5-OCH_3	Aspergillaceae (*Aspergillus, Penicillium*)
Terreic acid	2-CH_3, 3-OH, 5-6 > 0, 5,6- or 2,3-dihydro	Aspergillaceae (*Aspergillus*)
Gliorosein	2,3-CH_3, 5,6-OCH_3, 2,3-dehydro	Moniliales (*Gliocladium*)

Phoenicin (R=H)
Oosporin (R=OH)

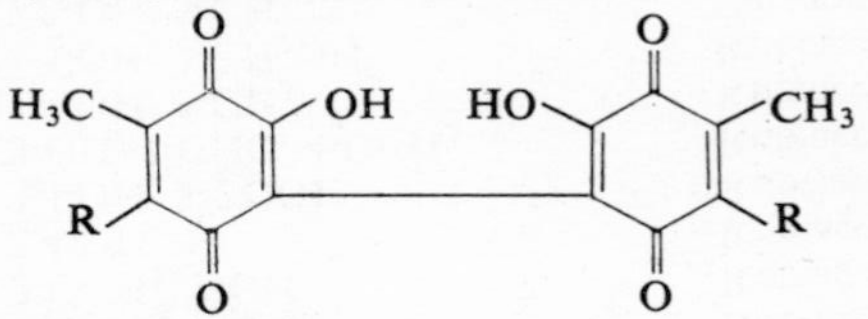

Aspergillaceae (*Penicillium*)
Sphaeriaceae (*Chaetomium*)
Moniliales (*Oospora, Verticillium, Acremonium*)

Compound	Substituents	Family (genus)
Volucrisporin	2,5-C_6H_4-OH (meta)	Volucrispora
Polyporic acid	2,5-C_6H_4, 3,6-OH	Polyporaceae (*Polyporus*) / Thelephoraceae (*Peniophora*) / Lichens (*Sticta*)
Atromentin	2,5-C_6H_4-OH (para), 3,6-OH	Agaricaceae (*Paxillus*)
Leucomelone	2-C_6H_4-OH (para), 3,6-OH; 5-$C_6H_3(OH)_2$(meta-para)	Polyporaceae (*Polyporus*)
Aurantiacin	3,6-O-CO-C_6H_5, 2,5-C_6H_4-OH (para)	Hydnaceae (*Hydnum*)
Muscarufin	3-CH=CH—CH=CH—CO_2H, 6-OH; 2,5-C_6H_4-CO_2H (ortho)	Agaricaceae (*Amanita*)

Thelephoric acid

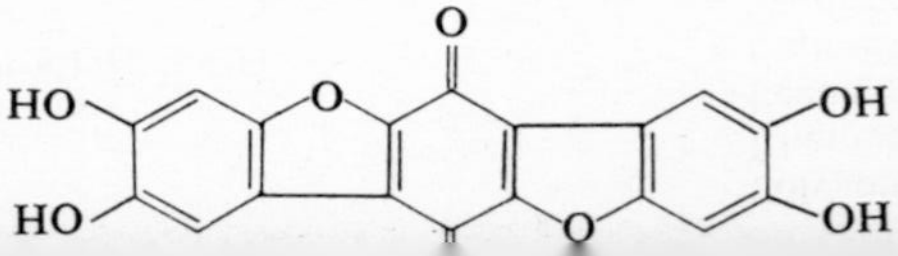

Thelephoraceae (*Thelephora*)
Hydnaceae (*Hydnum*)
Agaricaceae (*Cantharellus*)
Polyporaceae (*Polystictus*)

D. FUNGI AND LICHENS

The fungi and lichens, like the flowering plants, contain a large number of benzo-, naphtho-, anthra- and dianthra-quinones of varying structure (Tables 3 and 4). Only a few quinones, however, occur both in fungi and in flowering plants (e.g. chrysophanol, emodin, physcion: see Table 4). The quinones which have so far been isolated from fungi come from three main groups: the Euascomycetes, mainly from the families Aspergillaceae and Sphaeriaceae; the Fungi imperfecti, mainly Moniliales and Sphaeropsidales; and the Basidiomycetes, mainly the large family of Agaricaceae.

Our present scattered knowledge of the systematic distribution of these quinones coupled with difficulties in the classification of the fungi prevent any exact taxonomic interpretations being made. We can, nevertheless, make some generalizations: with the exception of volucrisporin (Table 3), all the known terphenylquinones have been found among the Agaricaceae (Basidiomycetes). The same quinone has never been found in both the lower fungi and in Basidiomycetes, although the two groups contain related quinones. Anthraquinones and dianthraquinones (called skyrins) often occur in the same species; with the exception of boletol, they occur only in lower fungi and in lichens.

E. FLOWERING PLANTS

The green parts of most, if not all, higher plants contain one of the plastoquinones (IV) which are, therefore, of no taxonomic value. But as in the fungi, we find various quinones with a very limited range of distribution particular to one family or even to one genus or part of a genus. The taxonomic importance of such quinones is dealt with in Section IV.

III. BIOGENESIS OF NATURALLY OCCURRING QUINONES

Several mechanisms have been proposed to explain the biogenesis of quinones. The first of these is the shikimic acid route (Fig. 1). This hypothesis has not yet been rigorously confirmed by experiment, but it seems a plausible route for the biosynthesis of the terphenylquinones. In fact, if the fungus *Volucrispora aurantiaca* is grown in a nutrient medium containing ^{14}C labelled shikimic acid, the terphenylquinone, volucrisporin (Table 3), produced is highly radioactive. The series of transformations proposed are indicated in Fig. 2. This mechanism, however, does not explain the formation of any other group of quinones.

The second pathway to quinones is the polyacetate route, and this biogenetic scheme can be applied to all members of the class. The route proceeds initially in the same way as that for the fatty acids, but little or no reduction of the carbonyl group takes place and a polyketo acid is formed instead. This hypothesis has been shown to be true in several cases for the quinones occurring in fungi: e.g. the biosynthesis of 6-methylsalicylic acid (Fig. 3). The initial products formed may undergo secondary transformation. For example,

TABLE 4. Fungal and lichen anthraquinones

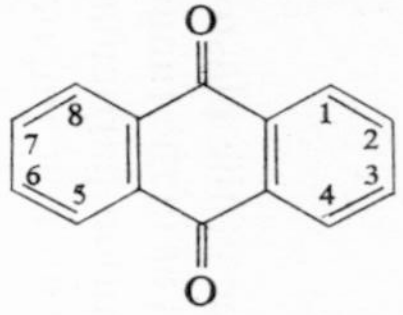

Name	Substituents	Occurrence
Pachybasin	1-OH, 3-CH$_3$	Moniliales (*Pachybasium*)
Chrysophanol	1,8-OH, 3-CH$_3$	Moniliales (*Pachybasium, Sepedonium*) Sphaeriaceae (*Chaetomium*) Aspergillaceae (*Penicillium*)
Islandicin	1,4,8-OH, 3-CH$_3$	Aspergillaceae (*Penicillium*)
Helminthosporin	1,5,8-OH, 3-CH$_3$	Moniliales (*Helminthosporium*) Moniliales (*Cladosporium*)
Emodin	1,6,8-OH, 3-CH$_3$	Aspergillaceae (*Penicilliopsis*) Sphaeriaceae (*Chaetomium*) Agaricaceae (*Cortinarius*) Polyporaceae (*Polystictus*)
Citreo-rosein	1,6,8-OH, 3-CH$_2$OH	Aspergillaceae (*Penicillium*)
Emodic acid	1,6,8-OH, 3-CO$_2$H	Aspergillaceae (*Penicillium*)
Physcion	1,8-OH, 3-CH$_3$, 6-OCH$_3$	Aspergillaceae (*Penicillium, Aspergillus*) Lichens (*Xanthoria, Teloschistes, Caloplaca, Gasparrinia*)
Fallacinol	1,8-OH, 3-CH$_2$OH, 6-OCH$_3$	Lichens (*Xanthoria, Teloschistes*)
Fallacinal	1,8-OH, 3-CHO, 6-OCH$_3$	Lichens (*Xanthoria*)
Parietinic acid	1,8-OH, 3-CO$_2$H, 6-OCH$_3$	Lichens (*Xanthoria*)
Roseopurpurin	1-OCH$_3$, 3-CH$_2$OH, 6,8-OH	Aspergillaceae (*Penicillium*)
Nalgiovensin	1,8-OH, 3-CH$_2$—CHOH—CH$_3$, 6-OCH$_3$	Aspergillaceae (*Penicillium*)
Nalgiolaxin	5- or 8-chloronalgiovensin	Aspergillaceae (*Aspergillus, Penicillium*)
Endocrocin	1,6,8-OH, 2-CO$_2$H, 3-CH$_3$	Hypocreaceae (*Claviceps*) Lichens (*Nephromopsis*)
Versicolorin	1,6,8-OH, 2-CH$_2$OH (?)	Aspergillaceae (*Penicillium*)
Cynodontin	1,4,5,8-OH, 3-CH$_3$	Moniliales (*Helminthosporium*) Sphaeropsidales (*Phoma, Deuterophoma*) Sphaeriaceae (*Pyrenophora*)

Catenarin   1,4,6,8-OH, 3-CH$_3$

Tritisporin   1,4,6,8-OH, 3-CH$_2$OH

Erythroglaucin   1,4,8-OH, 3-CH$_3$, 6-OCH$_3$

Solorinic acid   {1,3,8-OH, 2—CO·(CH$_2$)$_4$-CH$_3$ 6-OCH$_3$

Rhodocladonic acid   1,3,6,8-OH, 2-CH$_2$OH, 7-COOCH$_3$

Asperthecin   1,2,5,6,8-OH, 3-CH$_2$OH

Dermocybin   pentahydroxyanthraquinone

Moniliales (*Helminthosporium*)
Sphaeropsidales (*Phoma, Deuterophoma*)
Aspergillaceae (*Aspergillus, Penicillium*)
Moniliales (*Helminthosporium*)
Aspergillaceae (*Penicillium*)
Aspergillaceae (*Aspergillus*)
Lichens (*Solornia*)
Lichens (*Cladonia*)

Aspergillaceae (*Aspergillus*)
Agaricaceae (*Cortinarius*)

Phomazarin   Sphaeropsidales (*Phoma*)

Skyrins

All present in (*Penicillium* species.)

Rugulosin: R = H
Luteoskyrin: R = OH

Skyrin: R$_1$ = R$'_1$ = OH; R$_2$ = R$'_2$ = CH$_3$
Hydroxyskyrin: R$_1$ = R$'_1$ = OH; R$_2$ = CH$_3$; R$'_2$ = CH$_2$OH
Iridoskyrin: R$_1$ = R$'_1$ = H; R$_2$ = R$'_2$ = CH$_3$.

hydroxy groups can be introduced: e.g. in the biosynthesis of flaviolin; *C*-methyl groups can be formed: e.g. in the biosynthesis of aurantiogliocladin which was recently demonstrated by the use of ^{14}C labelled acetate and malonate; and rearrangement of the carbon skeleton also seems to take place and would explain certain anomalous exceptions to the polyacetate theory (Fig. 3). Both theory and experiment have shown that the polyacetate mechanism is the one most generally used by plants to synthesize a wide variety of

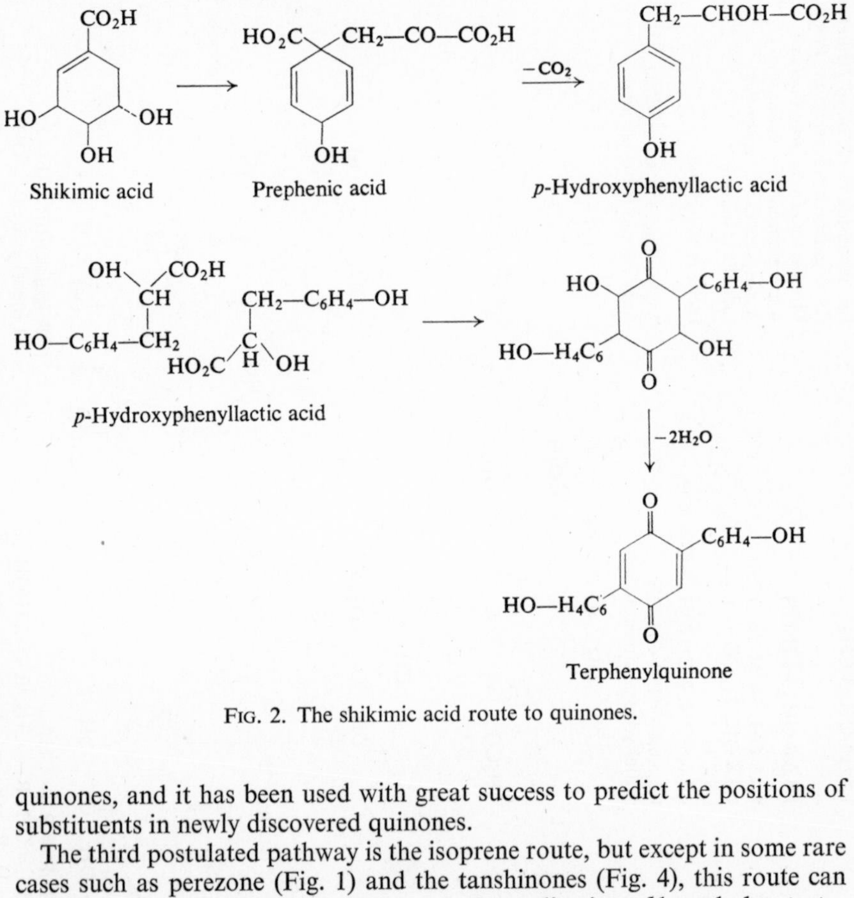

Shikimic acid   Prephenic acid   *p*-Hydroxyphenyllactic acid

p-Hydroxyphenyllactic acid

Terphenylquinone

FIG. 2. The shikimic acid route to quinones.

quinones, and it has been used with great success to predict the positions of substituents in newly discovered quinones.

The third postulated pathway is the isoprene route, but except in some rare cases such as perezone (Fig. 1) and the tanshinones (Fig. 4), this route can explain only part of the biosynthesis: e.g. the cyclization of lapachol to tectoquinone, or to β-lapachone which occurs in the same plant (see Fig. 5). It is also interesting to note that although certain benzo- and naphtho-quinones are known to have isopentenyl side-chains which may be cyclized to give another ring, no anthraquinones of this type have yet been found.

Finally, some naphthodianthrones may be formed from emodin anthrone: e.g. biogenesis of hypericin in *Hypericum* species (see Fig. 6). With the exception

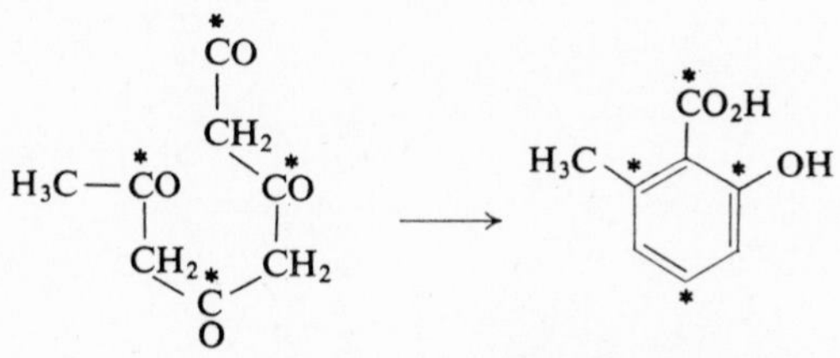

6-Methylsalicylic acid

(a) Normal mode of biosynthesis, e.g. 6-methylsalicylic acid

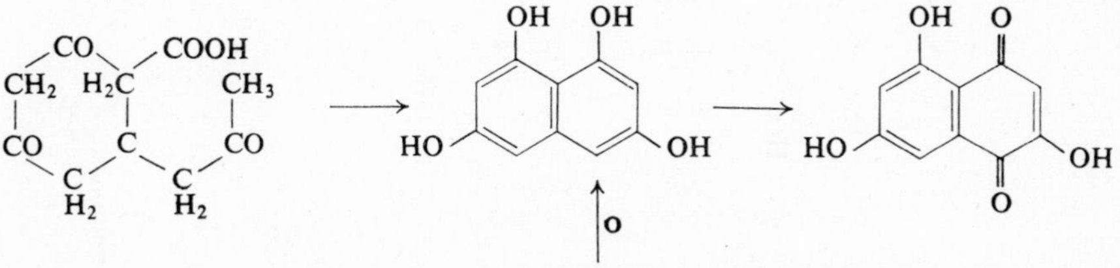

(b) Introduction of hydroxy groups, e.g. Flaviolin

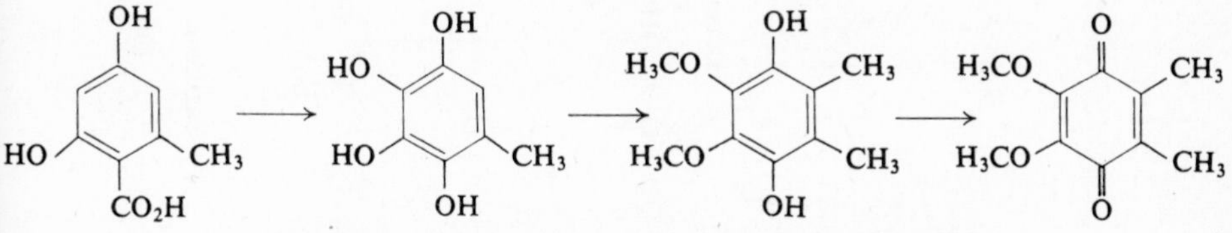

(c) Introduction of C-methyl groups, e.g. Aurantiogliocladin

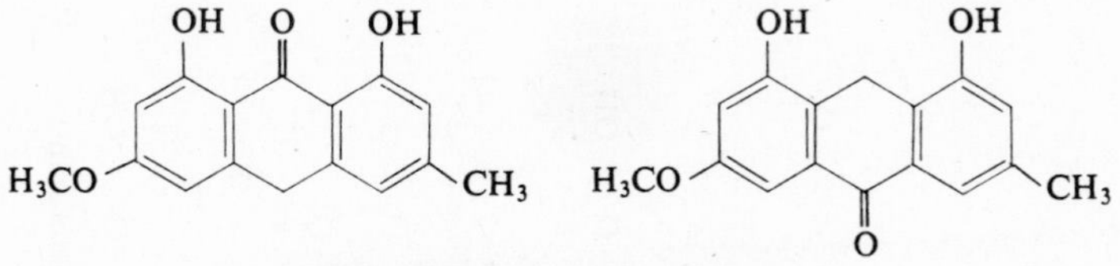

(d) Rearrangement of the carbon skeleton

FIG. 3. The polyacetate route to quinones showing secondary reactions.

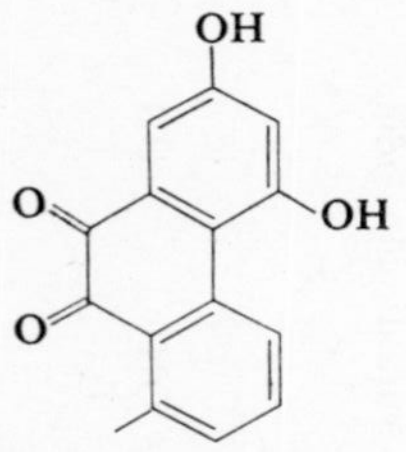

C. Mathis

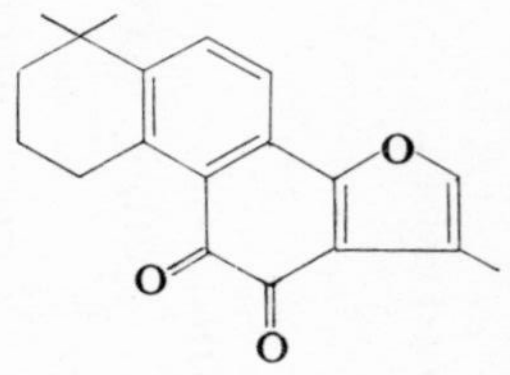

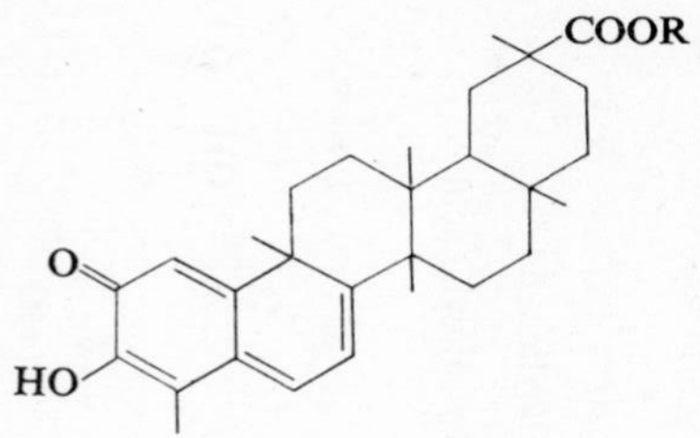

FIG. 4. Complex quinones.

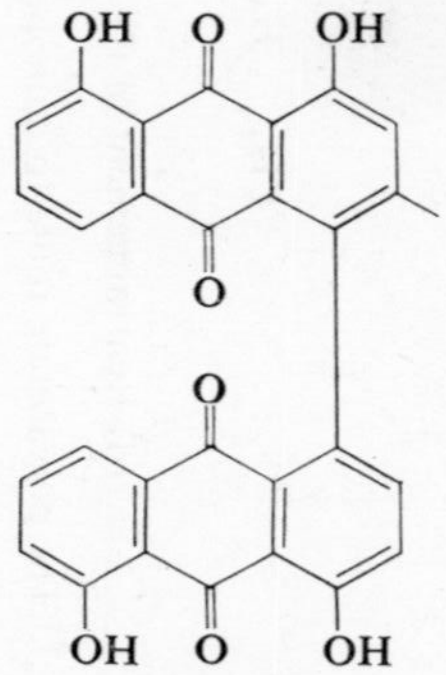

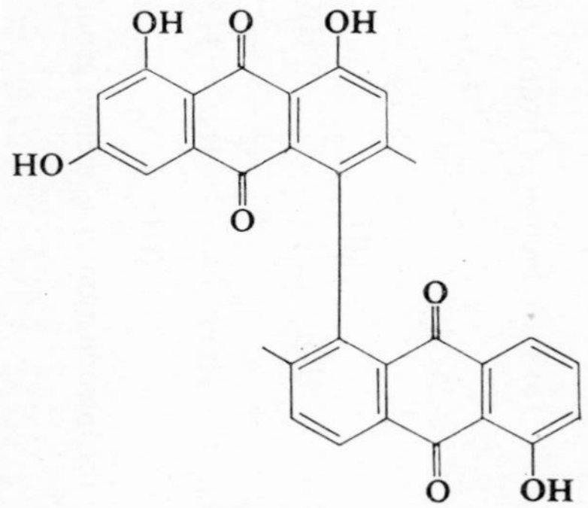

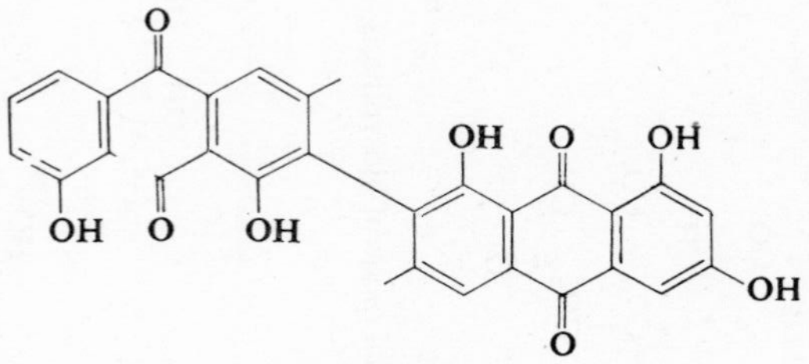

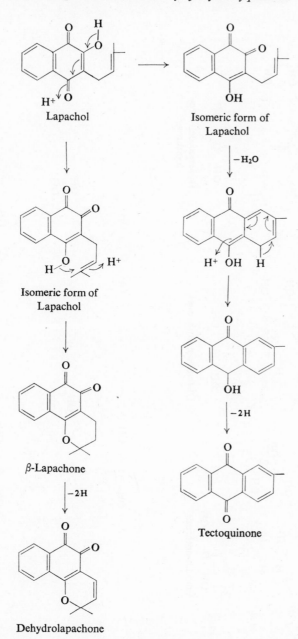

Fig. 5. Biogenetic relationships of quinones containing isoprenoid substituents

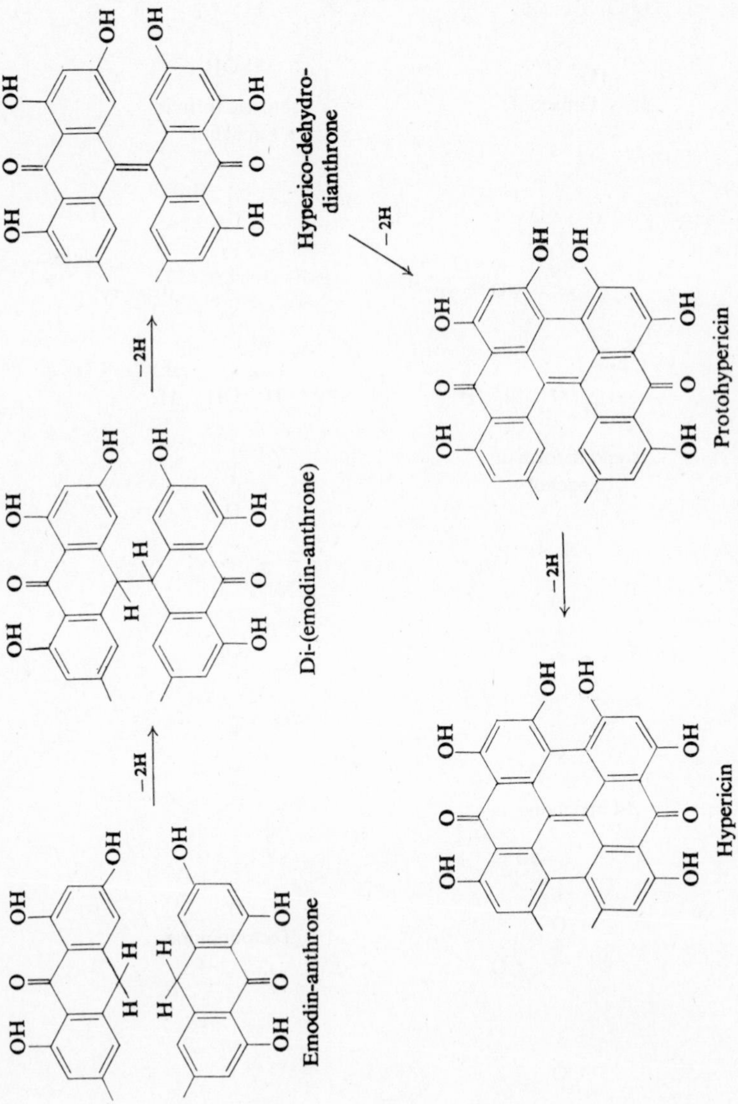

C. Mathis

Fig. 6. Biosynthesis of hypericin.

of di- (emodin—anthrone), all the intermediates shown in Fig. 6 have been isolated from the same plant.

The biogenetic mechanisms which we have discussed above appear to be the most likely, but it is possible that other routes leading to the formation of natural quinones exist and will be uncovered in the future.

IV. TAXONOMIC SIGNIFICANCE OF QUINONES IN HIGHER PLANTS

A. GYMNOSPERMS AND MONOCOTYLEDONS

In gymnosperms, quinones have been isolated in too few cases to be useful for taxonomic purposes. In monocotyledons, the presence of eleutherin (Table 1) in the genus *Eleuthera*, of dianellinone (Table 1) and stypanthone which were recently discovered in the genera *Dianella* and *Stypanthus* respectively, also are of little use as taxonomic criteria. Prof. Hegnauer has shown, however, that the distribution of anthraquinone derivatives in the Liliaceae is of interest to taxonomy; the most typical of these compounds is aloin, the *C*-glucoside of 1,8-dihydroxy-3-hydroxymethyl-anthrone, localized in the so-called aloin cells. Three phytochemical characters are common to the tribe Aloinae and part of the tribe Asphodelae: the presence of 1,8-dihydroxy-anthraquinones and calcium oxalate raphides, and the absence of steroid saponins. Such saponins are present in other members of the Asphodelae.

B. DICOTYLEDONS

The distribution of quinones in the dicotyledons is outlined in Table 5. Each order (following mainly Engler's classification) will be discussed briefly to emphasize the points shown.

Juglandales. Juglone (which is 5-hydroxynaphthoquinone, Table 1) and dihydrojuglone glucoside are found in three genera of the small family of Juglandaceae; these substances constitute, therefore, a very valuable taxonomic criterion for this family or at least for the genera in which they occur.

Proteales. Lomatiol (Table 1) forms a coloured layer on the seeds of five Australian *Lomatia* species, but is absent in three Chilean species of the same genus; thus, the presence of this compound may be characteristic of the Australian members of the genus.

Polygonales. Various classes of quinones have been found in the family of Polygonaceae, e.g. polygonaquinone (Fig. 1), a benzoquinone found in *Polygonum*; fagopyrin, a naphthodianthrone found in *Fagopyrum*; denticulatol (Fig. 4), a phenanthraquinone found in *Rumex*; anthraquinones, mainly 1,8-dihydroxyanthraquinones. These compounds are mainly found in the sub-family, Polygonoideae, especially in *Rumex*, *Rheum* and *Polygonum*. The nature of the anthraquinone derivatives may perhaps allow one to differentiate between these genera or even between certain species of the Polygonoideae: e.g. aloe-emodin and rhein (Table 2) have only been found as yet in the genus *Rheum*.

TABLE 5. Distribution of quinones in flowering plants

Order-family-genus	Anthraquinones	Naphthoquinones	Benzoquinones
Juglandales Juglandaceae { Juglans, Carya, Pterocarya		Juglone (and dihydrojuglone glucoside)	
Proteales Proteaceae { Lomatia, Protea, Grevillea		Lomatiol	Arbutin Arbutin
Polygonales Polygonaceae { Polygonum Polygonae { Fagopyrum	Chrysophanol Emodin Physcion Fagopyrin (naphthodianthrone)		Polygonaquinone
Rumiceae { Rumex	Chrysophanol Rhein Emodin Physcion Denticulatol (phenanthraquinone.)		
Rumiceae { Rheum	Chrysophanol Aloe-emodin Rhein Emodin Physcion Chrysarone		
Ranunculales Ranunculaceae Adonis			2,6-Dimethoxybenzoquinone
Guttiferales Guttiferae Hypericum	Naphthodianthrones: Hypericin Pseudohypericin (and precursors)		

Order / Family	Genus			
Sarraceniales Droseraceae	Drosera		Droserone Hydroxydroserone Plumbagin	Arbutin Rapanone
Rosales Saxifragaceae	Saxifraga	Chrysophanol		
Rosaceae	Pyrus			
Leguminosales Connaraceae	Connarus			
Mimosaceae	Acacia	Anthraquinone		
Cesalpiniaceae	Cassia*	{ Chrysophanol Aloe-emodin Rhein Emodin Physcion Obtusifolin Obtusin Aurantio-obtusin Chryso-obtusin Sennosides A, B (dianthrones)		
	Cordeauxia		Cordeauxiaquinone	
Papilionaceae	Dalbergia	Chrysophanol		Dalbergiones Rapanone
Geraniales Oxalidaceae	Oxalis			2,6-Dimethoxybenzoquinone
Euphorbiaceae	Cluytia			
Rutales Simarubaceae, Meliaceae				$2\text{-}n\text{-}C_{19}H_{39}$-benzoquinone
Sapindales Anacardiaceae	{ Campnosperma Quebrachia	Anthraquinone	2-Methoxy; 2,3-Dimethoxy naphthoquinone	
Balsaminaceae	Impatiens			
Celastrales Celastraceae				{ Celastrol Pristimerin (triterpenoids) Tingenone

* Also cassiamin, cassianin, siameanin (dianthraquinones).

TABLE 5—*continued*

Order-family-genus	Anthraquinones	Naphthoquinones	Benzoquinones
Rhamnales			
Rhamnaceae: *Rhamnus* and various genera	Chrysophanol, Aloe-emodin, Emodin, Physcion, Alaternin		
Myrtiflorales			
Lythraceae: *Lawsonia*		Lawsone	
Lythraceae: *Sonneratia*	Chrysophanol, Emodin		
Umbelliferales			
Umbelliferae			Thymoquinone, thymoquinol
Ericales			
Pyrolaceae: *Chimaphila*		Chimaphilin	
Pyrolaceae: *Moneses*		Chimaphilin	
Pyrolaceae: *Pyrola*		Chimaphilin	Pyrolatin, Arbutin
Ericaceae: *Monotropa*			Arbutin
Ericaceae: *Kalmia*			Arbutin
Ericaceae: *Azalea*			Arbutin
Ericaceae: *Vaccinium*			Arbutin
Ericaceae: *Arctostaphylos*			Arbutin
Ericaceae: *Arbutus*			Arbutin

Taxon	Compound
Primulales	
Myrsinaceae — *Myrsine*	Embelin
Embelia	Embelin
Rapanea	Embelin
	Rapanone
Maesa	Maesaquinone
Plumbaginales	
Plumbaginaceae — *Plumbago*	Plumbagin (2-methyljuglone)
Ebenales	
Ebenaceae — *Diospyros*	Diosquinone (orthoquinone)
	2-Methyljuglone
	2-Methyl dihydrojuglone
	7-Methyl juglone
Asterales	
Compositae — *Perezia*	Perezone
Trixis	Perezone
	2,6-Dimethoxybenzoquinone
Gentianales	
Apocynaceae — *Rauwolfia*	
Rubiales	
Rubiaceae	
Cinchonoideae — *Oldenlandia*	2-Hydroxy anthraquinone Alizarin (and its 1-methyl ether)
	Hystazarin 3-methyl ether

TABLE 5—continued

Order-family-genus	Anthraquinones	Naphthoquinones	Benzoquinones
Coprosma	2-Hydroxy 3-methyl anthraquinone Rubiadin (and its 1-methyl ether) Lucidin Soranjidiol Anthragallol (and its methyl-ethers) Morindone, copareolatin Alizarin Rubiadin } 1-Methyl ethers 6-Methyl xanthopurpurin		
Morinda	Damnacanthol Nordamnacanthal Soranjidiol Morindone Damnacanthol Nordamnacanthal		
Damnacanthus	Damnacanthal Juzunal		
Coffeoideae { Coelospermum	Coelulatin Alizarin (and its 1-methyl ether) Xanthopurpurin (and its methyl-ethers)		
Rubia	Rubiadin Munjistin Purpurin Pseudopurpurin		
Relbunium	Purpurin Pseudopurpurin Alizarin		
Galium	Rubiadin Pseudopurpurin Alizarin		
Crucianella	Pseudopurpurin Alizarin		
Asperula	Alizarin		

Taxon		Description	Compounds
Tubiflorales			
Borraginaceae	Alkanna		Alkannan
	Various genera		Shikonin
	Lithospermum		Lapachol
Verbenaceae	Avicennia		Lapachol
	Tectona	Tectoquinone 2-Hydroxy-3-methylanthraquinone	
Labiatae	Monarda		Thymoquinol / Thymoquinone
	Salvia	Tanshinones (phenanthraquinone) Digitolutein	
Scrophulariaceae	Digitalis		
	Capraria		Biflorin Dunnione
Gesneriaceae	Streptocarpus		
	Didymocarpus		Pedicinin / Methylpedicinin
Bignoniaceae	Tecoma		Lapachol
	Bignonia		Lapachol
	Tabebuia		Lapachol
	Stereospermum		Lapachol β-lapachone Dehydrolapachone
		Anthraquinone derivatives	
Iridales			
Iridaceae	Eleutherine		Eleutherins
Liliales			
Liliaceae			
Asphodeloideae			
Dianella	Dianella Stypandra		Dianellinone Stypanthone
Asphodelae			
Aloineae		Anthraquinones derivatives (aloin is the most typical)	
Loandreae			

Guttiferales. Hypericin (Fig. 6) and pseudohypericin, although related to fagopyrin, are very characteristic substances of the genus *Hypericum*, about half of the 300 species of which contain these pigments; in addition, they are only found in anatomically well-defined glands whose distribution, depending on the organ, is of equal taxonomic interest.

Sarraceniales. The isolation of three nearly related naphthoquinones (droserone, hydroxydroserone and plumbagin, Table 1) in various *Drosera* species is of undoubted taxonomic value in defining the genus or perhaps a part of it. The presence of plumbagin in both the Sarraceniales and the Plumbaginales, however, is not sufficient in itself to show a closer relationship between the two orders.

Leguminosales. Only the anthraquinone derivatives of the genus *Cassia* are, at present, of taxonomic importance and probably constitute a general criterion for this genus. Perhaps the more specific dianthraquinones (cassiamin, cassianin, siameanin, Fig. 4) and pentahydroxyanthraquinones (obtusins, etc.), which have more recently been found in the genus, will be of greater value when more is known about their distribution.

Celastrales. In recent years, the existence of triterpene quinones has been demonstrated in various genera of the family of Celastraceae (*Celastrus, Evonymus, Maytenus, Denhamia, Tripterygium*). These substances (tingenone, celastrol, pristimerin, Fig. 4) should certainly be useful in defining either the whole or at least part of this family.

Rhamnales. Anthraquinone derivatives have been found in the principal genera of the subfamilies of Rhamnaceae: Rhamnae, Ziziphae, Ventilaginae and Colletiae; these compounds seem to be present in the greater part of the vast genus, *Rhamnus* (155 species), but the experimental results must be interpreted with care because of the existence of colourless glycosidic derivatives. Until now the tetrahydroxyanthraquinone, alaternin, has been found only in Rhamnaceae.

Ericales. There is an obvious biogenetic relationship between the naphthoquinone, chimaphilin (Table 1) and the benzoquinone, pyrolatin (Fig. 1). Within the limits of present investigations pyrolatin and chimaphilin are peculiar to the family of Pyrolaceae and are thus taxonomically interesting. On the other hand, arbutin, which has been reported in Pyrolaceae, Ericaceae and other families, seems to be less useful.

Plumbaginales and *Ebenales.* The presence of plumbagin which is a naphthoquinone connects the Plumbaginaceae on the one hand with Ericales (since chimaphilin and plumbagin have very similar structures) and on the other with Ebenales in which four naphthoquinones of the juglone-type have been isolated.

Primulales. The presence of benzoquinones (embelin, rapanone, maesaquinone, Fig. 1) in four genera of Myrsinaceae differentiates this family from the neighbouring family of Primulaceae; it is not likely, however, that the nature of the quinones will allow a distinction to be made between the genera in the Myrsinaceae.

Rubiales. The anthraquinones (Table 2) from the Rubiaceae are characterized by the following points: with the exception of coelulatin, recently isolated from the genus *Coelospermum*, no 1,8-dihydroxyanthraquinone derivatives have been isolated from the Rubiaceae. The majority of the quinones have so far been isolated from the roots, and this may explain why some authors working on dried specimens (sometimes without roots) have reported no quinones. Usually all the hydroxyl groups are substituted on only one of the benzene rings, but more and more exceptions to this generalization are being discovered. These quinones clearly characterize the Rubiaceae, but it is difficult to use them for distinguishing subfamilies or genera. It is, however, interesting to note that xanthropurpurin (Table 2) and its methyl esters, purpurin and pseudopurpurin, have only been found so far in the tribe of Rubiae.

Asterales. Perezone (Fig. 1) is a sesquiterpenoid derivative, isolated from various *Trixis* species and closely related to the sesquiterpenes of other Compositae.

Tubiflorales. This order contains a great variety of quinones which are often quite unrelated. We will only deal here with the quinones from the family of Borraginaceae and from the group, Bignoniaceae–Verbenaceae.

The family of Borraginaceae and particularly the subfamily of Borraginoideae is clearly characterized by the presence of the naphthoquinone, alkannin and its reduction product, alkannan (Table 1), which have not been found up to the present in any other plant family. Hundreds of species in about 20 genera have been shown to contain alkannin, and certain species which apparently do not, probably contain colourless leucoderivatives.

The Verbenaceae and the Bignoniaceae both contain lapachol and tecto-quinone (Fig. 5). The pedicinins (Fig. 1) are benzoquinones found in the Gesneriaceae; they perhaps show that this family is related to the other two, but since the two groups of quinones have clearly different structures, their relationship is obviously not close. Moreover, in the Gesneriaceae, the quinones are present as a deposit on the leaves and flowers, whereas the quinones of the two other families are mainly found in the cells of the bark and phloem.

V. CONCLUSION

Our chemotaxonomic studies of the hydroxyquinones have shown that in this field few of the problems have been partially, let alone completely, solved. We hope, however, that in this brief survey, we have brought to light the fundamental problems which remain, and that this will stimulate further research into the distribution of this interesting group of compounds.

REFERENCES

Florkin, M. and Stotz, E. H. (1963). "Comprehensive Biochemistry", Vol. 9, pp. 158, 200, 213, Elsevier Publishing Co., New York.
Harborne, J. B. (1964). "Biochemistry of Phenolic Compounds". Academic Press, London and New York.

270   *C. Mathis*

Hegnauer, R. (1959). *Planta Medica* **7**, 344.
Hegnauer, R. (1962). "Chemotaxonomie der Pflanzen", Vol. I (1962), Vol. II (1963) and Vol. III (1964), Birkhäuser Verlag, Basel.
Thomson, R. H. (1957). "Naturally Occurring Quinones". Academic Press, New York and London.
Thomson, R. H. (1962). *In* "Comparative Biochemistry", Vol. 3 (M. Florkin, and H. S. Mason, eds.), p. 631. Academic Press, New York and London.
Zechmeister, L. (1950). *Fortschr. Chem. org. Naturst.* **6**, (1950), **20**, (1962) and **21**, (1963).

CHAPTER 16

The Evolution of Flavonoid Pigments in Plants

John Innes Institute, Hertford, Herts, England

I. INTRODUCTION

As an aid to plant systematics, the study of flavonoids has much to offer since these substances show much variation in structure; they are not essential metabolites but can be found in almost every higher plant. The potential importance of these pigments to plant taxonomy has, in fact, been discussed in some detail in recent years (see e.g. Harborne and Simmonds, 1964; Davis and Heywood, 1963) and several chapters in "Chemical Plant Taxonomy" (Swain, 1963) deal *inter alia* with this subject. The drawback is that total ascertainment has only been achieved so far in one or two small plant groups so that the present contribution of flavonoids to taxonomy is slight. Many more surveys at the generic and family level are required before chemical information of this type can be incorporated with confidence into plant classification.

A more fruitful approach, at the present time, to the natural distribution of flavonoids is, perhaps, to consider their relationship to phylogeny. While the evolution of phenolic constituents of the leaf in the angiosperms has been related to lignification in plants (Bate-Smith, 1962), little consideration has been given to the petal constituents. Bate-Smith (1962) rejected consideration of petal pigments as taxonomic characters because of their susceptibility to genetic variation. However, from the evolutionary point of view, flower pigments provide one of the most interesting examples of natural selection in operation. Furthermore, his criticism of flower pigments as unstable taxonomic

---

* Present address: Dept. of Botany, University of Liverpool.

271

characters is not valid as long as "wild type" material is chosen for examination. Flower colour is no more variable than many morphological characters (e.g. leaf shape) and only a relatively few species growing in natural conditions exhibit much colour variation; the variations that do occur more often involve changes in concentration, pH, distribution of co-pigment, etc. than alterations in chemical structure. An observation of Parks (1965), who studied flavonol glycosides in petals of three *Gossypium* species, is pertinent here; he found that species could still be identified in colour mutant forms by the residual array of pigments not affected by the mutant alleles.

In the present paper, emphasis will therefore be given to petal constituents, although flavonoids occurring in the leaf and other organs will also be considered. The natural distribution of flavonoids will thus be discussed from an evolutionary point of view and the opportunity will also be taken to record phytochemical data obtained since the last review (Harborne, 1963a). In particular, the results of surveys being carried out in this laboratory on the Plumbaginaceae and Gesneriaceae will be presented, since these prove the value of concentrating efforts on limited but interesting plant groups.

For the present purpose, the term flavonoid pigments is taken to include the commonly occurring anthocyanins and flavones, the less common chalcones and aurones and the biosynthetically related colourless substances such as flavanones and leucoanthocyanidins. While no completely up-to-date and comprehensive listing of flavonoid pigments is available, there are a number of recent reviews which report discoveries made after the publication in 1957 of W. Karrer's dictionary of plant constituents. These include those by Venkataraman (1959), Harborne and Simmonds (1964), Swain and Bate-Smith (1962) and Hegnauer (1962, 1963, 1964). Much use has also been made here of the publication by Bate-Smith (1962) of his leaf survey in the dicotyledons.

II. FLAVONOIDS IN LOWER PLANTS

There are no satisfactory records of flavonoid pigments occurring in bacteria, fungi or algae. Various reports have appeared from time to time of the occurrence of anthocyanins in bacteria and fungi (e.g. Peterson *et al.*, 1961; Avadhani and Lim, 1964; see also Erikson *et al.*, 1938) but in no case has the bacterial or fungal pigment been compared directly with authentic flavylium salts. In addition, phenazines and quinones, pigments known to occur in these organisms, are similar in colour and solubility to anthocyanins. By contrast, the occurrence of flavonoids in mosses and ferns is now well established and there is a clear indication, in the case of the anthocyanins, of a relationship to phylogeny. The pigments that have been isolated are biogenetically primitive in that they lack the 3-hydroxyl group of the common anthocyanins, and are derivatives of apigeninidin (I) or luteolinidin (II).

Only one genus of moss has so far been studied; the 5-mono- and -diglucoside of luteolinidin have been identified in *Bryum cryophyllum*, *B. rutilans* and *B. weigelii* (Benz *et al.*, 1962; Benz and Martensson, 1963). A number of other red-coloured mosses are known and it will be of considerable interest to see

I. Apigeninidin   II. Luteolinidin

whether some of these contain similar pigments. Other types of red pigments are probably present in mosses; the cell wall pigments in *Sphagnum magellanicum* and *S. rubellum*, which have been examined by Rudolph (1965), do not appear to be anthocyanin in character, as was once thought (Paul, 1908). A larger number of ferns have been studied and glycosides of apigeninidin and luteolinidin have been found in the juvenile fronds of *Adiantum veitchianum*, *Blechnum brasiliense* var. *corcoradense*, *Osmunda regalis*, *Pteris longipinnula* and *P. quadriaurita* (Harborne, 1965a, and unpublished results). In an earlier report, it was suggested that these fern pigments were derived from 6-hydroxy-pelargonidin or 6-hydroxycyanidin (Price *et al.*, 1938) but this possibility has now been excluded. Although ordinary anthocyanins (derivatives of pelargonidin and cyanidin) have been reported in *Davallia divaricata* (Price *et al.*, 1938) and *Dryopteris erythrosora* (Hayashi and Abe, 1955) a re-examination of the pigment in *D. erythrosora* indicated that luteolinidin and apigeninidin, but not cyanidin, were present. However, reinvestigation of *Davallia divaricata* showed that it did contain pelargonidin (as the glycoside, monardein).

Turning to the flavones, there is again much less information available about moss pigments than about those in ferns. The early report (Molisch, 1911) that the glycoflavone, saponarin, occurs in the moss *Madotheca platyphylla* has been partly substantiated by current work in this laboratory. In a survey of some thirty mosses, flavones were detected in *Mnium cuspidatum*, *M. undulatum*, in *Plagiochila asplenoides* and in a *Cleridium* species. That these substances are glycoflavones follows from the similarity of their spectral maxima and R_f values to those of vitexin and lutexin and their resistance to acid hydrolysis. However, none of the substances examined corresponded exactly with a known glycoflavone and more work is clearly needed to establish the structure of these moss constituents.

By contrast with the situation in mosses, there is abundant evidence that flavones, flavonols and leucoanthocyanidins of the type found in angiosperms are present in the Pteridophyta. For example, luteolin, apigenin, quercetin or kaempferol have been detected in acid hydrolysed extracts of some twenty fern genera (cf. Hegnauer, 1962) and Bate-Smith (1954) has recorded leucocyanidin and/or leucodelphinidin in fifteen of some twenty species surveyed. Where isolated, the flavonols have been found to occur as the 3-glucosides, 3-rhamnosides or 3-rutinosides. Two more complex kaempferol 3,7-diglycosides have been found, however, in the marsh horsetail, *Equisetum palustris* (Beckmann and Geiger, 1963). A number of other flavonoid types have also been isolated from ferns. Thus, the glycoflavone vitexin (III) has been found in *Cyathea*

III. Vitexin   IV

and *Sphenomeris* spp. (Ueno, 1963). Of even more phylogenetic interest is the report of the chalcone (IV) in *Pityrogramma chrysophylla* (Polypodiaceae); it is accompanied by the corresponding dihydrochalcone and by the related chalcone lacking a methoxyl group in the 4-position (Nilsson, 1961). In addition, the flavanones, matteucinol and demethoxymatteucinol, have been known for some time to be present in the Polypodiaceae (in *Matteuccia*) (Fujise, 1929). The presence of these three biogenetically simple flavonoid types, chalcones, dihydrochalcones and flavanones, in ferns certainly suggests that they are primitive characters in plants and what is known of their distribution in higher plants generally supports this view (see also Section V).

Only one attempt has been made so far to use flavonoid characters in evolutionary studies within the ferns; this is the work of Smith and Levin (1963) on reticulate evolution in the Appalachian *Asplenium* complex. Chromatographic studies of the phenolics in a range of diploids and tetraploids confirmed the concept of reticulate evolution already advanced on the basis of morphology, hybridization and cytology. Unfortunately, although flavonols are known to occur regularly in *Asplenium*, the pigments were not identified in these particular plants so that it is not possible to consider the chemical changes produced in this evolutionary sequence.

III. FLAVONOIDS IN GYMNOSPERMS

The range of flavonoid types found in the gymnosperms is generally similar to that in the ferns. Thus, the occurrence of flavonols, flavones, flavanones and leucoanthocyanidins is well recorded (cf. Hegnauer, 1962) and, recently, a glycoflavone related to vitexin has been found in *Ephedra americana* (Harborne, unpublished work). Another link between the ferns and the gymnosperms is *C*-methylation, since flavanones with this modification (e.g. strobopinin (V) from *Pinus strobus*) are known both in *Matteuccia* (Polypodiaceae) and in *Pinus* (Pinaceae). In *Pinus*, 6-*C*-methylquercetin (VI) and 6-*C*-methylmyricetin have also been found.

There is still some question as to whether anthocyanins occur in gymnosperms. The report, based on the old colour tests, of cyanidin 3-glucoside in *Picea obovata* needles (Beale *et al.*, 1941) clearly needs confirmation by modern methods, particularly since reddish colours may be formed in this group of plants by autoxidation of simple phenols. A preliminary re-investigation in this laboratory of reddish-tinged tissue from several Gymnospermae failed to reveal any anthocyanin pigments.

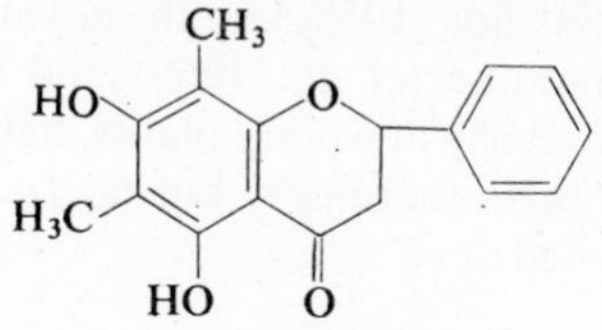

V. Strobopinin

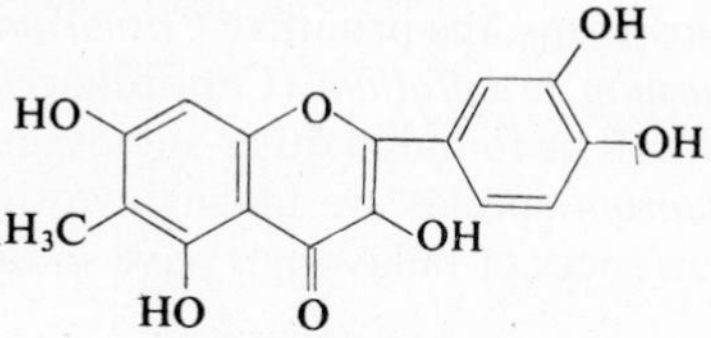

VI. Pinoquercetin

The main chemical feature distinguishing the gymnosperms from both lower plants and the angiosperms is biflavonyl formation. Biflavonyls such as amentoflavone (VII), formed from two molecules of apigenin by carbon-carbon coupling at the 8- and 3'-positions, are widely distributed in gymnosperms and their general taxonomic significance has already been widely discussed (see e.g. Harborne and Simmonds, 1964). Only two points need mentioning here. The first is to report that a new type of biflavonyl in which

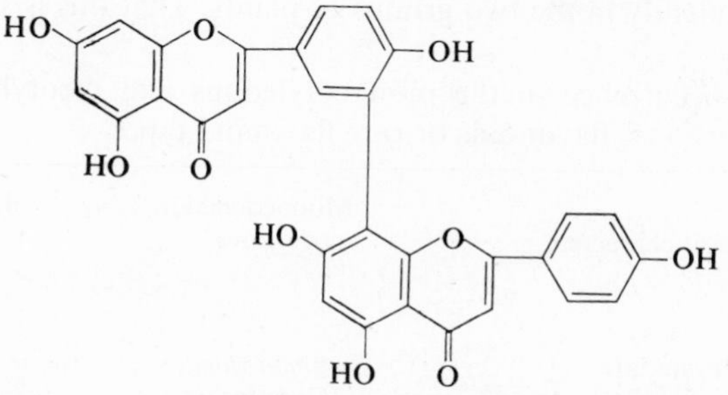

VII. Amentoflavone

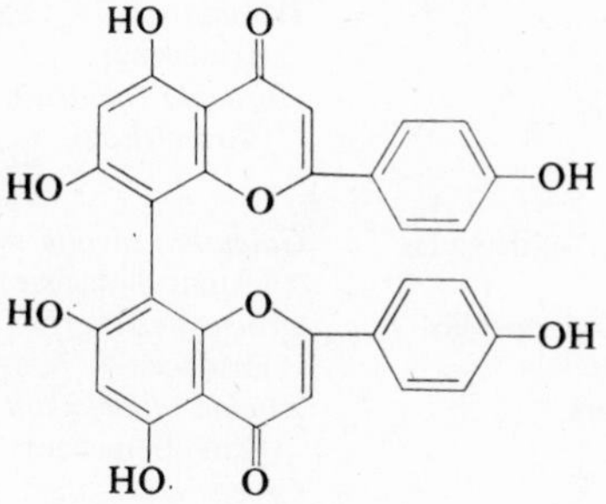

VIII. Cupressoflavone

carbon-carbon coupling has taken place in the 8- and 8'-positions has been found in *Cupressus torulosa* and *C. sempervirens* by Murti *et al.* (1964); this is the pigment cupressoflavone (VIII). The second is to stress the importance of surveying a range of angiosperms in order to confirm their general absence from most higher plants. Biflavonyls have already been discovered in two

10

angiosperms, the primitive *Casuarina stricta* (Sawada, 1958) and the advanced *Viburnum prunifolium* (Caprifoliaceae) (Hörhammer *et al.*, 1965) and they may well be found in other angiosperms, particularly in woody plants. Several *Casuarina* species are under investigation in this laboratory, but no further occurrences of biflavonyls have so far been noted.

IV. Flavonoids in Monocotyledons

The pattern of flavonoid pigments in the monocotyledons is very similar to that in the dicotyledons and since the difference between primitive and advanced families is much clearer in the latter group, the evolution of flavonoids in the angiosperms will not be considered here but in the following section. The flavonoids of the monocotyledons are more interesting from another point of view, that of parallel evolution. If, as is generally accepted, the mono- and dicotyledons arose separately from a common seed-bearing stock, then many chemical modifications in flavonoid synthesis must have arisen independently in the two groups of plants. That this is so is very clear

TABLE 1. Co-occurrence in the monocotyledons and dicotyledons of rare flavonoids or rare flavonoid types

Chemical character	Monocotyledon source	Dicotyledon source
Anthocyanins		
3-Deoxyanthocyanidins (Luteolinidin, Apigeninidin)	*Sorghum vulgare* (Gramineae)	*Gesneria cuneifolia* (Gesneriaceae)
Pelargonidin 3-sophoroside-7-glucoside	*Watsonia* sp. (Iridaceae)	*Papaver* sp. (Papaveraceae)
Pelargonidin 3-gentiobioside	*Tritonia* sp. (Iridaceae)	*Primula sinensis* (Primulaceae)
Cyanidin 3-α-arabinoside	*Hordeum vulgare* (Gramineae)	*Rhododendron* (Ericaceae)
Flavones		
Flavonol 3-sophoroside-7-glucosides	*Galanthus nivalis* (Amaryllidaceae)	*Helleborus foetidus* (Ranunculaceae)
Flavonol 3-rutinoside-7-glucosides	*Crocus fleischeri* (Iridaceae)	*Baptisia* sp. (Leguminosae)
Flavonol 4'-methyl ethers	*Alpinia officinarum* (Zingiberaceae)	*Tamarix* sp. (Tamaricaceae)
Acacetin	*Crocus laevigatus* (Iridaceae)	*Linaria vulgaris* (Scrophularicaceae)
Tricin	*Triticum* sp. (Gramineae)	*Orobanche* sp. (Orobanchaceae)
Other Flavonoids		
Isoflavones	*Iris* sp. (Iridaceae)	Papilionatae (Leguminosae)
Phloretin	*Smilax glycyphylla* (Liliaceae)	*Malus* sp. (Rosaceae)

from Table 1, which lists some examples of the same type of flavonoid or the same flavonoid being found under quite different circumstances in the mono-cotyledons and in the dicotyledons.

One of the most remarkable chemical links is that between the very disparate families, the Iridaceae (monocotyledons) and the Leguminosae (dicotyledons). Not only do they have the unusual isoflavones in common, but they are also both rich in glycoflavones and in addition, both synthesize the glycoxanthone, mangiferin (Bate-Smith and Harborne, 1963). Though obviously very different morphologically, these two families might be considered to have independently reached about the same stage in evolution. Whether this is actually so can only be judged when evidence from other sources becomes available. There are also some chemical similarities between the Gramineae (monocotyledons) and families in the order Tubiflorae. For example, the rare methylated flavone tricin is only known in grasses and in seeds of the *Orobanche* parasite (Oroban-chaceae) and in lucerne (Leguminosae). A more recently discovered link is the presence of apigeninidin and luteolinidin, first found in the Gesneriaceae (Robinson *et al.*, 1934), in the leaf of *Sorghum* (Stafford, 1965).

At least, the above results (Table 1) illustrate the similarity in flavonoid pattern of mono- and dicotyledons in terms of unusual modifications. In fact, the only two types of dicotyledon flavonoid not so far recorded in monocoty-ledons are aurones and 6- or 8-hydroxylated flavones, such as scutellarein and quercetagetin. That these substances will be found in monocotyledons is very likely, since chalcones are already known (in *Xanthorrheia*) as are isoflavones with the 5,6,7-trihydroxylation pattern (in *Iris*). The common flavonoids found in the monocotyledons are exactly the same as those in dicotyledons. The fact that the percentage distribution of flavonols and leucoanthocyanidins is lower in monocotyledons than in dicotyledons (Swain and Bate-Smith, 1962) is perhaps only a reflection of their predominantly herbaceous character.

V. FLAVONOIDS OF THE DICOTYLEDONS

The distribution of the commonly occurring flavonoids (e.g. quercetin and cyanidin) in higher plants is very well documented (Swain and Bate-Smith, 1962; Bate-Smith, 1962) but much less is known about those flavonoids which differ from the common types by one or more structural modifications and which are of most interest to the phytochemist. They are assumed to be of limited distribution because they have only been isolated from a few plants. In plant surveys, these rarer substances may have either been mistaken on occasion for common flavonoids (e.g. 7-*O*-methylmyricetin has the same R_f value as quercetin on paper chromatograms using the Forestal solvent) or may not have been detected at all (this applies particularly to isoflavones, flavanones, biflavonyls and highly methylated flavones). In future surveys of leaf and petal constituents, it is to be hoped that thin layer and gas chromatography will be used to complement the widely used paper chro-matographic techniques so that the absence as well as the presence of the rarer flavonoids can be reported with confidence.

While it is clearly important for the above reason not to place too much reliance on existing information, nevertheless it seems unlikely that the general distribution patterns will be radically altered by further surveys. Indeed, there is already such a vast body of information on the flavonoids in dicotyledons that it is not possible to summarize the situation in the present compass. It is proposed here to concentrate on recent studies, particularly those which have a bearing on plant systematics and then consider the evolutionary aspects.

A. ANTHOCYANINS

The general distribution of these pigments has been described in detail earlier (Harborne, 1963a). Since then, further work has been done on glycosidic patterns but attention has been mainly devoted to the distribution of rare anthocyanidins, i.e. those lacking a 3-hydroxyl group or those having an *O*-methyl group in the 5- or 7-position.

1. 3-*Deoxyanthocyanins*

3-Deoxyanthocyanins (e.g. gesnerin, IX) were first isolated from plants of the Gesneriaceae by Robinson *et al.*, in 1934 and later studies (Table 2) have shown that they are of regular occurrence in this highly evolved family. There are only three other records of their presence in the dicotyledons; in the related Bignoniaceae (*Arrabidaea chica*), and in the unrelated Theaceae (*Camellia sinensis*) and Sterculiaceae (*Chiranthodendron pentadactylon*).

The distribution of these rare anthocyanins in the Gesneriaceae has recently been examined in this laboratory as part of a collaborative study of the chemistry of the family with B. L. Burtt of Edinburgh. The results to date are presented in Table 2; the tribes and genera are arranged here according to Burtt's reclassification (1962) of the family into two sub-families on the basis

TABLE 2. Anthocyanins of the Gesneriaceae

Tribe, genus and species	Flower anthocyanins[1]	Leaf anthocyanins	3-Deoxyantho-cyanidin present or absent
New World species (*sub-family* Gesnerioideae)			
Columneae			
Alloplectus vittatus	Luteolinidin 5G	Columnin	+
Columnea × *banksii*	Columnin	Columnin	+
C. Columnea c.v. 'Stavenger'	Columnin	Columnin	+
C. microphylla	Columnin	Columnin	+
C. affinis	New type (without 3OH)	—	+ ?
C. kucyniakii	Carotenoid	Columnin	+
Episcia reptans	Pelargonidin 3RG	Columnin	+
Nautilocalyx lynchii	—	Columnin	+
Trichantha minor	Columnin	—	+

TABLE 2—*continued*

Tribe, genus and species	Flower anthocyanins[1]	Leaf anthocyanins	3-Deoxyantho-cyanidin present or absent
New World species (sub-family Gesnerioideae*)—continued*			
Gloxinieae			
Achimenes cvs.	Pelargonidin or malvidin 3RG5G	Columnin	+
Koellikeria erinoides	—	Columnin (stem)	+
Kohlerieae			
Kohleria eriantha	Gesnerin	Luteolinidin 5G	+
× *Kohleria*	Pelargonidin 3RG		
	Malvidin 3RG5G	Columnin	+
Gesnerieae			
Gesneria cuneifolia	Gesnerin, luteolinidin 5G	—	+
G. ventricosa	Pelargonidin 3RG	—	—
Sinningieae			
Rechsteineria cardinalis	Gesnerin, luteolinidin 5G	—	+
R. macropoda	Gesnerin	Green	+
Sinningia speciosa	Malvidin 3RG	—	—
Sinningia cv. (≡*gloriosa*?)	Pelargonidin and cyanidin 3RG	—	—
Sinningia barbata	—	Columnin	+
Old World species (sub-family Cyrtandroideae*)*			
Trichosporeae			
Aeschynanthus obconicus	Pelargonidin 3XG	Green	—
A. parvifolius	Pelargonidin 3XG	Cyanidin 3XG (sepal)	—
Didymocarpeae			
Chirita lacunosa	Malvidin glyc.	Green	—
C. micromuca	Cernuoside (aurone)	Green	—
Boea hygroscopica	Delph. and cyanidin glyc.	Green	—
Dichiloboea speciosa	Delph., pet. and malvidin glyc.	Green	—
Ornithoboea wildeania	Pet. and malvidin glyc.	Green	—
Saintpaulia ionanthe	Malvidin 3RG5G	Cyanidin 3XG	—
Streptocarpus, 9 spp.	Malvidin 3RG5G	Cyanidin 3XG	—
Streptocarpus dunnii	Cyanidin 3XG	Cyanidin 3XG	—
Streptocarpus cyanandrus	Cyanidin 3XG	Cyanidin 3XG	—

[1] Abbreviations: 5G, 5-glucoside; 3XG, 3-sambubioside; 3RG5G, 3-rutinoside-5-glucoside; 3RG, 3-rutinoside; glyc., glycoside unidentified; delph = delphinidin; pet. = petunidin; —, material not available for examination.

of geographical distribution and presence or absence of anisocotyly. While New World species contain 3-deoxyanthocyanins with great frequency (17 out of 20 species examined were positive), Old World species appear to have only the normal type of anthocyanin pigments. The results thus support the validity of Burtt's reclassification.

A case in point is the genus *Columnea*, which was placed by Fritsch (1893–94), on the basis of the superior position of the ovary, into the sub-family Cyrtandroideae. The presence in all but one of the *Columnea* species examined of a new 3-deoxyanthocyanin, columnin, indicates that *Columnea* properly belongs to the sub-family Gesnerioideae where Burtt (1962) has placed it. The aglycone, columnidin, of this new 3-deoxyanthocyanin, columnin, has been provisionally identified as 5,7,8,3′,4′-pentahydroxyflavylium (X). The structure of another new 3-deoxyanthocyanidin in *Columnea affinis* is being actively pursued.

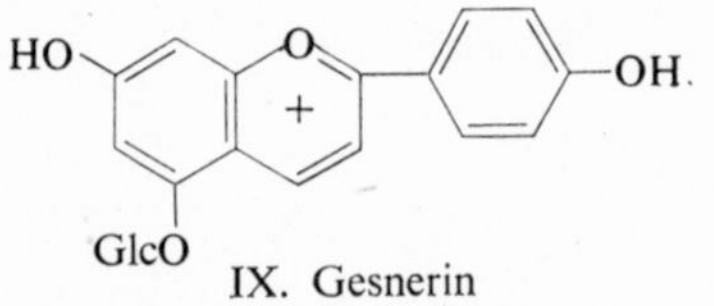

IX. Gesnerin 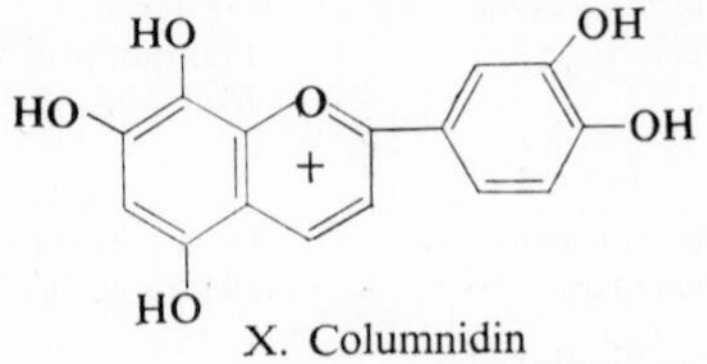   X. Columnidin

A measure of the taxonomic interest in the Gesneriaceae is the fact that none of the three species originally examined by Robinson and his co-workers in 1934 still bears the same name. *Gesneria cardinalis* is now *Rechsteineria cardinalis* and *Isoloma hirsutum* is *Kohleria eriantha*. The identity of the third plant "*Gesnera fulgens*" can no longer be traced, but it may possibly be a species of *Smithiantha* (C. V. Morton, private communication). Fortunately, the name gesnerin for apigeninidin 5-glucoside is still appropriate as this pigment has recently been found in one true species of this genus, *Gesneria cuneifolia*.

2. *Anthocyanins with A-ring methylation*

Methylation of the hydroxyl groups in the A-ring of anthocyanidins, unlike methylation of the B-ring, is a rare feature in plants and pigments of this type are only known in three sympetalous families. Hirsutidin (XI, $R = OCH_3$) and rosinidin (XI, $R = H$) are known in the Primulaceae (*Primula*) and Apocynaceae (*Lochnera*) and capensinidin (XII, $R = CH_3$) in the Plumbaginaceae (*Plumbago capensis*) (cf. Harborne, 1963a). Further surveys of these three families are in progress and two new anthocyanidins have recently been found in the latter family (Harborne, unpublished results).

The first, pulchellidin, occurs in flowers of *Plumbago pulchella* and has been provisionally characterized as 5-*O*-methyldelphinidin (XII, $R = H$). The second, europinidin, occurring in petals of *Plumbago europea* and *Ceratostigma plumbaginioides*, appears to be a new dimethyl ether of delphinidin since its

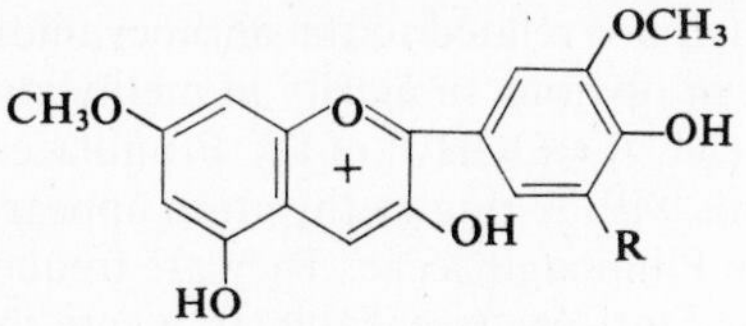

XI. $R = CH_3$; Hirsutidin
    $R = H$; Rosinidin

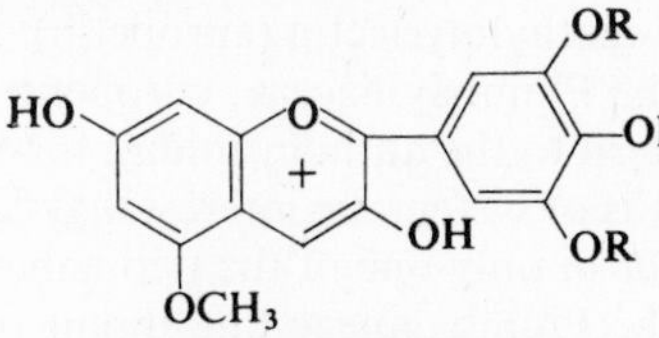

XII. $R = CH_3$; Capensinidin
     $R = H$; Pulchellidin?

R_f value is close to but different from malvidin (0·69 and 0·62 on paper chromatograms in Forestal solvent respectively); it may be the 5,3'-dimethyl ether. A major problem in identification is that neither pigment can be obtained in quantity; for example, *P. pulchella* does not flower profusely and its corollas, which are less than 1 cm long, do not yield much colouring matter.

TABLE 3. Distribution of flavonoids in the Plumbaginaceae

	Flavonoids[1] present in	
Organ	Tribe Plumbagineae (20–30 spp.)	Tribe Staticeae (150–300 spp.)
Root	Plumbagin (8/8)[2]	Plumbagin absent (0/28)
Leaf	Europetin (1/6) 5-OMe Myricetin (1/6) Leucodelphinidin (5/6)	Myricetin (8/9) Leucodelphinidin (4/4)
Petal	Pulchellidin (1/7) Europinidin (2/7) Capensinidin (1/7) Azaleatin (5/7) Gallotannins (1/7)	Delphinidin (1/4) Malvidin (3/4) Quercetin (1/4) Kaempferol (1/4) Cernuoside (1/4)

[1] Figures in brackets show numbers of occurrences and number of species examined. Glycosidic forms of the flavonoids have been ignored for simplicity, but there are differences in the anthocyanidin glycosidic patterns (3-rhamnoside or 3-glucoside in Plumbagineae; 3,5-diglucoside and 3-rhamnoside-5-glucoside in Staticeae).

[2] This pigment was first isolated from *Plumbago europea* roots in 1828 by Dulong and was found in three other *Plumbago* species prior to the present investigation (see Thomson, 1957). Plumbagin occurs in a colourless combined form and is liberated from root tissue by acid treatment.

Flavonols with the same type of methylation pattern as these anthocyanidins also occur in the Plumbaginaceae. Azaleatin (5-*O*-methylquercetin) occurs as a petal constituent in most *Plumbago* and *Ceratostigma* species examined and traces of 5-*O*-methylmyricetin are present in the leaf of *Plumbago europea*. The major flavonol constituent of this latter plant, however, is a new flavonol,

J. B. Harborne

7-*O*-methylmyricetin (europetin). Although not related to the anthocyanidins in the Plumbaginaceae, europetin bears an obvious similarity in methylation pattern to the anthocyanidin, hirsutidin (XI, R = OCH$_3$), of the Primulaceae.

It is of systematic interest that flavonoids with A-ring methylation appear to occur in only one of the two tribes of the Plumbaginaceae. They are frequent in the Plumbagineae but absent from the Staticeae (see Table 3), where they are replaced by the more usual flavonoids, myricetin, delphinidin and malvidin. This distribution is correlated with that of another, but unrelated, chemical character, the yellow naphthoquinone pigment plumbagin (XIII). This occurs in the roots of all members of the Plumbagineae examined (8/8) but is absent

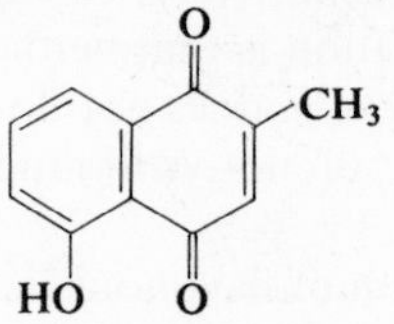

XIII. Plumbagin

from those of the Staticeae (0/28). It is satisfying to find that studies by Baker (1948) of the pollen morphology also divide the Plumbaginaceae in the same way; the Plumbagineae are monomorphic, the Staticeae, dimorphic. The only exception is the plant *Aegialitis annulata*, a member of the Staticeae but monomorphic. It will be interesting to examine the chemistry of this key relict species. This will shortly be possible, since seeds have recently been obtained from wild material growing in the Port Moresby area of New Guinea.

TABLE 4. Flavonoid constituents in the genus *Plumbago*

Flavonoids		Distribution in *Plumbago* species				
Organ	Substance	*Capensis*	*Europea*	*Pulchella*	*Rosea*	*Zeylanica*
Leaf constituents	Europetin	—	+	+	—	—
	Myricetin/quercetin	—	—	+	—	—
	Leucodelphinidin	+	—	+	+	+
	Plumbagin[1]	—	+	+	—	—
Petal anthocyanidins	Capensinidin	+	—	—	—	—
	Europinidin	—	+	—	—	—
	Pulchellidin	—	—	+	—	—
	Delph./cyan./ pelargonidin	—	—	—	+	—
Other petal constituents	Azaleatin	+	—	+	—	+
	Plumbagin	—	+	—	—	—
	Gallotannins	—	—	—	+	—

[1] Plumbagin is uniformly present in root of all spp. (see Table 3).

At the species level, it is interesting that four of the five *Plumbago* species so far studied contain 5-*O*-methylated flavonoids. The fifth, *Plumbago rosea*, is very distinct, having glycosides of kaempferol, quercetin, pelargonidin, cyanidin and delphinidin in its petals and also two additional rare phenolics, galloylglucose derivatives. *P. rosea* is also morphologically different, and is the only species with red (instead of blue or white) flowers. All five species, in fact, are very readily distinguished by their phenolic constituents in leaf and flower (Table 4) so that the genus *Plumbago* appears to be a very favourable plant group in which to apply chemical data to problems of species relationships.

3. *Glycosidic patterns of the anthocyanidins*

The relationship between anthocyanidin glycosidic patterns and plant systematics has been reviewed recently (Harborne, 1963a). While the general distribution of the various glycosidic types is fairly clear, few detailed studies at the family and genus level have been carried out. To illustrate the type of information that can be obtained, the results of a survey of pigments in the Scrophulariaceae are given (Table 5). As in most chemotaxonomic work, the coverage is limited, confined as it is to 22 species from 13 genera in a family of 3000 species and 200 genera. The survey was a random one and the results obtained so far must to some degree be representative.

TABLE 5. Anthocyanins of the Scrophulariaceae

Genus and species	Petal anthocyanin[1]
Antirrhinum	
Section Antirrhinum (5 spp.)	Cy 3-rutinoside
Section Saerorhinum (2 spp.)	Dp 3,5-diglucoside
Asarina procumbens	Cy 3-rutinoside
Digitalis purpurea	Cy and Pn 3,5-diglucoside, Cy 3-glucoside
D. fulvia	Cy 3-rutinoside
Gambelia speciosa	Pg 3-rutinoside
Lathreae clandestina	Cy 3-rutinoside, Dp 3-rutinoside
Linaria maroccana	Cy 3,5-diglucoside (in red forms); Dp 3,5-diglucoside (in blue forms)
Maurandia speciosa	Dp 3,5-diglucoside
Mimulus cardinalis	Pg and Cy 3-rutinoside, Pg and Cy 3-glucoside
M. glutinosus	Pg and Cy 3-rutinoside
M. lewisii	Cy 3-glucoside
Misopates orontium	Cy 3-rutinoside
Nemesia strumosa	Cy 3-sambubioside
Penstemon heterophyllus	Dp 3,5-diglucoside
Phygelius capensis	Cy 3-rutinoside
Torenia fourreri	Dp, Pt and Mv 3,5-diglucoside

[1] Abbreviations: Dp, delphinidin; Cy, cyanidin; Pg, pelargonidin; Mv, malvidin; Pt, petunidin; Pn, peonidin. Data from unpublished work, except for *Torenia*, which was studied by Endo (1962).

The following points may be noted. (1) There are two common glycosidic patterns in the family: 3,5-diglucoside and 3-rutinoside, which rarely appear together in the same genus. When they do occur together, as in *Antirrhinum*, they are present characteristically in different sections. (2) While all six common anthocyanidins have been detected, methylation is rare; cyanidin and delphinidin are the most common aglycones. (3) Acylation, a feature of the anthocyanins in several related families in the Tubiflorae, is apparently absent.

In studying the variation in anthocyanin pigmentation in the Scrophulariaceae, the distribution of several unit processes or chemical characters in the family is being uncovered. While at present the results have little relevance to systematics, they may well be used eventually, together with other chemical characters (such as presence or absence of aurone pigment), to solve taxonomic problems at the generic level.

B. FLAVONOLS AND FLAVONES

Each year many identifications of flavones and flavonols in plants are reported by phytochemists, mostly of known compounds from new sources. These are frequently carried out because of the pharmaceutical interest and unfortunately, systematic surveys of related groups of plants are almost never attempted. By contrast, the systematic surveys that are carried out by botanists are usually lacking in any report on the chemistry of the plants examined. It seems a dangerous practice to place reliance on the presence or absence of unidentified phenolic spots on two-dimensional chromatograms, as has been done in many recent investigations (e.g. Alston and Irwin, 1961; Stebbins *et al.*, 1963; Torres and Levin, 1964). At least, an attempt should be made to identify the major constituents, as has been done by Harney and Grant (1964, 1965) in their chemical survey of the genus *Lotus*. Even here, use of simple chromatographic procedures (Bate-Smith, 1962) might fail to indicate the presence of a novel or systematically interesting constituent. For example, Pecket (1959, 1960) surveying leaf and flower of *Lathyrus* for flavonols, did not observe a novel flavonol, syringetin (myricetin 3',5'-dimethyl ether) as a major constituent of *Lathyrus pratensis*. The time has surely come for collaborative studies, in which both botanical and chemical aspects are thoroughly examined.

In considering the distribution of flavonols and flavones in the dicotyledons, only three of the many biologically interesting aspects can be considered.

1. *Replacement of flavonols by flavones*

As long ago as 1954, Bate-Smith noted that flavonols mainly occur in woody species of plants, whereas flavones and flavanones predominate in herbaceous plants. This conclusion was based on observing the presence or absence of flavonol in plant leaves, as his methods did not specifically reveal occurrences of flavones and flavanones. More recently, surveys of both leaf and flower have been carried out in a number of sympetalous families, the results of which (Table 6) considerably strengthen the evidence supporting this generalization.

TABLE 6. Occurrence of flavonols and flavones in the Tubiflorae

Family	Records of flavonols (Bate-Smith, 1962)	Records of flavones
Acanthaceae	0/9	Luteolin in leaves and flowers (11/11) (Nair et al., 1965)
Bignoniaceae	2/8	Chrysin, baicalein and oroxylon A in Oroxylon
Gesneriaceae	0/4	Flavonols absent from leaf and flower (0/40); luteolin and/or apigenin in 5 genera J. B. (Harborne, unpublished)
Labiatae	3/15	Luteolin chief flavone in 7 genera; kaempferol and quercetin only detected in Lamium and Prunella (Hörhammer and Wagner, 1962).
Scrophulariaceae	3/15	Luteolin and/or apigenin common in flowers (Harborne, 1963 and unpublished)
Solanaceae	19/25	Kaempferol and quercetin common in Solanum and Nicotiana (90 spp. examd.) (Hörhammer and Wagner, 1962). Luteolin in Solanum stoloniferum (Harborne, 1963a)
Verbenaceae	0/10	Luteolin and apigenin in garden Verbena; vitexin (glucosylapigenin) in Vitex

The idea that flavonols are replaced by flavones in more highly evolved plants is of considerable phylogenetic interest and will be discussed in more detail later. For the present, some comment on the evidence supporting this hypothesis is necessary.

The results given in Table 6 refer to seven families of the Tubiflorae, and, as expected, in six out of the seven, flavones occur more readily than do flavonols. The view that flavone synthesis is dominant to flavonol synthesis in these highly evolved plants is apparent from chemical studies of colour mutants in *Streptocarpus* (Gesneriaceae) and *Antirrhinum* (Scrophulariaceae). In *Streptocarpus hybrida*, the wild type and most mutants contain apigenin and luteolin along with the usual anthocyanidins. The flavonol, kaempferol, is found only in the bottom recessive pale pink form and is presumably formed here as a "by-product" of pelargonidin synthesis. This constitutes the only record of a flavonol in a survey of leaf and flower of over forty gesneriads. Similarly, in *Antirrhinum*, all wild species have only flavones and so little kaempferol and quercetin are produced, compared to apigenin and luteolin, in mutant forms of *A. majus* that their identification has presented considerable difficulties (see Sherratt, 1959).

The seventh family, the Solanaceae, in Table 6 is atypical in having more records of flavonol than flavone. This is probably a reflection of the woody nature of many of its species, but it is gratifying that this family is recognized to be a primitive member of the Tubiflorae (see e.g. Lawrence, 1961). Comparable values for flavonol occurrences in primitive woody families (taken from Bate-Smith, 1962) are: Betulaceae (6/6), Magnoliaceae (10/11), Fagaceae (5/7) and Nyctaginaceae (5/5). Interesting phylogenetically is the presence of luteolin in only one of 70-tuber-bearing *Solanum* species examined (Harborne, 1962). This species, *S. stoloniferum*, is an allotetraploid and must be considered one of the more recently evolved species in the group. A contrasting case is the primitive Mexican species, *S. pinnatisectum*, which retains the ability to synthesize coumarins in the leaf and petal, while nearly all the other species, including *S. stoloniferum*, have lost what appears to be a primitive chemical character (Harborne, 1960).

2. *Hydroxylation and methylation patterns*

There are many rare flavonols and flavones which differ from the common structures by having an extra hydroxyl group or by having an *O*-methyl substituent. These additional structural features are probably formed as unit processes occurring late in biosynthesis. One example is quercetagetin (XIV), presumably formed by 6-hydroxylation of quercetin, which occurs in *Papaver* (Papaceraceae), *Primula* (Primulaceae), *Rhododendron* (Ericaceae) and *Lotus* (Leguminosae) as well as in *Tagetes* (Compositae), from which genus it received its name. Another is azaleatin, 5-*O*-methylquercetin (XV), the distribution of which in the Plumbaginaceae has already been described, but which was first found in *Rhododendron* (Ericaceae) (Wada, 1956).

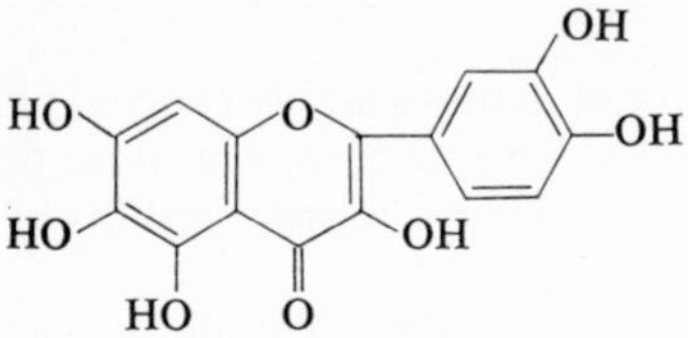

XIV. Quercetagetin 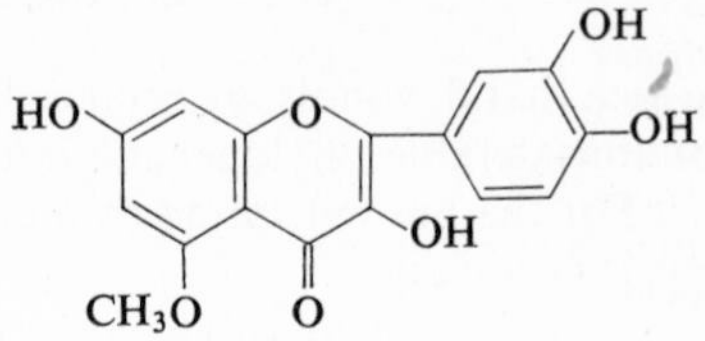　　　　　　　　　XV. Azaleatin

The distribution of these substances in the families from which they have been isolated (Table 7) has hardly been studied, except in the case of azaleatin (see p. 281), but is almost bound to yield results of systematic interest. Their general distribution in plants is of phylogenetic interest, since they appear to occur for the most part in highly evolved plants (see Table 7). This is particularly true of flavones with 6- or 8-hydroxylation which have been recorded in six sympetalous families and only once in the Archichlamydeae.

These substances occur with particular abundance in the Compositae and in families in the Tubiflorae and Contortae. Their occurrences in families of these two orders are listed in Table 8, since their presence or absence may be of considerable interest in interrelating the families in these orders. The opportunity has been taken to include distribution data on other known chemical characters in these orders; the families in Table 8 are given according to Engler's original classification (see Willis, 1960).

Only two points about the data in Table 8 can be mentioned here. The first is the close chemical relationship between families which are difficult to separate on morphological grounds. As Fritsch (1893–4) says, "the relationships of the Gesneriaceae to allied orders, especially Scrophulariaceae,

Orobanchaceae and Bignoniaceae, are so close that it is almost impossible to draw the dividing line." With these four families, one might include the Verbenaceae on the chemical data. The Solanaceae, Convolvulaceae and Hydrophyllaceae form a contrasting chemical group.

The second point is concerned with the position of the Loganaceae and particularly *Buddleia* in the Contortae. In the recent edition (1964) of Engler's "Syllabus", *Buddleia* is separated from the Loganaceae, raised to family rank

TABLE 7. Distribution in dicotyledons of flavonols and flavones of systematic interest

Class of flavonoid	Present in families
Flavonols with 6- or 8-hydroxylation, e.g. quercetagetin, gossypetin	Archichlam: Chenopodiaceae, Combretaceae, Leguminosae (5 genera), Papaveraceae, Rutaceae, Saxifragaceae Sympetalae: Compositae (6 genera), Ericaceae, Malvaceae, Primulaceae, Rubiaceae, Scrophulariaceae, Verbenaceae
Flavonols with *O*-methylation, e.g. azaleatin, rhamnetin	Archichlam: Begoniaceae, Combretaceae, Leguminosae, Polygonaceae, Phytolaccaceae, Rutaceae, Rhamnaceae Sympetalae: Compositae, Ericaceae, Plumbaginaceae, Scrophulariaceae, Solanaceae
Flavones with 6- or 8-hydroxylation, e.g. scuttelarein, wogonin	Archichlam: Leguminosae Sympetalae: Bignoniaceae, Compositae, Labiatae, Pedaliaceae, Plantaginaceae, Scrophulariaceae
Flavones with *O*-methylation, e.g. diosmetin, tricin	Archichlam: Leguminosae, Rosaceae, Rutaceae, Thymelaeaceae Sympetalae: Gesneriaceae, Hydrophyllaceae, Labiatae, Loganaceae, Scrophulariaceae
Flavanones with 6- or 8-hydroxylation, e.g. isopedicin, isookanin	Archichlam: Leguminosae, Rosaceae Sympetalae: Compositae, Gesneriaceae
Flavanones with *O*-methylation, e.g. hesperitin, sakuranetin	Archichlam: Rosaceae, Rutaceae Sympetalae: Ericaceae, Gesneriaceae, Hydrophyllaceae

and placed in the Tubiflorae close to the Scrophulariaceae. This is very satisfactory from the chemical point of view, since the data in Table 8 for Loganaceae mostly refer to studies on *Buddleia* and clearly point to such a rearrangement. The water soluble yellow carotenoid gentiobioside, crocein, provides yet another link between *Buddleia* and the Scrophulariaceae. This rare pigment of *Crocus* pollen was found in petals of *Verbascum phlomoides* by Schmid and Kotter in 1932, but has recently been discovered in another Scrophulariaceae, *Nemesia strumosa*, and, remarkably enough, also in *Buddleia variabilis* (Harborne, 1965b and unpublished results).

Of the many other interesting aspects of hydroxylation and methylation patterns in flavones, two of phylogenetic interest must be mentioned. Hydroxylation in the 2'-position of flavonoids is a rare event and only occurs with

TABLE 8. Distribution of chemical characters in the orders Contortae and Tubiflorae

Order and family	Distribution of				
	Methylated flavones	Hydroxylated flavones	Flavone replacing flavonol	Orobanchin[2]	Iridoids
CONTORTAE					
Oleaceae	—	—	+	+	—
Loganaceae	+	—	+	+	+
Gentianaceae	—	—	+	—	—
Apocynaceae	—	—	—	—	+
Asclepiadaceae	—	—	—	—	—
TUBIFLORAE[1]					
Convolvulaceae	+	—	—	—	—
Hydrophyllaceae	+	—	—	—	—
Verbenaceae	—	+	+	+	+
Labiatae	+	+	+	—	—
Solanaceae	+	—	—	—	—
Scrophulariaceae	+	+	+	+	+
Bignoniaceae	—	+	+	+	+
Orobanchaceae	+	—	+	+	+
Gesneriaceae	+	—	+	+	—
Acanthaceae	—	—	+	+	—

[1] Some families are omitted, since data is not available about their chemical constituents.

[2] This complex caffeic-sugar derivative was first isolated from *Orobanche minor* by Bridel and Charaux (1924) and similar substances have been detected in *Buddleia* (W. D. Ollis, unpublished results), *Catalpa* and *Syringa* (Birkofer *et al.*, 1961). It may be identical to verbascoside from *Verbascum* (Scarpati and Monache, 1963). Other records are from unpublished results of the author.

flavones and isoflavones. 2'-Hydroxyflavones are found in the Rutaceae, 2'-hydroxyflavonols (e.g. morin (XVI)) in three genera of the Moraceae, and in the Anacardiaceae, Datiscaceae and Leguminosae, and 2'-hydroxyiso-flavones in the Amarantaceae and Leguminosae. All these families are included in Engler's "Archichlamydeae" and it is significant that the only other occurrence of a 2'-hydroxyflavonoid in nature is in the gymnosperm, *Podocarpus* (Coniferales).

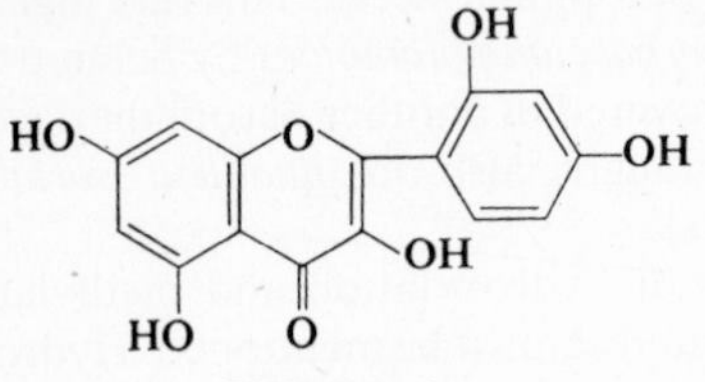

XVI. Morin

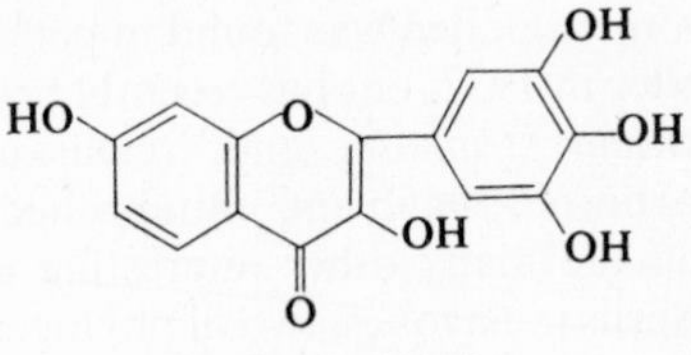

XVII. Robinetin

The other modification of special interest in the absence of a 5-hydroxyl group. Flavonoids of this type, e.g. robinetin (XVII) are found mostly in the Leguminosae, but also in Anacardiaceae and Celastraceae (both in the order Sapindales). Related chalcones occur in the Leguminosae and also in the Compositae, but their presence in the latter family is almost certainly related to aurone production. Thus, absence of 5-hydroxylation, like presence of 2′-hydroxylation, is a characteristic feature of certain families of the Archichlamydeae, and notably of the Leguminosae.

3. *Glycosidic patterns*

Over forty classes of flavonol and flavone glycosides are known and the systematic significance of some of the rarer sugar combinations has been discussed recently (Harborne, 1964). Two newer examples will be described

TABLE 9. Distribution of flavonoids in the genus *Baptisia*[1]

Group	Flavonoids present in leaf and flower
White flowered species (e.g. *B. alba*, *B. pendula*)	Flavonols[2] (predominantly 3 mono- and diglycosides) isoflavones[3]
B. leucophaea species group	Luteolin 7-glucoside, apigenin 7-glucoside Isoflavones Traces of flavonols
B. sphaerocarpa species group	Luteolin 7-rutinoside, apigenin 7-rutinoside Flavonols (in flowers only) Isoflavones Flavanones[4]

[1] Date from R. E. Alston, H. Rösler, J. Kagan and T. J. Mabry (unpublished results).
[2] Flavonols are kaempferol and quercetin, present as the 3-glucoside, 3-rutinoside, 7-glucoside, 3,7-diglucoside, 3-glucoside-7-rutinoside and 3-rutinoside-7-glucoside.
[3] Isoflavones are biochanin A, tectoridin, genistein, genistin, orobol, oroboside and genistein 7-rutinoside.
[4] Eridictyol, naringenin and their glycosides plus possibly others.

here. The first refers to the distribution of flavone glycosides in the genus *Baptisia* (Leguminosae). The twenty or so species of *Baptisia* can be divided arbitrarily into three groups on the basis of their flavonoid content (Table 9). As can be seen, the main difference between the *leucophaea* and *sphaerocarpa* groups is in the flavone glycosidic pattern. On the basis of this chemical evidence and assuming that flavones are more highly evolved than flavonols (see an earlier section), the three groups fall into the evolutionary sequence: flavones absent→flavone 7-glucosides→flavone 7-rutinosides. The application of these data to the systematics of the group has still to be worked out, but it is to be hoped that these chemical results will illuminate species relationships in this plant genus.

The second example is also taken from work on legume flavonoids, this time on the genus *Pisum*. Many of the flavonol glycosides present in the three main groups of pea plant have been identified (Table 10). The only true representative available of a wild pea is *P. fulvum* but there are primitive cultivated forms growing today in Nepal and Ethiopia which differ from the modern European cultivars. These three groups, on the chemical evidence, fall into a natural

TABLE 10. Flavonoids of *Pisum*

Pea material	Major[2] flavonol glycoside in leaf
Pisum fulvum	Quercetin 3-glucoside
Primitive Asian and African cultivars (*P. nepalensis, P. abbysinicum*)	Kaempferol and quercetin 3-sophoroside
Modern European varieties (*P. sativum, P. elatior, P. jomardii*)[1]	Kaempferol and quercetin 3-(*p*-coumaroylglucosylsophoroside)

[1] The field pea, *P. arvense*, contains kaempferol 3-(*p*-feruloylgalactosylsophoroside).
[2] Other flavonol glycosides occurring in minor amounts have not yet been fully identified. The acylated pigments in modern cultivars are usually accompanied by the corresponding unacylated glycosides.

evolutionary sequence: flavonol monoglucoside→flavonol diglucoside→ flavonol triglucoside. This result does not solve the problem of the origin of the garden pea, but it does suggest that chemical data may be worth considering in evolutionary studies of this kind.

C. OTHER FLAVONOIDS

The distribution of five other flavonoid classes in dicotyledons is shown in Table 11. The data are obviously incomplete, since no systematic search for any of these classes of substance has been carried out. The main point to note from the available data is that flavanones, chalcones and dihydrochalcones, three classes known to occur in lower plants (see earlier, p. 274) are randomly distributed in the dicotyledons. Thus, they are present in primitive (e.g. Ranunculaceae, Salicaceae) as well as advanced (e.g. Compositae) families. By contrast, the two flavonoid classes not found in lower plants, isoflavones and aurones, are of much more restricted distribution. Isoflavones are found mainly in the Leguminosae and the other isoflavone families are all in the Archichlamydeae and are not too distantly related to the legumes; indeed, the Rosaceae is in the same order. Similarly, aurones, present in four sympetalous families and two advanced families in the Archichlamydeae, clearly represent an advanced character in higher plants.

TABLE 11. Distribution in dicotyledons of minor flavonoids

Class	Families present in
Flavanones	Archichlam: Leguminosae, Rosaceae, Rutaceae Sympet: Ericaceae, Hydrophyllaceae, Compositae
Chalcones	Archichlam: Caryophyllaceae, Ranunculaceae, Rosaceae, Leguminosae, Piperaceae Sympet: Acanthaceae, Gesneriaceae, Compositae
Dihydrochalcones[1]	Archichlam: Rosaceae, Salicaceae Sympet: Ericaceae
Isoflavones	Archichlam: Amarantaceae, Leguminosae (sub-family: Papilionatae), Moraceae and Rosaceae
Aurones	Archichlam: Leguminosae, Oxalidaceae Sympet: Compositae, Gesneriaceae, Plumbaginaceae and Scrophulariaceae

[1] Distribution of dihydrochalcones is discussed in detail by A. H. Williams in Chapter 17.

The different distribution patterns of these rarer flavonoids are of considerable evolutionary interest and they will be considered in the following section.

VI. EVOLUTION OF FLAVONOIDS

All the simple flavonoids are derived biosynthetically from a common C_{15}-precursor, probably a chalcone or flavanone intermediate, which is then modified in a variety of ways in a series of unit processes by oxidation, reduction, ring closure, glycosylation, methylation, etc. to yield the many substances of this type found in plants. From studying the known distribution patterns in plants, it is possible to suggest that some structural modifications are primitive and others are advanced characters (Table 12). It should be noted that elimination of trihydroxylation is a loss mutation *only in the leaf flavonoids* and is correlated with the replacement of flavonols by flavones (Bate-Smith, 1962). It explains the complete absence in nature of a flavone corresponding to myricetin, i.e. 5,7,3′,4′,5′-pentahydroxyflavone (tricetin). In the petal anthocyanidins, trihydroxylation (cyanidin→delphinidin) is a gain mutation associated with evolution towards a blue flower colour.

A few modifications do not appear to be either exactly primitive or precisely advanced and are classified here as "isolated characters", since they are found almost exclusively in a few related families in the Archichlamydeae, but do not appear to any extent elsewhere. Isoflavone formation and 2′-hydroxylation, for example, both occur characteristically in the Leguminosae; the combination of these two characters in the same family leads to the production of a more complex group of flavonoids, the insecticidal principles, the rotenoids (Grisebach and Ollis, 1961). Another "isolated character" studied by Bate-Smith

TABLE 12. Evolution of flavonoids in plants

Primitive characters

1. 3-Deoxyanthocyanidins
2. Flavonols
3. Leucoanthocyanidins
4. Chalcones, flavanones and dihydrochalcones
5. C-substitution (C-methylation, C-prenylation, C-glycosylation, biflavonyl formation)

Advanced characters

Gain mutations

1. Complex O-glycosylation (including acylation of sugars)
2. 6- or 8-Hydroxylation
3. O-Methylation
4. Oxidation of chalcones→aurones

Loss mutations

1. Replacement of flavonols by flavones
2. Elimination of leucoanthocyanidins
3. Elimination of trihydroxylation

Isolated characters

1. Replacement of anthocyanin by betacyanin
2. Shift of flavonoid B ring to 3-position: isoflavone formation (Leguminosae)
3. 2′-Hydroxylation
4. Elimination of 5-hydroxyl group

(1962, and unpublished results) is ellagic acid, which is present in the Rosales and a large area of the Englerian Archichlamydeae, but is absent from the Sympetalae (except Ericaceae) and from lower plants. It is, like aurones and 6-hydroxylated flavones, one of the few dicotyledon polyphenols not known in the monocotyledons.

The biological significance of an evolving pattern of flavonoid synthesis is not entirely clear and may be an indirect response to changes in primary metabolic pathways. It must be related to some extent to the plant's need to produce a fruit or flower colour attractive to insect vectors. Many modifications such as anthocyanidin hydroxylation, O-glycosylation, O-methylation are presumably related to the search for a stable blue petal colour. Similarly, aurone formation in highly evolved families must be advantageous compared to chalcone formation, since the aurone is a more stable and brighter yellow pigment.

Two questions of particular phylogenetic interest arise when considering the distribution of anthocyanins: (1) how is it that primitive 3-deoxyantho-cyanidins are found anomalously in the highly evolved Gesneriaceae? and (2) why do betacyanins replace anthocyanins only in the order Centrospermae?

There are three possible answers to the first question. Production of the orange and scarlet 3-deoxyanthocyanidins in gesneriads may represent the retention of a primitive moss and fern character throughout evolution; a rather unlikely situation. More probably, it represents a response to selection for red flower colour, favoured by bird pollinators, in place of the more common blue colour preferred by bees. There is, however, no clear evidence that gesneriads are bird-pollinated, though this phenomenon is known in the closely related Bignoniaceae; the 3-deoxyanthocyanidins are, however, found in species which are most likely to be bird-pollinated, i.e. the sub-tropical New World subfamily, the Gesnerioideae. A third possibility is that 3-deoxyanthocyanidin production is related to the replacement of flavonol by flavone in the leaf. Synthesis of flavonoids in leaf and in petal are known to be closely related processes and since flavonols are completely absent from the family, an analogous loss of an enzyme for 3-hydroxylation of anthocyanidins might well take place in some species.

Betacyanin production is puzzling because it is completely restricted to one order of plants, the Centrospermae, which are generally acknowledged to be a fairly primitive group. The reason why betacyanins have not replaced anthocyanins in most other higher plants must be because they represent a biosynthetic cul-de-sac. The colour range among the betacyanins is more restricted than with the anthocyanins, and it may be difficult for them to yield a pure blue tone. Betacyanins are certainly very labile pigments chemically and may be similarly labile *in vivo*. Finally, their synthesis requires amino acid nitrogen, an element usually in short supply in the plant and one better employed in protein synthesis.

A similar explanation to that employed above may be used to account for the restriction of isoflavone synthesis to the Leguminosae. Transfer of the B ring in the flavone nucleus from the 2- to 3-position reduces completely any potentiality for colour production; many flavones are pale yellow in colour but no coloured isoflavones have ever been found.

Another major factor controlling evolution of flavonoids is lignification, since lignins and flavonoids are derived, in part, from a common C_9 precursor. Bate-Smith (1962) has already noted that the reduction in lignification, caused by a changeover from woody to herbaceous habit, alters the pattern of leaf flavonoids. Leucoanthocyanidins and flavonols disappear to be replaced by flavones and methoxycinnamic acids. The relative wealth of flavone production in some herbaceous plants, e.g. members of the Compositae, may represent a means of avoiding a build-up of lignin precursors. Herbaceous plants, because they lack woody lignins and astringent leucoanthocyanidins, are much more edible to animals and, in order to survive, may have to produce some other distasteful constituent. Certainly, many herbaceous plants are rich in alkaloids and saponins (Bate-Smith, 1954). Flavonoid evolution is no doubt affected in some way by considerations of plant edibility and plant survival.

VII. Conclusion

The attempt has been made to outline an evolutionary scheme for flavonoids in plants. This will no doubt require modification as new occurrences of key flavonoids are reported. Further surveys of flavonoids in lower plants are badly needed. The scheme (Table 12) is based mainly on present ideas of evolutionary status (see Davis and Heywood, 1963) and on biogenetic considerations. It cannot therefore be used to prove or disprove any new evolutionary schemes that are proposed on morphological or other considerations. If it has any use, it will be in studies of narrow ranges of plants and only then in conjunction with all the usual biological criteria.

A major handicap in discussing flavonol distribution and evolution is the paucity of our knowledge regarding their function. It is not possible to conclude that leaf flavonoids have no biological function since their complete elimination would be expected in the most highly advanced plants which is not the situation. The fact that flavonoids are not primary metabolites is, of course, strongly in their favour as taxonomic markers.

There is much to be said for concentrating future efforts in this field on a few plant families: the Leguminosae, the Compositae and the Tubiflorae group in the dicotyledons and the Gramineae in the monocotyledons. It is a happy coincidence that the two plant groups of prime economic importance— the grasses and the legumes—are among the most richly versatile in their biosynthetic equipment. Flavonoid chemistry has certainly provided new data for studying plant relationships and may well play a part in solving evolutionary problems in some of our crop plants.

REFERENCES

Alston, R. E. and Irwin, H. S. (1961). *Amer. J. Bot.* **48**, 35.
Avadhani, P. N. and Lim, G. (1964). *Abstr. X Int. Bot. Congr.* 326.
Baker, H. G. (1948). *Ann. Bot.* N.S. **12**, 207.
Bate-Smith, E. C. (1954). *Biochem. J.* **58**, 122.
Bate-Smith, E. C. (1962). *J. Linn. Soc.* (*Bot.*) **58**, 95.
Beale, G. H., Price, J. R. and Sturgess, V. C. (1941). *Proc. roy. Soc.* B **130**, 113.
Beckmann, S. and Geiger, H. (1963). *Phytochemistry* **2**, 281.
Benz, G., Martensson, O. and Terenius, L. (1962). *Acta chem. scand.* **16**, 1183.
Benz, G. and Martensson, O. (1963). *Acta chem. scand.* **17**, 266.
Birkofer, L., Kaiser, C., Nouvertné, W. and Thomas, U. (1961). *Z. Naturf.* **16b**, 249.
Bridel, M. and Charaux, C. (1924). *C.R. Acad. Sci., Paris* **178**, 1839.
Burtt, B. L. (1962). *Notes Roy. Botan. Gardens Edinb.* **24**, 205.
Davis, P. H. and Heywood, V. H. (1963). "Principles of Angiosperm Taxonomy." Oliver and Boyd, Edinburgh.
Endo, T. (1962). *Jap. J. Genet.* **37**, 284.
Engler, A. (1964). *In* "Syllabus der Pflanzenfamilien" (H. Melchior, ed.), Vol. II. Springer-Verlag, Basle.
Erikson, D., Oxford, A. E. and Robinson, R. (1938). *Nature, Lond.* **142**, 211.
Fritsch, K. (1893–4). *In* "Die Naturliche Pflanzenfamilien" (Engler, A. and Prantl, K., eds.), IV (3B), p. 133. Leipzig.
Fujise, S. (1929). *Sci. Papers Inst. Phys. Chem. Res.* (*Tokyo*) **11**, 111.
Grisebach, H. and Ollis, W. D. (1961). *Experientia* **11**, 4.
Harborne, J. B. (1960). *Biochem. J.* **74**, 270.
Harborne, J. B. (1962). *Biochem. J.* **84**, 100.
Harborne, J. B. (1963a). *In* "Chemical Plant Taxonomy", (T. Swain, ed.), p. 359. Academic Press, London and New York.

Harborne, J. B. (1963b). *Phytochemistry* **2**, 327.

Harborne, J. B. (ed.) (1964). "Biochemistry of Phenolic Compounds", p. 129. Academic Press, London.

Harborne, J. B. and Simmonds, N. W. (1964). *In* "Biochemistry of Phenolic Compounds", (J. B. Harborne, ed.), p. 77. Academic Press, London.

Harborne, J. B. (1965a). *Nature, Lond.* **207**, 984.

Harborne, J. B. (1965b). *Phytochemistry.* **4**, 647.

Harney, P. M. and Grant, W. F. (1964). *Amer. J. Bot.* **51**, 621.

Harney, P. M. and Grant, W. F. (1965). *Canad. J. Genet. Cyt.* **7**, 40.

Hayashi, K. and Abe, Y. (1955). *Bot. Mag., Tokyo* **68**, 299.

Hegnauer, R. (1962–1964). "Chemotaxonomie der Pflanzen", Vols. I–III. Birkhäuser Verlag, Basel.

Hörhammer, L. and Wagner, H. (1962). *In* "Chemistry of Natural and Synthetic Colouring Matters", (T. S. Gore *et al.*, eds.), p. 315. Academic Press, New York and London.

Hörhammer, L., Wagner, H. and Reinhardt, H. (1965). *Naturwissenschaften* **52**, 161.

Lawrence, G. H. M. (1961). "Taxonomy of Vascular Plants". Macmillan Co., New York.

Molisch, C. (1911). *Ber. dtsch. Bot. Ges.* **29**, 487.

Murti, V. V. S., Raman, P. V. and Seshadri, R. T. (1964). *Tetrahedron Letters*, 2995.

Nair, A. G. R., Nagarajan, S. and Subramanian, S. S. (1965). *Curr. Sci. (India)* **34**, 79.

Nilsson, M. (1961). *Acta chem. scand.* **15**, 154, 211.

Parks, C. R. (1965). *Amer. J. Bot.* **52**, 309.

Paul, H. (1908). *Mitt. Bayer. MoorkAnst*, Heft 2, 63.

Pecket, R. C. (1959). *New Phytol.* **58**, 182.

Pecket, R. C. (1960). *New Phytol.* **59**, 138.

Peterson, G. E., Livesay, R. and Futch, H. (1961). *Chem. Abstr.* **55**, 17738.

Price, J. R., Sturgess, V. C. and Robinson, R. (1938). *Nature, Lond.* **142**, 356.

Robinson, G. M., Robinson, R. and Todd, A. R. (1934). *J. chem. Soc.* 809.

Rudolph, H. (1965). *Planta* **64**, 178.

Sawada, T. (1958). *J. Pharm. Soc. Japan* **78**, 1023.

Scarpati, M. L. and Monache, F. D. (1963). *Ann. Chim. (Rome)* **53**, 356.

Schmid, L. and Kotter, E. (1932). *Mh. Chem.* **59**, 346.

Sherratt, H. S. A. (1959). *J. Genet.* **56**, 28.

Smith, D. M. and Levin, D. A. (1963). *Amer. J. Bot.* **50**, 952.

Stafford, H. A. (1965). *Plant Physiol.* **40**, 130.

Stebbins, G. L., Harvey, B. L., Cox, E. L., Rutger, J. N., Jelencovic, G. and Yagil, E. (1963). *Amer. J. Bot.* **50**, 830.

Swain, T. (ed.) (1963). "Chemical Plant Taxonomy". Academic Press, London. and New York.

Swain, T. and Bate-Smith, E. C. (1962). *In* "Comparative Biochemistry", (M. Florkin and H. S. Mason, eds.), Vol. III, p. 755. Academic Press, New York and London.

Thomson, R. H. (1957). "Naturally Occurring Quinones". Butterworth, London.

Torres, A. M. and Levin, D. A. (1964). *Amer. J. Bot.* **51**, 639.

Ueno, A. (1963). *Chem. Abstr.* **59**, 736.

Venkataraman, K. (1959). *Fortschr. Chem. org. Naturst.* **17**, 1.

Wada, E. (1956). *J. Amer. chem. Soc.* **78**, 4725.

Willis, J. C. (1960). "Dictionary of Flowering Plants and Ferns". 6th Ed. Cambridge University Press.

CHAPTER 17

Dihydrochalcones

A. H. WILLIAMS

Long Ashton Research Station, University of Bristol, England

I. INTRODUCTION

To be of use in comparative phytochemistry, a compound clearly must have a limited range of distribution in plants. If the range is very wide, as with the derivatives of quercetin and caffeic acid, very little distinction can be made between species, genera or families; on the other hand, if its reported occurrences are few, a compound may be of use only as a marker for a single species. Between these limits lie the most useful compounds, which may mark off whole families or orders as do, for example, the betacyanins (see Chapter 14).

The dihydrochalcones are compounds of very limited distribution, and a consideration of their occurrence illustrates both the uses and limitations of such compounds in taxonomic problems.

II. DIHYDROCHALCONES IN *MALUS* SPECIES

The best known dihydrochalcone, phloridzin (I), should not perhaps be described as a rare compound; it occurs in quantity in all parts of the apple tree except the flesh and juice of the mature fruit, ranging from over 1% of the fresh weight in leaf to over 12% of the dry weight in root bark. In view of the annual production of apple leaf in the orchards of the world, the amount of phloridzin synthesized is quite considerable.

Phloridzin was the first dihydrochalcone to be isolated, by de Koninck in 1835, and his first publication of his discovery made a claim that has bedevilled the comparative phytochemistry of phloridzin ever since. In his first paper (de Koninck, 1835a) he said that phloridzin occurs in the bark of apple, cherry, pear and plum; while in the second paper (de Koninck, 1835b) he describes in some detail its isolation from apple root bark only. Neither de Koninck nor anyone else produced further evidence of the occurrence of phloridzin in pear, cherry or plum, and the original statement was corrected in the case of the pear tree by Rivière and Bailhache (1904), followed by Bourquelot and Fichtenholz (1911) and Lincoln (1926). Nevertheless writers of surveys and textbooks have continued to reproduce the mistake down to the present day. To mention two recent examples, it is quoted in a chapter in "The Chemistry of Flavonoid Compounds" (Shimokoriyama, 1962) and in "Chemical Plant Taxonomy" (Paris, 1963). From my own work it can be stated definitely that phloridzin is confined to the apple (*Malus*) among the fruit trees and does not occur in pear, cherry or plum, nor in any of their relatives among the *Rosaceae*. Perhaps this error concerning the occurrence of phloridzin may at last be laid to rest.

Phloridzin (I) is, in chemical terms, both a phloroglucinol derivative and an aromatic *o*-hydroxyphenylketone. This makes it very suitable for detection on paper chromatograms, since the phloroglucinol nucleus gives a strong orange-red colour with diazotized *p*-nitroaniline while the *o*-hydroxyphenyl-ketone structure chelates with aluminium to give a very typical ultraviolet

I. Phloridzin   II. Sieboldin

fluorescence. Chromatographic examination of leaf extracts of different varieties of cultivated apple, whether dessert, culinary or cider varieties, showed the presence of phloridzin in all. But these varieties all derive from a single species, *Malus pumila*, whereas botanically speaking there are many other *Malus* species; Rehder (1954) lists twenty-five in his classification. Examination of leaf extracts from these species by paper chromatography showed that while the majority still have phloridzin as the major phenolic, it is in a few species either replaced or accompanied by another substance which gives a dark brown spot with the diazo reagent. When separated and subjected to alkaline fission, this substance gave phlorin (phloroglucinol glucoside) and β-(3,4-dihydroxyphenyl)propionic acid; phloridzin under the same conditions gives phlorin and *p*-hydroxyphenylpropionic acid. So the new material at first was thought to be 3-hydroxyphloridzin; but methylation and hydrolysis to confirm this structure gave not the expected 2′-hydroxy-3,4,4′,6′-tetramethoxydihydrochalcone but the isomeric 4′-hydroxy-3,4,2′,6′-tetramethoxydihydrochalcone. The new compound, *sieboldin*, must therefore differ from phloridzin in having its glucose molecule attached at position 4′ instead of at 2′, besides having an additional phenolic hydroxyl at C-3 (Williams, 1961) and thus has the structure (II).

The name sieboldin for the new dihydrochalcone derives from the fact that it occurs in the majority of specimens of species 8–11 in Rehder's classification, the series Sieboldianae; the species are *M. floribunda, M. zumi, M. sieboldii* and *M. sargenti*. It would be more satisfying to be able to report the unfailing occurrence of sieboldin in these species, but unfortunately specimens from some sources do not contain it. Probably some of the supposed specimens are seedlings rather than trees vegetatively propagated from the type. Specimens of the Sieboldianae series from the collection at the John Innes Institute which had been checked for their botanical characteristics all contained sieboldin except for *M. zumi*. This species is particularly variable in its chemistry; it is in fact regarded as a probable hybrid from *M. sieboldii* by Rehder, so variability of seedlings would be expected. In a test, 256 open-pollinated seedlings were raised from seed from a specimen of *M. zumi* (from Kew) which contained sieboldin; 44% contained only phloridzin in their leaf while 56% contained sieboldin as well, 20% containing more sieboldin than phloridzin.

Sieboldin has not been found outside the series Sieboldianae or their hybrids except in one variety of *M. prunifolia*, Rinki; this may thus well be a hybrid with a Sieboldianae species in its pedigree.

Further examination of chromatograms of leaf extracts of apple species showed that the main spot from *M. trilobata* differed slightly from phloridzin in its diazo colour and aluminium fluorescence reactions, although its R_f values were very similar. Acid hydrolysis gave glucose and phloretin in equivalent amounts, alkaline fission phlorin and *p*-hydroxyphenylpropionic acid; the compound, trilobatin, was clearly the 4'-glucoside of phloretin, isomeric with phloridzin. Methylation experiments confirmed this structure (III) (Williams, 1961).

III. Trilobatin   IV. Asebotin

This third dihydrochalcone from *Malus* leaf, trilobatin (III), has been found only in *M. trilobata* among the *Malus* species. A point of special interest is that *M. trilobata* seems to contain no phloridzin; the change of dihydrochalcone glycosidation pattern is complete, and there is no co-occurrence of the two types as happens with phloridzin and sieboldin in the Sieboldianae. This poses an interesting question about glucosylating enzymes; if glucosylation is a late stage in the synthesis of the dihydrochalcone glucosides, why should *M. trilobata* put glucose onto phloretin or its precursor exclusively in position 4' while the majority of *Malus* species do so only in the 2' position? Again, in the Sieboldianae series glucose appears to be attached to phloretin exclusively in the 2' position simultaneously with the glucosylation of 3-hydroxyphloretin in the 4' position. Certainly glucosylation seems to follow the establishment of the hydroxylation pattern in the C_{15} component; there is no evidence to suggest the possibility of direct hydroxylation of already formed dihydrochalcone glucosides; sieboldin (II) does not occur in the same species as trilobatin, or vice versa.

Trilobatin (III) has been found only in the one species *M. trilobata*. No hybrids of this species seem to have been recorded, and in this country at least the tree, which is not common, seems to be a shy fruiter; seed has not been available from Kew over the several years since trilobatin was isolated, so it has not been possible to investigate the dihydrochalcones of its progeny. By sheer chance, however, we did record trilobatin as the exclusive dihydro-chalcone of one seedling supposedly from *M. sieboldii arborescens*; the significance or otherwise of this observation cannot even be guessed at on the basis of a single isolated case.

The geographical background of the occurrence of the new dihydrochalcone glucosides is of interest. All the Sieboldianae group of species, containing sieboldin (II), come from Japan; only one, *M. sieboldii arborescens* is men-tioned as occurring on the mainland of Asia as well, in Korea close to Japan. The only other sieboldin-containing variety, *M. prunifolia* Rinki comes from East Asia too. It seems therefore that the appropriate gene-combination for the production of sieboldin originated in the area of Japan. Many hybrids of the Sieboldianae group have been raised, both in Japan and elsewhere; most of them seem to contain some sieboldin.

M. trilobata, by contrast, comes from West Asia. No hybrids have been reported, and it seems to be an isolated species which has arisen independently of the Sieboldianae.

To complete the present survey of the *Malus* dihydrochalcones, it is of interest to note that probably three other dihydrochalcones, all phloretin polyglycosides, occur in the common apple tree, *M. pumila*, along with phloridzin but in much smaller amounts. The only one nearly fully charac-terized, occurring mainly in very young leaf is phloretin 2'-xylosyl-glucoside, essentially phloridzin with a xylose molecule linked to the glucose. Another is probably a similar compound with arabinose attached to the phloridzin glucose; the third, judging by its chromatographic behaviour, probably has three sugar molecules attached to phloretin (A. H. Williams, unpublished results). As these compounds have been investigated only in *M. Pumila*, they are of no taxonomic interest.

So far my emphasis has been on the dihydrochalcones of the genus *Malus*, because it contains the greatest variety and largest amounts of these compounds. No other occurrences have been recorded in the Rosaceae apart from the apple-pear hybrids raised by M. B. Crane at the (then) John Innes Horticultural Institution, which contain the full range of phenolics from both their parents (Williams, 1955).

III. DIHYDROCHALCONES IN THE ERICACEAE

The next well authenticated occurrence of dihydrochalcones is in the family Ericaceae, in the genera *Pieris* and *Kalmia*. Here the previously published work appeared somewhat confusing; the presence of both phloridzin and asebotin was claimed, sometimes in apparent contradiction. Asebotin (IV), the 4'-methyl ether of phloridzin was noted long ago by Eykman (1883) in what he called *Andromeda japonica* (now *Pieris japonica*), and later by Bour-quelot and Fichtenholz (1912) in *Kalmia latifolia*. Then Bridel and Kramer (1931) claimed that the *Kalmia latifolia* glucoside was phloridzin; Murakami and Fukuda (1955) reported similarly of the *Pieris* compound. This situation was first brought to my notice indirectly by Dr. Bate-Smith, who had examined several *Pieris* samples from Wisley in the course of his survey of the phenolic aglycones of the dicotyledons; he noted clear differences in their phenolic contents, and a search of the literature led me to carry out chromatographic surveys of both *Pieris* and *Kalmia*.

Referring once more to Rehder's Manual (1954), he lists three main species of *Kalmia*; *latifolia*, *polifolia* and *angustifolia*. Examination of leaf extracts of a number of specimens from each group showed quite clear-cut distinctions in terms of the dihydrochalcones. *K. polifolia* has no dihydrochalcones, *K. latifolia* has phloridzin (I) and *K. angustifolia* asebotin (IV). *K. latifolia* contains in addition small amounts of another dihydrochalcone (not asebotin), probably a phloretin polyglycoside.

With *Pieris* the situation is not quite so clear-cut. Rehder lists only two main species, *P. floribunda* and *P. japonica*; here the distinction is quite clear since *P. floribunda* has no dihydrochalcones while *P. japonica* has. It is with his subdivision of *P. japonica* that the complexities arise. He lists four "related species" with *P. japonica*; *P. formosa*, *P. forrestii*, *P. taiwanensis* and *P. nana*. Other authorities list *P. forrestii* as a variety of *P. formosa*, and *P. taiwanensis* as a distinct species. Leaf samples of several specimens of all except *P. nana* have been examined, and the results are equivocal. Most samples of *P. japonica* contain only phloridzin (I), and most *P. taiwanensis* only asebotin (IV); *P. formosa* and *P. forrestii* are very variable, some containing only phloridzin (I) or only asebotin (IV), others having both. It seems clear that Rehder is right in separating off *P. floribunda* as a distinct species and regarding all the rest as more closely related. Geographically this makes sense too; *P. floribunda* is from North America while the *P. japonica* group come from Asia and its nearby islands. It is probable too, of course, that some of the specimens examined are seedlings raised in nurseries where the *P. japonica* sub-species have been grown near together and so may be hybrids.

The occurrence of dihydrochalcones in the family Ericaceae is, so far as my experience goes, confined to *Kalmia* and *Pieris*. Of the thirty-five genera listed by Rehder, chromatographic examination of leaf extracts of varying numbers of species from twenty-four of these genera has failed to show dihydrochalcones in any others; nor are there references in the literature to any other occur-rences; dihydrochalcones seem to be almost as rare in Ericaceae as in Rosaceae.

IV. GLYCYPHYLLIN IN *SMILAX* SPECIES

The only other long-standing record of a dihydrochalcone in the literature refers to a plant far removed from those already discussed; a monocotyledon from Australia, *Smilax glycyphylla* (Liliaceae). In the 1880's E. H. Rennie, Professor of Chemistry at Adelaide, investigated the leaf of this plant and isolated the substance that he called glycyphyllin. He recognized its close similarity to phloridzin and showed that it was in fact a rhamnoside of phloretin (V) (Rennie, 1886). When in the present survey of dihydrochalcones an attempt was made to check Rennie's observations a curious situation developed. Since *S. glycyphylla* is a Australian plant not grown in this country even at Kew, dried material collected by other people had to be used. The first sample, obtained through the kindness of Mr. W. E. Hillis, was found to contain no dihydrochalcone and in fact only traces of phenolic material of any sort. An examination of several other species of *Smilax* available at Kew (*S. excelsa, glauca, megalantha*, and *rotundifolia*) gave similarly negative results. It began to look as though Rennie's plant might have been wrongly identified, and on the suggestion of Dr. Melville of Kew, an examination was made of a number of plants from the same area that might possibly be confused with *S. glycyphylla*; several species of *Rhipogonum, Eustrephus latifolius, Geitonoplesium cymosum* and *Parsonia lanceolata*. None showed any trace of

V. Glycyphyllin   VI. Mangiferin

a phloretin derivative. The solution to the problem came as a result of a casual mention to Dr. W. Bottomley, who then offered some material of *S. glycyphylla* gathered from the same locality as Rennie's original sample in New South Wales; Hillis' sample had come from Queensland. A single paper chromatographic test on the new sample at once revealed a strong spot very like phloridzin in its reactions, and the compound was quickly isolated and proved to be phloretin-2'-α-L-rhamnoside.

Further samples of the two forms of *S. glycyphylla* were obtained and tested. Glycyphyllin (V), was confirmed in the New South Wales samples, likewise its absence from the Queensland samples; but this time the Queensland samples contained another phenolic compound giving unusual colour and fluorescence reactions and ultraviolet absorption. Its nature remained obscure, as did that of an apparently identical compound noted by Drs. Bate-Smith and Harborne in certain *Iris* species. Two years ago at the Oxford meeting of the Plant Phenolics Group, Dr. Wagner provided the key by pointing out the similarity of the spectrum of the unknown to that of a xanthone derivative; Dr. Harborne quickly showed it to be mangiferin (VI) (Bate-Smith and Harborne, 1963).

This is a very peculiar situation; two specimens of a single species of plant containing two completely different phenolic compounds which have no close chemical relationship; as will be seen, the compounds do not even share the same carbon skeleton, and one is a *C*-glucoside, the other an *O*-rhamnoside. Australian botanists seem to be satisfied that both forms of the plant are *S. glycyphylla* by the classical criteria. While the absence or presence of a single compound might be explained as due to seasonal, environmental or nutritional differences, one hesitates to attribute the appearance of a completely different compound to such causes. Possibly this is a case of a mutation of limited geographical distribution which gives rise to a chemical difference without causing morphological differences. One Australian botanist has observed that the leaf of the New South Wales form has a strong odour or taste of sarsaparilla which is absent from the Queensland form.

V. Other Occurrences of Dihydrochalcones

My personal experience of the dihydrochalcones ends here, but to complete the survey two other occurrences and two false reports should be mentioned. So far, the dihydrochalcones discussed occurred as glycosides; the remaining examples are occurrences of the free aglycones. Nilsson (1961a) has isolated

VII   VIII

two aglycones (VII and VIII) structurally similar to phloretin from the fronds of the silver fern *Pityrogramma chrysophylla* var. marginata; they may be described as the 4-deoxy (VII) and 4-*O*-methyl (VIII) derivatives of 4′-*O*-methylphloretin. The first compound (VII) was also found by Goris and Canal (1935) in the buds of the balsam poplar.

The false reports of the occurrence of dihydrochalcones concern phloridzin in *Prunus virginiana* var. demissa (Harvey, 1929) and *Micromelum tephrocarpum* (Heilbron 1937). Both these plants have been re-examined and no trace of phloridzin could be found.

VI. The Comparative Phytochemistry of Dihydrochalcones

We can now survey the somewhat erratic comparative phytochemistry of the dihydrochalcones (Table 1). Phloridzin, by its presence, clearly marks off the genus *Malus* in the sub-family Pomoideae of the family Rosaceae; it is absent from the other Pomoideae and has not been found in any other Rosaceae examined. Within the genus *Malus*, deviations from the phloridzin structure provide distinction between some species. The other occurrences of dihydrochalcones seem quite unrelated; in the family Ericaceae they occur in the genera *Kalmia* and *Pieris*, but only in some species and with variations due to the presence or absence of methylation. Then in the monocotyledons we have the isolated example of glycyphyllin in some samples of one species of *Smilax* in the Liliaceae; and lastly the dihydrochalcone aglycones in *Populus* (Salicaceae) and the fern, *Pityrogramma*. Other occurrences of dihydrochalcones may of course yet be found, but it seems clear that their distribution is both limited and highly erratic, in contrast to that of the common phenolics already mentioned such as derivatives of quercetin and caffeic acid. Neither the very rare nor the common types are of much help in a chemical approach to taxonomy over a wide range of plants; nevertheless within limited fields rare classes such as dihydrochalcones are perhaps the more useful.

TABLE 1. Naturally occurring dihydrochalcones

Compound	Substituents[1]	Occurrence
Phloridzin	2′,4,4′,6′-Tetrahydroxy 2′-β-D-glucoside	Most species of *Malus*, *Kalmia latifolia*, various species of *Pieris*.
Trilobatin	2′,4,4′,6′-Tetrahydroxy 4′-β-D-glucoside	*Malus trilobata*.
Sieboldin	2′,3,4,4′,6′-Pentahydroxy 4′-β-D-glucoside	Most species and hybrids of the Sieboldianae series of *Malus* species.
Glycyphyllin	2′,4,4′,6′-Tetrahydroxy 2′-α-L-rhamnoside	Some specimens of *Smilax glycyphylla*.
Asebotin	2′,4,6′-Trihydroxy-4′-methoxy 2′-β-D-glucoside	*Kalmia angustifolia*, various species of *Pieris*.
Poplar and fern dihydrochalcone (VII)	2′,6′-Dihydroxy-4′-methoxy	Balsam poplar, silver fern.
Fern dihydrochalcone (VIII)	2′,6′-Dihydroxy-4,4′-dimethoxy	Silver fern

[1] Numbering follows that for phloridzin (I).

Finally the relationship of the dihydrochalcones to the other flavonoid compounds may be considered. All share the basic C_6-C_3-C_6 skeleton; their most obvious relatives are the chalcones and flavanones. The dihydrochalcones differ from the chalcones only by having a saturated $C\alpha$-$C\beta$ double bond, and it is remarkable that no co-occurrence of the two types ever seems to have been recorded. The closest connection is in the ferns, where there is alternative occurrence; the silver variety marginata of *Pityrogramma chrysophylla* has two dihydrochalcone aglycones, while the golden variety heyderi has the corresponding chalcones (Nilsson 1961b); even here there is no co-occurrence. With the dihydrochalcone glycosides, there is not even the occurrence of chalcone glycosides in any closely related plants. Chalcone glycosides are most common in the Compositae and Leguminosae, families in which no dihydrochalcones have been found; and the most common hydroxylation pattern of these chalcones (resorcinol type in the A-ring) differs from that of the dihydrochalcones which all have a phloroglucinol type A-ring. The only chalcones corresponding exactly in structural pattern to dihydrochalcones are isosalipurposide, occurring in *Salix purpurea* (Charaux and Rabaté, 1931), *S. daphnoides* (Williams, 1962) and *Helichrysum arenarium* (Jerzmanowska, 1956, Hänsel and Heise, 1959), the precise analogue of phloridzin (I), and neosakuranin in *Prunus cerasoides* (Puri and Seshadri, 1954) corresponding similarly to asebotin (IV). No dihydrochalcones have been found in these genera. The chalcones show much more variation in hydroxylation pattern than do the dihydrochalcones; beside having the resorcinol or phloroglucinol pattern, they may be additionally hydroxylated at the 3' (okanin, lanceolin) or 5' (stillopsin) positions of the A-ring. Sugars are most commonly attached at positions 4 or 4', sometimes at 2' or 3, whereas with the dihydrochalcones glycosidation occurs only at 2' or 4'.

Flavanones may be regarded as ring-closed chalcones, and it is therefore reasonable to look at their co-occurrence or otherwise with dihydrochalcones. Only those with a phloroglucinol-type A-ring corresponding to the dihydrochalcones will be considered. Eriodictyol 7-glucoside (IX) is found with

IX. R=OH; Eriodictyol-7-glucoside
X. R=H; Naringenin-7-glucoside

XI. Angolensin

phloridzin in apple bark (Williams, 1962), but the structures do not correspond; the flavanone 7-glucoside is the equivalent of a 4' dihydrochalcone glucoside, and the 3' hydroxyl group of the flavanone is extra, so the relationship is hardly close. Eriodictyol 7-glucoside has in fact the exact hydroxylation and glucosylation pattern of sieboldin (II), one of the dihydrochalcone glucosides of the Japanese *Malus* species described earlier; but it was found as a minor phenolic of *M. pumila* which does not contain sieboldin. The flavanones corresponding to phloridzin (I), and trilobatin (III), i.e. the 5- and 7-glucosides of naringenin (X), occur in *Salix purpurea* (Charaux and Rabaté, 1931; Williams, 1962), but without any sign of the dihydrochalcones; conversely, no derivatives of naringenin have been found in *Malus* in the course of our work. Henke (1963) claimed to detect naringenin derivatives in practically all species and hybrids of *Malus*, but we have not so far been able to confirm his results. Naringenin 7-glucoside and its 4'-O-methyl derivative sakuranin occur in many species of *Prunus* (Shimokoriyama, 1962), but the corresponding dihydrochalcones trilobatin (III) and asebotin (IV) have not been found with them here either.

The dihydrochalcones thus do not fit into any scheme of coexistence with what seem to be chemically their nearest relatives; they remain a small and isolated group of phenolic compounds. But perhaps not the smallest group of all; by analogy with the flavone—isoflavone relationship one can arrive at the isodihydrochalcones, so far consisting of a single compound, angolensin (XI).

Little is known about the intermediates on the biosynthetic route to dihydro-chalcones. As with other flavonoids, the B-ring (cf. I) appears to derive from the shikimic acid pathway; ^{14}C-labelled phenylalanine and cinnamic acid are incorporated into the B-ring of the phloridzin molecule by *Malus* leaf, while the A-ring is built up from acetate units. Hydroxylation of the B-ring seems to take place at a later rather than at an early stage in the synthesis; tyrosine is much less effective than phenylalanine as a precursor. (Hutchinson, *et al.*, 1959). The synthesis of phloridzin (I) in *Malus* leaf takes place only in young, growing leaf and ceases when the leaf is mature; labelled phenylalanine gives labelled phloridzin only in young leaf, while in old leaf the main products are glucose esters of cinnamic and *p*-coumaric acids. (Avadhani and Towers, 1961). The synthesis is in fact probably interrupted at an even earlier stage in old leaf, for the feeding of ^{14}C-labelled sugars or sorbitol to old leaf does not give labelled derivatives of cinnamic or *p*-coumaric acids (Williams, 1963). The synthesis of sieboldin (II) seems to be analogous in all respects (Williams, unpublished observations).

Grisebach and Patschke (1962) have shown that labelled phloridzin is formed in young apple leaf from added ^{14}C-labelled 2′, 4, 4′, 6′-tetrahydroxy-chalcone 2′-glucoside (isosalipurposide). This could be the final step in phloridzin formation, but in the absence of any identified intermediates between *p*-coumaric acid derivatives and phloridzin in apple tissue the sugges-tion must be only tentative; especially in view of the evidence of breakdown of chalcones to *p*-coumaryl derivatives in plants at the time of addition (Patschke, *et al.*, 1964).

Note added in proof. B. H. Koeppen and D. G. Roux (*Tetrahedron Letters*, (1965), **39**, 3497) have recently found a dihydrochalcone which they call aspalathin in the leaf of *Aspalathus linearis*. Besides being the first recorded occurrence of a dihydrochalcone in the Leguminosae, this compound is also the first example of a *C*-glycosyl dihydrochalcone. It is 3′-*C*-β-D-glucopyranosyl-3,4,2′,4′,6′ pentahydroxydihydrochalcone.

REFERENCES

Avadhani, P. N. and Towers, G. H. N. (1961). *Canad. J. Biochem. Physiol.* **39**, 1605.
Bate-Smith, E. C. and Harborne, J. B. (1963). *Nature, Lond.* **198**, 1307.
Bourquelot, E. and Fichtenholz, A. (1911). *C. R. Acad. Sci., Paris* **153**, 468.
Bourquelot, E. and Fichtenholz, A. (1912). *C. R. Acad. Sci., Paris* **154**, 526.
Bridel, M. and Kramer, A. (1931). *C. R. Acad. Sci., Paris* **193**, 748.
Charaux, C. and Rabaté, J. (1931). *C. R. Acad. Sci., Paris* **192**, 1478.
Eykman, J. F. (1883). *Rec. Trav. chim.* **2**, 99.
Goris, A. and Canal, H. (1935). *C. R. Acad. Sci., Paris* **201**, 1435, 1520.
Grisebach, H. and Patschke, L. (1962). *Z. Naturf.* **17b**, 857.
Hänsel, R. and Heise, D. (1959). *Arch. Pharm.* **292/64**, 398.
Harvey, E. M. (1929). *Plant Physiol.* **4**, 357.
Heilbron, I. M. (1937). "Dictionary of Organic Compounds", Vol. 3, p. 471, Eyre and Spottiswoode, London.
Henke, O. (1963). *Flora* **153**, 358.

Hutchinson, A., Taper, C. D. and Towers, G. H. N. (1959). *Canad. J. Biochem. Physiol.* **37**, 901.

Jerzmanowska, Z. (1956). *Acta polon. pharmac.* **13**, 901.

Koninck, L. de (1835a). *Liebigs Ann.* **15**, 75.

Koninck, L. de (1835b). *Liebigs Ann.* **15**, 258.

Lincoln, F. B. (1926). *Proc. Amer. Soc. Hort. Sci.* **23**, 249.

Murakami, S. and Fukuda, M. (1955). *J. Pharm. Soc. Japan* **75**, 603.

Nilsson, M. (1961a). *Acta. chem. scand.* **15**, 154.

Nilsson, M. (1961b). *Acta. chem. scand.* **15**, 211.

Paris, R. (1963). *In* "Chemical Plant Taxonomy", (T. Swain, ed.), p. 344. Academic Press, London and New York.

Patschke, L., Hess, D. and Grisebach, H. (1964). *Z. Naturf.* **19b**, 1114.

Puri, B. and Seshadri, T. R. (1954). *J. Sci. Indian Res.* **13B**, 698.

Rehder, A. (1954). "Manual of Cultivated Trees and Shrubs", 2nd Ed. The Macmillan Co., New York.

Rennie, E. H. (1886). *J. chem. Soc.* 857.

Rivière, G. and Bailhache, G. (1904). *C. R. Acad. Sci., Paris* **139**, 81.

Shimokoriyama, M. (1962). *In* "The Chemistry of Flavonoid Compounds", (T. Geissman, ed.), Chapter 10, Pergamon Press, Oxford.

Williams, A. H. (1955). *Nature, Lond.* **175**, 213.

Williams, A. H. (1961). *J. chem. Soc.* 4133.

Williams, A. H. (1962). *Ann. Rep. Long Ashton Res. Sta.* p. 31.

Williams, A. H. (1963). *Ann. Rep. Long Ashton Res. Sta.* p. 33.

CHAPTER 18

Flavonoid *C*-Glycosides

H. WAGNER

Institut für Pharmazeutische Arzneimittellehre,
Munich, Germany

I. INTRODUCTION

Since the publication of the comprehensive reviews on flavonoid *C*-glycosides by Haynes (1963) and Siekel (1963), much further progress has been made in our own laboratories and by other research groups on the isolation, analysis, and structural elucidation of flavonoid *C*-glycosides, and the time has now come to present a new survey of this field.

On the basis of extensive periodate degradation studies, Evans *et al.* (1957) suggested that the polyhydroxylated side chain in vitexin (I), the first flavonoid *C*-glucoside to be isolated, had a tetrahydrofuran (2,5-anhydrohexahydroxyhexyl) structure (IIa). This structure was adopted by later authors for other newly isolated *C*-glycosides in most cases without carrying out any experiments

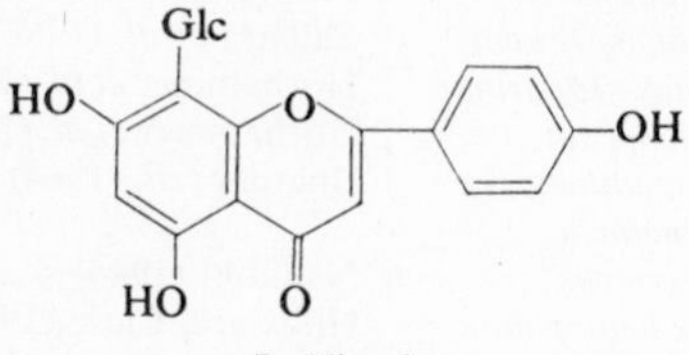

I. Vitexin

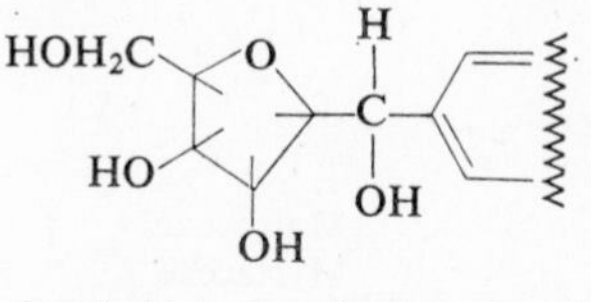

IIa. 2,5-Anhydrohexahydroxyhexyl

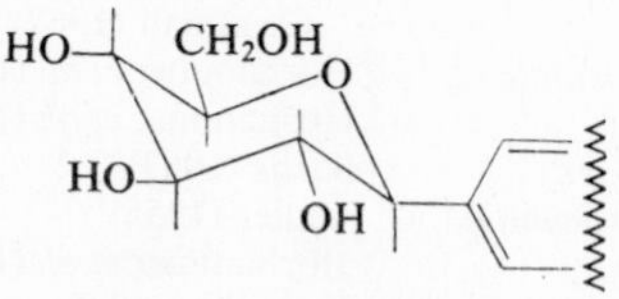

IIb. *C*-β-D-Glucosyl

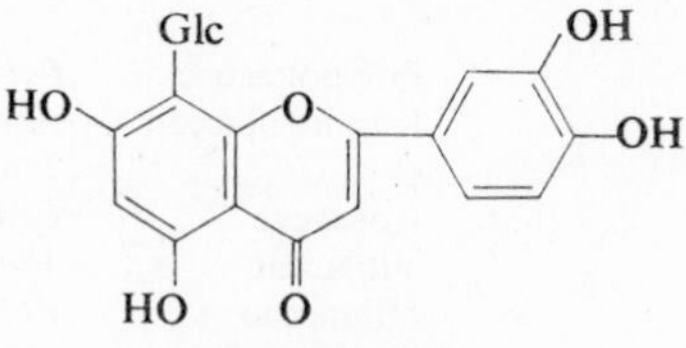

III. Orientin

of their own. Recent systematic NMR-spectroscopic studies on a number of
C-glycosides (Horowitz and Gentili, 1964), however, have shown that this
formulation of the sugar residue in vitexin could no longer be regarded as
correct. These authors believe that their NMR studies have demonstrated
that the side chain has a C-β-D-glucopyranosyl structure (IIb). This is in

TABLE 1. Occurrence of flavonoid *C*-glycosides in higher plants

Division	Family	Species	Reference
Ferns	Cyatheaceae	*Cyathea fauriei*	Ueno *et al.* (1963)
	Pteridaceae	*Sphenomeris chusana*	Ueno *et al.* (1963)
Monocotyledons	Gramineae	*Hordeum vulgare*	Seikel and Bushnell (1959), Seikel and Geissman (1957); Seikel *et al.* (1959)
		Spirodela oligorrhiza	Jurd *et al.* (1957)
Dicotyledons	Verbenaceae	*Vitex littoralis*	Perkin (1898)
		Vitex lucens	Briggs and Cambie (1958); Seikel and Mabry (1965), Seikel *et al.* (1959)
		Vitex peduncularis	Rao and Venkateswarlu (1956)
		Vitex agnus castus	Hänsel and Rimpler (1963)
	Caryophyllaceae	*Saponaria officinalis*	Barger (1906)
	Combretaceae	*Combretum micranthum*	Jentzsch *et al.* (1962)
	Cruciferae	*Alliaria officinalis*	Paris and Delaveau (1962)
	Fagaceae	*Castanospermum australe*	Eade *et al.* (1962)
	Leguminosae	*Aspalathus acuminatus*	Koeppen (1963) Koeppen *et al.* (1962)
		Cytisus laburnum	Paris, (1957) Stambouli and Paris (1961)
		Lespedeza capitata	Paris and Charles (1962)
		Pueraria thunbergiana	Shibata *et al.* (1962)
		Sarothamnus scoparium	Hörhammer *et al.* (1962)
		Spartium junceum	Hörhammer *et al.* (1960)
		Tamarindus indica	Bhatia *et al.* (1964)
		Baptisia tinctoria	
	Malvaceae	*Hibiscus syriacus*	Nakaoki (1944)
	Myrtaceae	*Eucalyptus hemophloia*	Hillis and Carle (1963)
	Rutaceae	*Citrus medica*	Chopin *et al.* (1964)
	Oxalidaceae	*Oxalis cernua*	Shimokariyama and Geissman (1962)
	Polygonaceae	*Polygonum orientale*	Hörhammer *et al.* (1962)
	Ranunculaceae	*Adonis vernalis*	Hörhammer *et al.* (1960)
		Trollius europaeus	Sachs (1963)
	Rosaceae	*Crataegus oxyacantha*	Fiedler (1955)
	Violaceae	*Viola tricolor*	Hörhammer *et al.* (1965)
	Ulmaceae	*Zelkova serrata*	Funaoka and Tanaka (1957, a, b)

agreement with the results of kinetic measurements on the periodic acid oxidation of vitexin, and of our own measurements on the periodic acid oxidation of orientin (III).

All the plants from which flavonoid C-glycosides have so far been isolated are listed in Table 1. It can be seen that some 30 different sources are now known. The compounds are not concentrated in any particular families as far as can be seen at present, although the highest percentages occur in the Leguminosae, in the tribe Papilionatae. The flavonoid C-glycosides have so far been found in all parts of plants (leaves, flowers, stems, bark, wood) except the roots. In general they are found together with their corresponding O-glycoside, and this may be of importance in relation to the as yet unexplained biosynthesis of the C-glycosides.

II. Types of Flavonoid C-Glycosides

About 25 different flavonoid C-glycosides known today are based on only three types of flavonoid compound, the flavones, isoflavones and flavanones.

A. FLAVONE TYPES

This group includes the apigenin (5,7,4'-trihydroxyflavone) derivatives, vitexin (I) and saponaretin (IV) (isovitexin or homovitexin) with the O-glyco-sides vitexin 4'-rhamnoside, vitexin 4'-xyloside, and saponarin (V) (saponaretin 7-mono-β-D-glucoside), as well as vicenin (VI) and violanthin (VII). The last two compounds are glycosides containing two C-C-linked sugar residues.

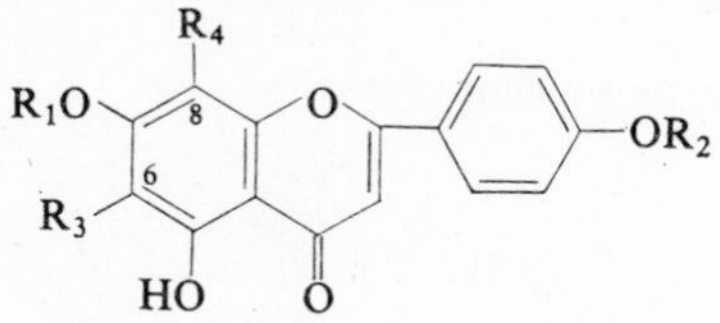

I. $R_1R_2R_3$=H, R_4=glucosyl: Vitexin
IV. $R_1R_2R_4$=H, R_3=glucosyl: Saponaretin
V. R_2R_4=H, R_1R_3=glucosyl: Saponarin
VI. R_1R_2=H, R_3R_4=hexosyl: Vicenin
VII. R_1R_2=H, R_3=rhamnosyl, R_4=glucosyl: Violanthin
VIII. R_1R_3=H, R_2=CH$_3$, R_4=glucosyl: Cytoside

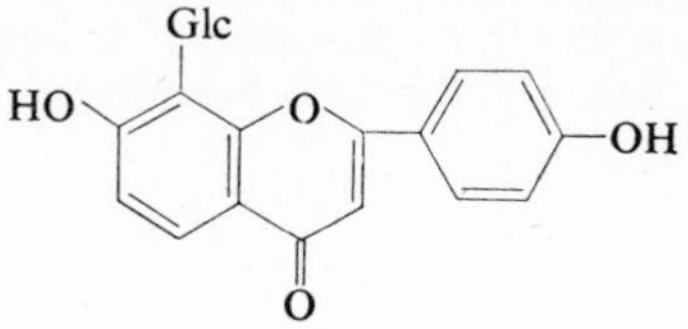

IX. Bayin

Stambouli and Paris (1961) isolated from *Cytisus laburnum*, a glycoside which they found to have the structure of a 8-C-glucosyl-5,7-dihydroxy-4'-methoxyflavone (cytoside) (VIII). This, however, was not the same as the second glycoside obtained by us, together with vitexin, from the same plant, which was found to be acacetin 7-rutinoside (linarin) Hörhammer *et al.*, (1962). The 5-deoxyvitexin, bayin (IX) (Eade *et al.*, 1962) may be regarded as a derivative of vitexin.

Luteolin (5,7,3′,4′-tetrahydroxyflavone) derivatives include orientin (lutexin) (III), and its 7-*O*-glycoside lutonarin, and the orientin *O*-xyloside adonivernith (X), the xylose being attached to the *C*-glucosyl moiety, as well as homo-orientin (lutonaretin = iso-orientin) (XI), scoparin (XII), and lucenin (XIII). Seikel and Mabry (1965) have adopted a structure similar to that of vicenin for lucenin. Another *C*-glycoside, which appears to be based on the diosmetin structure (XIV), was recently isolated from lemon rind by Chopin *et al.* (1964).

B. ISOFLAVONE TYPES

The only two known compounds of this group are both based on the 7,4′-dihydroxyisoflavone (daidzein) structure. These are puerarin (XV) and puerarin mono-*O*-xyloside, which are obtained from *Pueraria thunbergiana*. (Shibata *et al.*, 1959.)

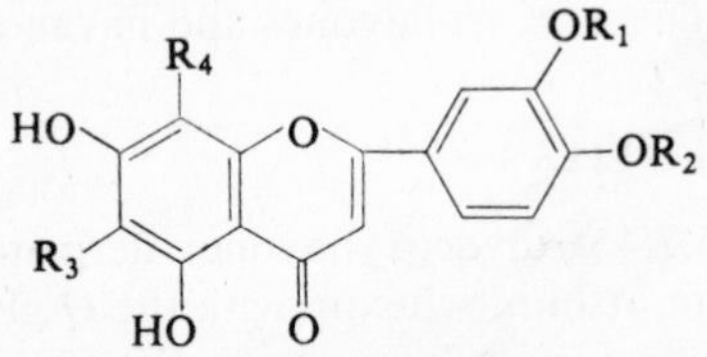

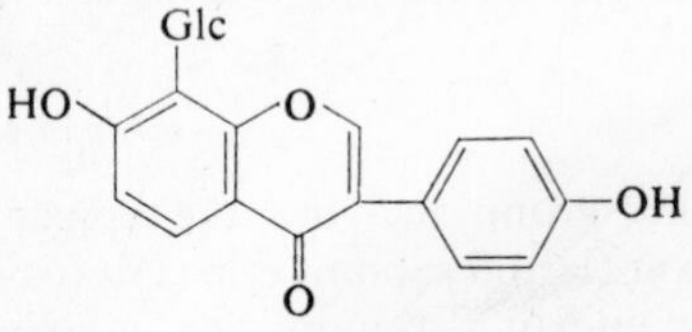

III. $R_1R_2R_3$=H; R_4=glucosyl: Orientin
X. $R_1R_2R_3$=H, R_4=xylosylglucosyl: Adonvernith
XI. $R_1R_2R_4$=H, R_3=glucosyl: Homo-orientin
XII. R_2R_3=H, R_1=CH_3, R_4=glucosyl: Scoparin
XIII. R_1R_2=H, R_3, R_4=glucosyl: Lucenin
XIV. R_1R_3=H, R_2=CH_3, R_4=glucosyl. Diosmetin-8-*C*-glucoside

XV. Puerarin

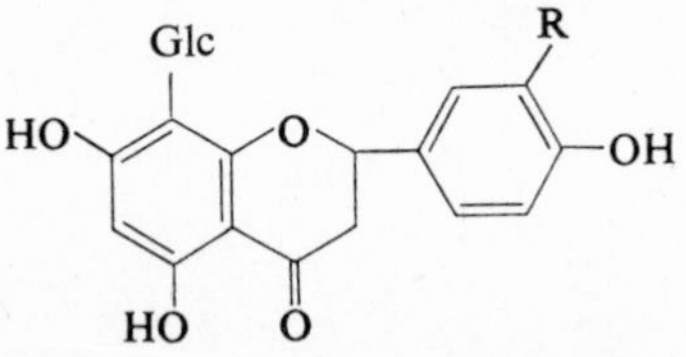

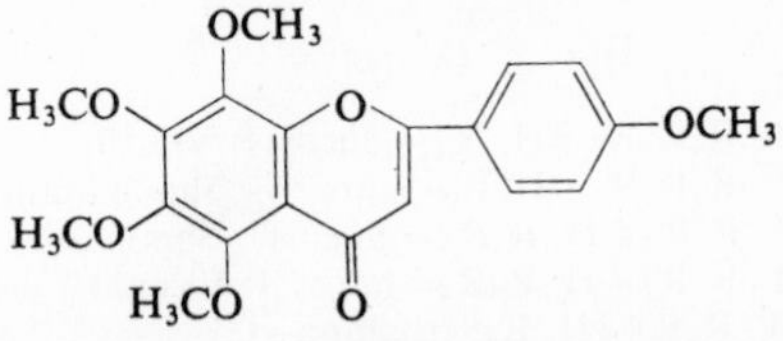

XVI. Aspalathin (or the corresponding C-6 compound)
XVII. Naringenin-8-*C*-glucoside (or the corresponding C-6 compound)

XVIII. Tangeritin

C. FLAVANONE TYPES

Koeppen (1962; Koeppen *et al.*, 1963) has reported the presence of aspala-thin (XVI), an eriodictyol-*C*-glycoside in *Aspalathus acuminatus*.* Hillis and

---

* *Note added in proof.* After this chapter was written, the aspalathin structure was revised to 5′-*C*-β-D-glucopyranosyl-3,4,2′,4′,6′-pentahydroxy-dihydrochalcone (B. H. Koeppen and D. G, Roux, *Tetrahedron Letters*, **39**, 3497, 1965).

Carle (1963) identified a glycoside from *Eucalyptus hemophloia* as a 8- (or -6)-*C*-glycosyl-5,7,4'-trihydroxyflavanone (XVII).

The flavonoids keyakinin and keyakinol, which were isolated from *Zelkowa serrata* by Funaoka *et al.* (1957a, b, and c) were originally reported by the authors to be *C*-glycosides of a flavanonol and a flavonol, respectively. While there is no doubt about the *C*-glycoside nature of the two flavonoids, the structure of the basic flavonoid skeletons requires re-examination. The fully methylated keyakinin gives tangeretin on degradation with periodic acid (Funakoa and Mitanaka, 1957a), and since tangeretin has now been assigned a 5,6,7,8,4'-pentamethoxyflavone structure (XVIII) (Goldsworthy and Robinson, 1957), keyakinin and keyakinol must contain a flavanone and a flavone nucleus, respectively.

The above survey shows that the *C*-glycosides are mostly flavone derivatives. Many also contain an additional *O*-glycoside linkage between a hydroxyl group in the 7- or 4'-positions and glucose, rhamnose, or xylose. An exception is adonivernith (X), which is obtained from *Adonis vernalis*, and in which the xylose is linked to the *C*-glycosyl residue.

III. IDENTIFICATION OF *C*-GLYCOSIDES

It is obviously of the utmost importance for systematic investigations to know how the flavone *C*-glycosides differ from the corresponding *O*-glycosides and how they can be detected in plant extracts without prior isolation and purification.

The only reasonably reliable indication is their behaviour towards acids. Flavone *C*-glycosides suffer little cleavage on prolonged treatment (about 24 hours) with 25% hydrochloric or sulfuric acid at 100°C. Flavonol-3-*O*-glycosides and 3,7-diglycosides, and both flavone and flavonol 7- and 4'-*O*-glycosides, on the other hand, are almost all readily hydrolysed in 30 minutes or so by 2N hydrochloric or sulfuric acid at 100°C. However, a number of flavone(ol) 7-glycosides are resistant and are hydrolysed only under much more drastic conditions (Bate-Smith and Swain, 1960; Egger, 1959), and so a hydrolysis time of 10 hours with 2N acid is recommended to ensure reliable differentiation. These conditions were recently subjected to critical testing by Harborne (1962).

We have found that the common *C*-glycosides of the apigenin and luteolin types can be identified chromatographically and give R_f values shown in Figure 1. In *n*-butanol-glacial acetic acid-water (4:1:5) all the hydrolysable *C*-glycosides except saponaretin (IV) lie in the R_f range 0·3–0·5. The R_f values of the few known di-*C*-glycosides (vicenin, violanthin (VI, VII) $= EG_2$, lucenin (XIII) $= EG_3$?) differ only slightly from these figures.

When a few milligrams of the substance are available for analysis, IR spectroscopy can also be used to aid identification. Comparison of the spectra of *C*-glycosides with those of corresponding normal *O*-glycosides reveals appreciable differences in the C–O stretching region between 1000 and 1100 cm^{-1}. *C*-Glycosides give only two weak bands at 1010 and 1035 cm^{-1}. This characteristic is no longer observed in the case of arabinosides or rhamnosides.

314 *H. Wagner*

In our experience, however, the best method of establishing the nature of the flavonoid is the reductive cleavage of the C-C bond with hydrogen iodide in glacial acetic acid or phenol, and subsequent paper-chromatographic detection of the "aglycone". Since in the case of the *C*-glycosides, as with the *O*-glycosides, the u.v. spectra are similar to those of the aglycones, u.v. spectroscopy in conjunction with elemental analysis and degradation with alkali may also help to elucidate the structural type.

Some difficulty may be experienced in the preparation of derivatives of *C*-glycosides. The most reliable method of obtaining pure fully acetylated

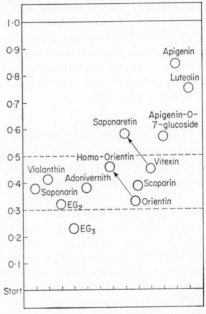

FIG. 1. Chromatography of glycoflavonoids.

C-glycoside, in our experience, is to heat the substance for a few minutes with 50 times as much acetic anhydride and a few drops of sulfuric acid, the crude acetate then being recrystallized from methanol. The undeca-acetate of violanthin has been obtained without difficulty by this method. The best method of methylating the phenolic hydroxyl groups is by treatment with dimethyl sulfate for 1 to 1½ hours, followed by the addition of 1N potassium hydroxide solution. Trimethylvitexin and trimethylsaponaretin can be obtained in this way. More drastic conditions are required for permethylation (Herzig and Tiring, 1918; Kuhn and Egge, 1963).

The greatest difficulties encountered so far have been in the identification of the glycosyl residue. Oxidation with periodic acid is unquestionably the best of the available methods, although the results obtained are comparable only if the conditions are standardized. We have carried out comparative periodic acid oxidations with 1% aqueous sodium periodate solution in 1N sulfuric acid (2 hours), with the object of studying the difference in behaviour between C-glycosides and O-glycosides. Under these vigorous conditions the polyhydroxyl system suffers total degradation. As is clear from Fig. 2(a), the degradation of an O-linked glucopyranosyl residue takes place in several stages:

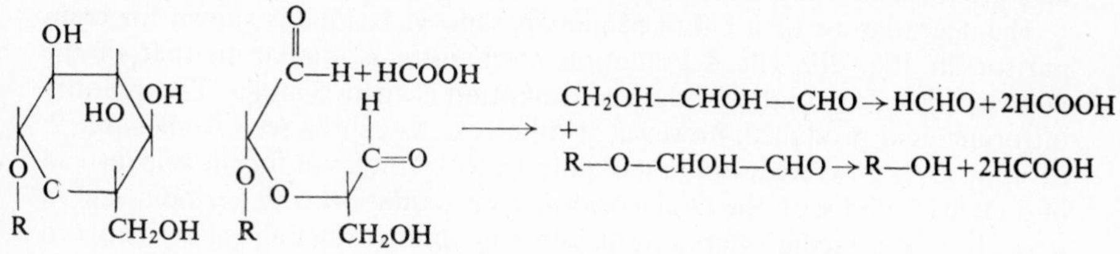

$$CH_2OH—CHOH—CHO \rightarrow HCHO + 2HCOOH$$
$$+$$
$$R—O—CHOH—CHO \rightarrow R—OH + 2HCOOH$$

$$R—O—C_6H_{11}O_5 + 5IO_4^- + H_2O \rightarrow R—OH + 5HCOOH + HCHO + 5IO_3^-$$

(a)

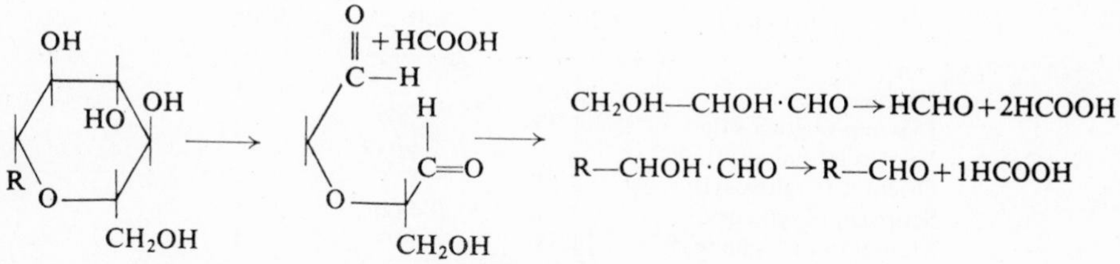

$$CH_2OH—CHOH \cdot CHO \rightarrow HCHO + 2HCOOH$$
$$R—CHOH \cdot CHO \rightarrow R—CHO + 1HCOOH$$

$$R—C_6H_{11}O_5IO_4^- \rightarrow R \cdot CHO + 4HCOOH + HCHO + 5IO_3^-$$

(b)

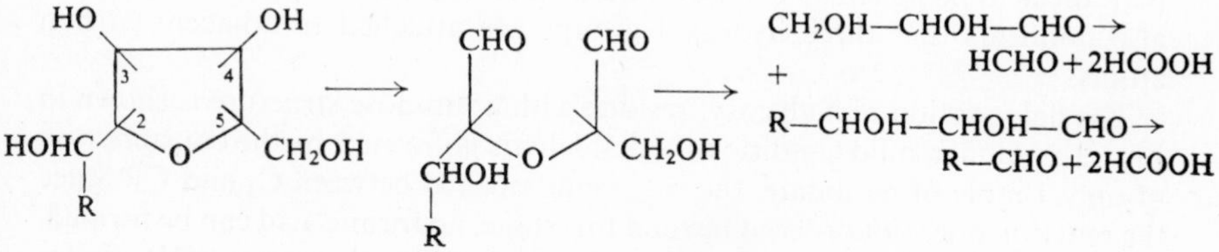

$$CH_2OH—CHOH—CHO \rightarrow$$
$$HCHO + 2HCOOH$$
$$+$$
$$R—CHOH—CHOH—CHO \rightarrow$$
$$R—CHO + 2HCOOH$$

$$R—C_6H_{11}O_5 + 5IO_4^- \rightarrow R—CHO + 4HCOOH + HCHO + 5IO_3^-$$

(c)

FIG. 2. Degradation of (a) O-linked and (b) C-linked glucopyranosyl units and of (c) C-linked glucofuranose.

(i) Ring-opening between C_2 and C_3, resulting in the consumption of 1 mole of periodate (Angyal and Klavins, 1961); (ii) Further C-C cleavage between C_3 and C_4, with consumption of a second mole of periodate, and leading to the formation of a dialdehyde and 1 mole of formic acid (Malaprade, 1934); (iii) Cleavage 2, of the oxygen bridge and further degradation in the manner indicated. A total of 5 moles of periodate is consumed, and 5 moles of formic acid are formed at the same time.

The degradation of a *C*-linked glucopyranosyl residue is shown for comparison in Fig. 2b. The degradation mechanism is similar to that of the *O*-glycosides. The total periodate consumption is again 5 moles. The quantity of formic acid produced, however, is different. As can be seen from Table 2, titration gives an average amount of only 3·9 to 4·0 moles of formic acid instead of 5 moles found with the *O*-glycosides. Thus oxidation with periodic acid in strongly acidic media offers one possible method of distinguishing between *O*-glycosides and *C*-glycosides.

TABLE 2. Formic acid produced from *O*- and *C*-linked glucosides

Substance	Mol HCOOH after		
	20′	120′	Theor.
Glucose	5·10	—	5
Flavone-*O*-glucoside	1·95	5·07	5
Vitexin (*C*-glucosyl)	—	4·05	4
Orientin (*C*-glucosyl)	2·14	3·97	4
Scoparin (*C*-glucosyl)	—	4·00	4
Mangiferin (*C*-glucosyl)	1·88	3·98	4

If, however, the oxidation is carried out under the mild conditions (0·01 M periodic acid at 15°C) used by Koeppen (1962), the degradation of the glucopyranosyl residue leads, after 23 hours, only to the dialdehyde, 2 moles of periodic acid being consumed and 1 mole of formic acid being produced. This also confirms that three hydroxyl groups are attached to adjacent carbon atoms.

The degradation of a glycosyl residue with a furanose structure is shown in Fig. 2(c). Under mild conditions, a dialdehyde is formed by the consumption of only 1 mole of periodate, the ring being opened between C_3 and C_4. Since the reaction does not proceed beyond this stage, no formic acid can be formed. Since formic acid is produced by both vitexin (I) and orientin (III), under mild conditions, neither of these compounds can contain a tetrahydrofuran structure. The result of the periodic acid oxidation of a furanose structure in strongly acidic media gives similar results to those with the *C*-glucopyranosyl residue. Thus information regarding the nature of the sugar ring system present can only be obtained by mild periodic acid oxidation.

Further indications as to the structure of the *C*-sugar residue can be obtained by the borohydride reduction of the dialdehyde formed on mild oxidation, followed by methanolysis (Viscontini *et al.*, 1955). The resulting glycols, which can be readily detected by paper chromatography, are characteristic of the type of sugar present (Fig. 3). Since we have obtained only glycerol from all the *C*-glycosides studied, with the exception of violanthin (VII), sugars other than glucose, galactose, fructose, mannose,and ribose can be ruled out.

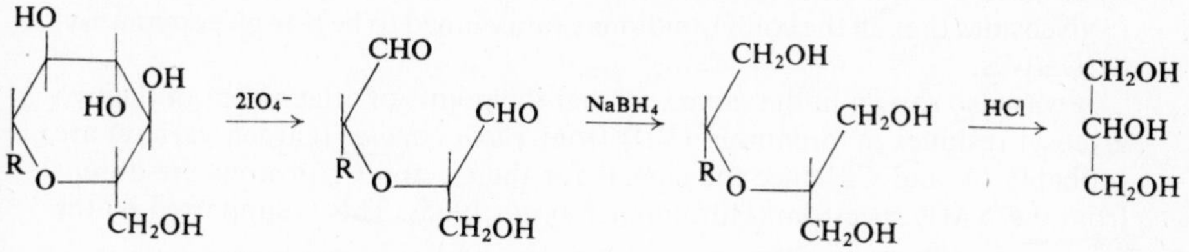

Glucose, Fructose, Ribose, Mannose and Galactose → Glycerol

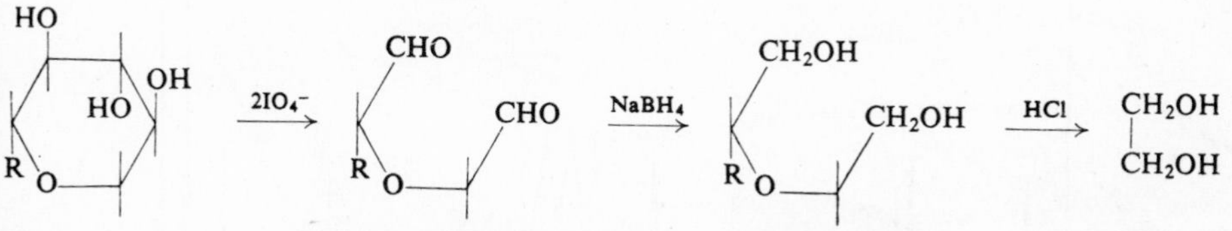

Sorbose, Arabinose and Xylose → Ethylene glycol

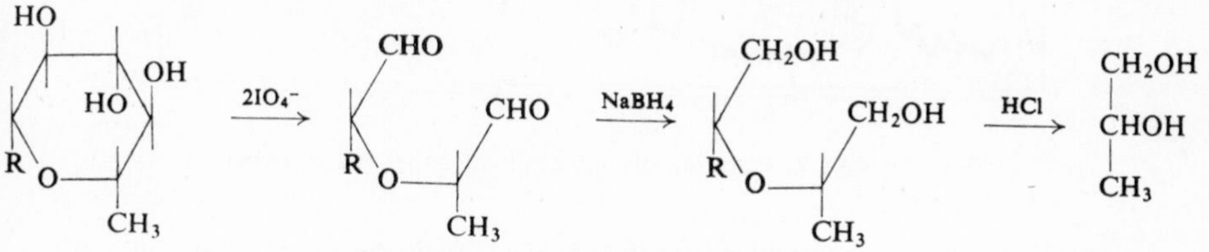

Rhamnose → 1,2-Propylene glycol

FIG. 3. Products obtained from different sugars by mild periodate oxidation followed by borohydride reduction.

In order to find the position of the C-linkage in the flavone molecule, the classic method was to oxidize the glycoside with periodic acid to the corresponding flavone aldehyde, which is then compared with a synthetic product. This method was used to show that the position of the linkage in vitexin (I) and puerarin (XV) is on C_8. In the meantime, NMR spectroscopy has proved to be useful (Waiss *et al.*, 1964; Mabry *et al.*, 1965; Hillis and Horn, 1965). Thus on the basis of a comparative NMR study, Horowitz and Gentili (1964) were able to show that saponaretin (IV) must have a 6-*C*-glucopyranosyl-apigenin structure, an alternative that had already been suggested by Koeppen (1964). This agrees with the fact that vitexin (I) and saponaretin (IV) are relatively easily interconvertable in acidic media, provided that the C_5-OH group is free. In co-operation with Professor Mabry, we have carried out NMR-spectroscopic investigations on trimethylsilyl ethers of a number of *C*-glycosides. We found that the position of the linkage in orientin (III) and scoparin (XII) must be assumed to be at C_8, while that in saponarin (V) appears to be at C_6 (Hörhammer *et al.*, 1965). It is fairly probable that homo-orientin (XI) (iso-orientin) is the luteolin analogue of saponaretin (IV). In a recent paper by Hillis and Horn (1965), it was shown from the NMR spectra of acetylated *C*-glycosides that all the compounds may be assumed to be β-D-glycopyranosyl derivatives.

It was also shown in the same way that the points of attachment of the two glycosyl residues in violanthin (VII) from *Viola tricolor* (garden variety) are probably C_8 and C_6, since the signals for the C_8 and C_6 protons are absent from the NMR spectrum (Hörhammer *et al.*, 1965). This is supported by the

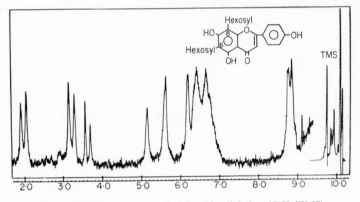

FIG. 4. N.M.R. spectrum of violanthin-silylether (CCl_4/TMS).

surprisingly large chemical shift of the $H_{2'6'}$ protons in comparison with apigenin-7-glucoside (see Fig. 4). Violanthin forms an undeca-acetate and consumes 10 moles of periodic acid on oxidation. Since the Viscontini degradation (Viscontini *et al.*, 1955) yields glycerol together with small quantities of propylene glycol, one of the two sugars must be a pentose, and probably rhamnose. This is confirmed by the NMR spectrum, and by the fact that violanthin (VII) is not identical with vicenin (6,8-di-*C*-glucosylapigenin (VI)) (Seikel, 1963). Including lucenin (XII), therefore, a total of three di-*C*-glycosyl compounds has been discovered in the plant kingdom. There is reason to believe, moreover, that derivatives of these new flavone types in which a third sugar residue is bound by an *O*-glycoside linkage (cf. adonivernith (X)) also exist in plants.

REFERENCES

Angyal, S. J. and Klavins, J. E. (1961). *Aust. J. Chem.* **14**/4, 577.
Barger, J. (1906). *J. chem. Soc.* **89**, 1210.
Bate-Smith, E. C. and Swain, T. (1960). *Chem. & Ind.* 1132.

Bhatia, V. K., Gupta, S. R. and Seshadri, T. R. (1964). *Curr. Sci. (India)* **33**, 581.
Briggs, L. H. and Cambie, R. C. (1958). *Tetrahedron* **3**, 269.
Chopin, J., Roux, B. and Durix, M. A. (1964). *C. R. Acad. Sci., Paris* **13**, 259.
Eade, R. E., Salasoo, I. and Simes, J. J. H. (1962). *Chem. & Ind.* 1720.
Egger, K. (1959). *Z. Naturf.* **14** B, 401.
Evans, W. H., McGookin, A., Jurd, L., Robertson, A. and Williamson, W. R. N. (1957). *J. chem. Soc.* 3510.
Fiedler, U. (1955). *Arzneimittelforsch.* **5**, 609.
Funaoka, K. and Tanaka, M. (1957a). *Nippon Kokuzai Gakkaishi* **3**, 144, 173 and 218.
Funaoka, K. and Tanaka, M. (1957b). *Chem. Abstr.* **51**, 18130 and **52**, 12395.
Funaoka, K. and Mitanaka (1957). *Nippon Kokuzai Gakkaishi* **3**, 218.
Funaoka, K. and Mitanaka (1958). *Chem. Abstr.* **52**, 12395.
Goldsworthy, L. and Robinson, R. (1957). *Chem. & Ind.* 47.
Hänsel, R. and Rimpler, H. (1963). *Arch. Pharm.* **296**, 598.
Harborne, J. B. (1962). *Chem. & Ind.* 222.
Haynes, L. J. (1963). *Advan. Carbohydrate Chem.* **18**, 227.
Herzig, J. and Tiring, G. (1918). *Mh. Chem.* **39**, 253.
Hillis, W. E. and Carle, H. (1963). *Aust. J. Chem.* **16**, 147.
Hillis, W. E. and Horn, D. H. S. (1965). *Aust. J. Chem.* **18**, 531.
Hörhammer, L., Wagner, H. and Gloggengiesser, F. (1958). *Arch. Pharm.* **291**/63, 126.
Hörhammer, L., Wagner, H. and Dhingra, H. S. (1959a). *Arch. Pharm.* **292**, 64, 83.
Hörhammer, L., Wagner, H., Nieschlag, H. and Wildi, G. (1959b). *Arch. Pharm.* **292**/64, 380.
Hörhammer, L., Wagner, H. and Leeb, W. (1960). *Arch. Pharm.* **293**/65, 264.
Hörhammer, L., Wagner, H. and Beyersdorff, P. (1962). *Naturwissenschaften* **49**, 392 B.
Hörhammer, L., Wagner, H., Rosprim, R., Mabry, T. J. and Rössler, H. (1965). *Tetrahedron Letters* **22**, 1707.
Horowitz, R. M. and Gentilli, B. (1964). *Chem. & Ind.* 498.
Jentzsch, K., Spiegel, P. and Fuchs, L. (1962). *Planta Medica* **10**, 1.
Jurd, L., Geissman, T. A. and Seikel, M. (1957). *Arch. biochem. Biophys.* **67**, 284.
Koeppen, B. H. (1962). *Chem. & Ind.* 2145.
Koeppen, B. H. (1963). *J. Lab. Clin. Med.* **9**, 141.
Koeppen, B. H. (1964). *Z. Naturf.* **19** B, 173.
Koeppen, B. H., Smit, C. J. B. and Roux, D. G. (1962). *Biochem. J.* **83**, 507.
Kuhn, R. and Egge, H. (1963). *Chem. Ber.* **96**, 3338.
Mabry, T. J., Kagan, J. and Rössler, H. (1965). *Phytochemistry* **4**, 177.
Malaprade, L. A. (1934). *Bull. Soc. chim. Fr.* **1**, 833.
Nakaoki, T. (1944). *J. Pharm. Japan* **64**, (11A) 57.
Nakaoki, T. (1952). *Chem. Abstr.* **46**, 108.
Paris, R. R. (1957). *C. R. Acad. Sci., Paris*, **245**, 443.
Paris, R. R. and Charles, A. (1962). *C. R. Acad. Sci., Paris* **254**, 352.
Paris, R. R. and Delaveau, P. (1962). *C. R. Acad. Sci., Paris* **254**, 928.
Perkin, A. G. (1898). *J. chem. Soc.* **73**, 1019.
Rao, C. B. and Venkateswarlu, V. (1956). *Curr. Sci. (India)* **25**, 328.
Sachs, H. (1963). Thesis, München.
Seikel, M. K. (1963). "Proceedings of the Third Annual Symposium P-P.G.N.A." (V. C. Runeckles, ed.), p. 31. Toronto.
Seikel, M. K. and Geissman, T. A. (1957). *Arch. biochem. Biophys.* **71**, 17.

Seikel, M. K. and Bushnell, A. J. (1959). *J. org. Chem.* **24**, 1995.

Seikel, M. K. and Mabry, T. J. (1965). *Tetrahedron Letters.* **16**, 1105.

Seikel, M. K., Bushnell, A. J. and Birzzalis, R. (1959a). *Arch. biochem. Biophys.* **85**, 272.

Seikel, M. K., Holder, D. J. and Birzzalis, R. (1959b). *Arch. biochem. Biophys.* **85** 272.

Shibata, S., Murakami, T., Nishikawa, Y. and Budidarmo, W. (1959). *Congr. Sci. Pharm.* **1959**, 214.

Shibata, S., Murakami, T., Nishikawa, Y. and Budidarmo, W. (1962). *Chem. Abstr.* **56**, 3564.

Shimokoriyama, M. and Geissman, T. A. (1962). *Recent Progress in Chemistry,* Nat. Synth. Color Matters related Fields 245.

Shimokoriyama, M. and Geissman, T. A. (1962). *Chem. Abstr.* **59**, 4020.

Stambouli, A. and Paris, R. (1961). *Ann. Pharm. Fr.* **19**, 435.

Ueno, A., Oguri, N., Hori, K., Saiki, Y. and Harada, T. (1963). *Yokugaku Zasshi* **82**, 1486; **83**, 420.

Viscontini, M., Hoch, D. and Karrer, P. (1955). *Helv. chim. acta* **38**, 642.

Waiss, A. C., Jr., Lundin, R. E. and Stern, D. J. (1964). *Tetrahedron Letters* **10**, 513.

Author Index

Numbers in italic refer to the reference pages at the end of the chapters where references appear.

A

Abe, Y., 273, *295*
Achmatowicz, O., 189, *194*
Adams, G. A., 143, 144, *155*, *157*, *158*
Adams, R., 112, *118*
Adiga, P. R., 198, *209*
Agranoff, B. W., 99, *119*
Ahluwalia, V. K., 29, *31*
Akiyoshi, S., 109, *118*
Albers-Schonberg, G., 164, *172*
Allen, M. B., 123, 126, *137*
Alston, R. E., 12, *18*, 38, 40, 45, *55*, 197, *209*, 284, *294*
Altman, R. F. A., 220, *230*
Anagnistopoulos, C., 152, *156*
Anderson, N. S., 147, *155*
Andrews, P., 153, 155, *155*
Angyal, S. J., 316, *318*
Aplin, R. T., 61, *74*
Araki, C., 146, *155*
Archer, B. L., 99, *118*
Archibald, A. R., 151, *155*
Argoud, S., 135, *137*
Arigoni, D., 28, *31*
Ariyoshi, H., 61, *75*
Arndt, C., 26, *31*, 88, *95*
Arndt, R. R., 226
Arni, P. C., 152, *155*
Arnold, W. N., 190, *194*
Aronoff, S., 46, 47, *55*
Arthur, H. R., 161, *172*
Arya, V. P., 114, *118*
Asahina, Y., 178, *185*
Aspinall, G. O., 139, 140, 141, 142, 143, 144, 149, 151, 152, *155*, *156*
Attaway, D. H., 61, 69, *75*
Avadhani, P. N., 272, *294*, 306, *306*
Ayrey, G., 99, *118*

B

Baarschers, W. H., 226
Bacon, J. S. D., 161, 166, 170, *172*
Bagby, M. O., 87, *95*

Bailey, G. F., 128, *137*
Bailhache, G., 298, *307*
Baillie, J., 149, *155*
Baker, E. A., 61, 63, *74*, *75*
Baker, E. G., 69, *74*
Baker, G. L., 66, 68, *74*
Baker, H. G., 282, *294*
Banaszek, H., 189, *194*
Banner, A. E., 64, *74*
Baptist, J. N., 59, *74*
Baraud, J., 128, 131, *137*
Barber, G. A., 154, *156*
Barbieri, J., 195, *209*
Barger, J., 310, *318*
Barghoorn, E. S., 69, 71, *74*, *77*
Barkemeyer, H., 161, *173*
Barrall, E. M., 63, *74*
Barrer, R. M., 63, *74*
Barrera, J. B., 61, 66, *74*
Barton, D. H. R., 23, *31*
Bartz, K. W., 64, *74*
Basset, E. W., 112, *120*
Bate-Smith, E. C., 11, *18*, 161, 169, *172*, 177, *188*, 271, 272, 273, 277, 284, 285, 291, 293, *294*, *295*, 302, *306*, 313, *318*
Bates, R. B., 111, *118*
Batt, R. F., 61, 63, *74*
Battaile, J., 106, *119*
Baumann, F., 63, *74*
Bazant, V., 71, *76*
Beale, G. H., 274, *294*
Bean, R. C., 145, *156*
Beattie, A., 151, *156*
Becher, P., 235, *244*
Beck, E., 241, *244*
Beckmann, S., 273, *294*
Beckurts, H., 178, *185*
Begbie, R., 142, *156*
Belardini, M., 114, 116, *119*
Bell, E. A., 52, 53, *55*, 197, 198, 199, 203, *209*
Bellen, Z., 189, *194*
Belsky, T., 63, 64, 66, 71, 72, *74*, *75*
Bencze, W., 164, *173*
Benjamin, C. R., 26, *31*
Benn, M. H., 192, *194*
Benson, L., 243, *244*

321

G

H

T

Genera and Species Index

A

Abronia, 239
Acacia, 263
  pycnantha, 144
  senegal, 144
Acanthaceae, 215, 285, 288, 291
Acanthosyris spinescens, 85
Aceraceae, 214
Acetabularia, 142
Achillea, 217
Achimenes, 279
Achyranthes, 239
Aconitum, 176, 178, 216, 227
  altaicum, 180
  anthora, 180
  bicolor, 180
  ferox, 180
  fischeri, 180
  lycoctonum, 180
  napellus, 180, 222
  ranunculifolius, 180
Acremonium, 252
Acrosiphonia centralis, 150
Actaea spicata, 180
Actinidia polygama, 165
Actinidiaceae, 214
Adenocarpus complicatus, 47
  hispanicus, 47
Adiantum veitchianum, 273
Adonis, 176, 178, 262
  aestivalis, 180
  amurensis, 180
  vernalis, 180, 310, 313
Aegialitis annulata, 282
Aerva, 239
Aeschynanthus nervifolius, 279
  obconicus, 279
Agardhiella tenera, 147·
Agaricaceae, 252, 254, 255

Agathis alba, 114
  australis, 113
  microstachya, 114
  robusta, 114
Agavaceae, 197
Agave vera-cruz, 152
Ahnfeltia, 146
Aizozaceae, 214, 232, 239, 240
Akaniaceae, 214
Alangiaceae, 215, 219
Albizzia lophanta, 188
Alchornea floribunda, 229
Algae, 61, 123, 247
Alisma lanceolatum, 184
Alkanna, 267
Alliaria officinalis, 310
Allium, 190, 207
  cepa, 188, 190
  sativum, 190
  schoenoprasum, 190
Alloplectus vittatus, 278
Alluandia, 239
Alluandiopsis, 239
Aloe, 61, 62, 63, 66, 67, 89, 251
Alpinia officinarum, 276
Alstonia constricta, 160
Alternanthera, 239
Amanita, 252
Amaranthaceae, 214, 232, 239, 240,
  291
Amaranthus, 239
  tricolor, 235
Amaryllidaceae, 197, 216, 276
Ammodendron, 47
Amorphophallus, 153
  konjak, 153
Anabasis, 47
  aphylla, 47, 222
Anacampseros, 239

U

Ulmaceae, 310
Ulva, 140, 149
  lactuca, 150, 151
Umbellales, 88
Umbelliferae, 80, 215, 218, 219, 264
Umbelliflorae, 80, 215
Urticales, 171, 213

V

Vaccinium, 264
Valerianaceae, 215
Veratreae, 217
Verbascum phlomoides, 287
Verbena, 170
Verbenaceae, 248, 250, 267, 285, 287, 288, 310
Verticillium, 252
Viburnum opulus, 130, 135
  prunifolium, 276
Vicia, 51, 198, 203
  articulata, 204, 205
  amphicarpa, 205
  augustifolia, 205
  aurantica, 205
  baicalensis, 205
  benghalensis, 204
  bithinica, 204
  cassubica, 204
  cordata, 205
  cornigera, 205
  cracca, 204
  cylindrica, 205
  dalmatica, 204
  disperma, 204
  dumetorum, 205
  ervilia, 204
  faba, 61, 205
  fulgens, 204, 205
  grandiflora, 205
  hirsuta, 204
  hybrida, 205

Vicia—cont.
  hyrcanica, 204
  lathyroides, 205
  ludoviciana, 205
  lutea, 204
  michauxii, 205
  narbonensis, 205
  onobrychoides, 204, 205
  orobus, 204
  pannonica, 204
  peregrina, 205
  picta, 205
  sativa, 205
  selloi, 204
  sepium, 205
  sicula, 205
  silvatica, 204
  tennuifolia, 204
  tennuisima, 204, 205
  tetrasperma, 204
  unijuga, 205
  villosa, 204, 205
Vinca, 33
Viola tricolor, 310, 318
Violaceae, 310
Violales, 171, 215
Virgilia araboides, 149
Vitex agnus castus, 310
  littoralis, 310
Vitex lucens, 310
  peduncularis, 310
Volucrispora, 252

W

Watsonia, 276
Whale, 68
Wolffia, 40
  arrhiza, 41
  columbiana, 41
  microscopica, 41
  papulifera, 41
  punctata, 41
  floridana, 41
  gladiata, 41

General Compounds Index

A

Abienol, 117
Abietic acid, 69, 118
Abietinal, 118
Abietinol, 118
Acacetin, 276
Acanthoidine, 217
Acanthoine, 217
Acetic acid, 24
N-Acetyl-L-djenkolic acid, 207
Acetylenes, 79
α-Acetylornithine, 197
Actinidine, 216
Actinorhodins, 247
Adenocarpine, 47
Adonivernith, 312, 313, 318
Agaropectin, 146
Agarose, 146, 147
Agathic acid, 114, 115
Agnuside, 166
Alaternin, 251, 264
Albizzine, 207, 208
Alginic acid, 140
Alizarin, 250, 265, 266
Alizarin 1-methyl ether, 250
Allosecurinine, 228, 229
Alkaloids, 171, 178, 211
N-Alkanes, 57
Alkannan, 249, 267, 269
Alkannin, 249, 269
Aloe-emodin, 251, 261, 262, 263, 264
Aloin, 261, 267
Amarantin, 232, 235
Amaryllidaceous alkaloids, 217
Amentoflavone, 43, 44, 275
Amino acids, 24, 51, 195
α-Amino-γ-oxalylaminobutyric acid, 52, 198, 199, 201

α-Amino-β-oxalylaminopropionic acid, 52, 198, 199, 201
β-Aminopropionitrile, 51
Ammodendrine, 47
Amylopectin, 151
Amylose, 151
Anabasine, 216, 217, 218
Anemonin, 179
Angelicin, 29
Angolensin, 305, 306
Annuloline, 217
Antheraxanthin, 122, 124, 125, 134, 135
Anthocyanins, 41, 278
Anthragallol, 251, 266
Anthragallol methyl ethers, 251
Anthraquinones, 250, 254, 262, 263
Aphylline, 50
Apigenin, 285, 289
Apigeninidin, 272
Aporphines, 223, 224
Arabinogalactans, 140, 143
Araucarenolone, 113, 115
Araucarol, 113, 115
Araucarolone, 113, 115
Araucarone, 113, 115
Arbutin, 246, 262, 263 264, 268
Arecoline, 217
Arginine, 52, 205
Aristolactam, 216
Aristolochic acid, 216
Arylbenzofurans, 29
Asebotin, 299, 301, 304
Asicine, 217
Aspalathin, 306, 312
Asparagine, 205
Asperocotillin, 169
Asperthecin, 255

351